MICHIO KAKU

PARALLEL WORLDS

Dr. Michio Kaku is professor of theoretical physics at the City University of New York and a cofounder of string field theory. He is the author of several widely acclaimed books, including *Visions*, *Beyond Einstein*, and *Hyperspace*, which was named one of the best science books of the year by *The New York Times* and *The Washington Post*. He hosts a nationally syndicated radio science program and has appeared on such national television shows as *Nightline*, *60 Minutes*, *Good Morning America*, and *Larry King Live*.

Also by Michio Kaku

Beyond Einstein

Einstein's Cosmos

Hyperspace

Visions

PARALLEL WORLDS

MICHIO KAKU

ANCHOR BOOKS
A Division of Random House, Inc.
New York

PARALLEL
WORLDS

A JOURNEY THROUGH
CREATION, HIGHER DIMENSIONS,
AND THE FUTURE OF THE COSMOS

FIRST ANCHOR BOOKS EDITION, FEBRUARY 2006

The Library of Congress has cataloged the Doubleday edition as follows:
Kaku, Michio.
    Parallel worlds : a journey through creation, higher dimensions, and the future of
the cosmos / Michio Kaku.—1st ed.
        p.   cm.
    Includes bibliographical references.
    1. Cosmology. 2. Big bang theory. 3. Superstring theories. 4. Supergravity. I. Title.
    QB981.K134 2004
    523.1—dc22

                                                                      2004056039

Anchor ISBN-13: 978-1-4000-3372-0

Book design by Nicola Ferguson

Illustrations by Hadel Studio

www.anchorbooks.com

Printed in the United States of America
20 19 18 17 16 15

This book is dedicated to my loving wife, Shizue.

CONTENTS

ACKNOWLEDGMENTS   xi

PREFACE   xv

PART I: THE UNIVERSE

CHAPTER ONE: Baby Pictures of the Universe   3

CHAPTER TWO: The Paradoxical Universe   22

CHAPTER THREE: The Big Bang   45

CHAPTER FOUR: Inflation and Parallel Universes   76

PART II: THE MULTIVERSE

CHAPTER FIVE: Dimensional Portals and Time Travel   111

CHAPTER SIX: Parallel Quantum Universes   146

CHAPTER SEVEN: M-Theory: The Mother of All Strings   181

CHAPTER EIGHT: A Designer Universe?   241

CHAPTER NINE: Searching for Echoes from
the Eleventh Dimension   256

PART III: ESCAPE INTO HYPERSPACE

CHAPTER TEN: The End of Everything   287

CHAPTER ELEVEN: Escaping the Universe   304

CHAPTER TWELVE: Beyond the Multiverse   343

NOTES   363

GLOSSARY   381

RECOMMENDED READING   403

INDEX   407

ACKNOWLEDGMENTS

I would like to thank the following scientists who were so gracious in donating their time to be interviewed. Their comments, observations, and ideas have greatly enriched this book and added to its depth and focus:

- Steven Weinberg, Nobel laureate, University of Texas at Austin
- Murray Gell-Mann, Nobel laureate, Santa Fe Institute and California Institute of Technology
- Leon Lederman, Nobel laureate, Illinois Institute of Technology
- Joseph Rotblat, Nobel laureate, St. Bartholomew's Hospital (retired)
- Walter Gilbert, Nobel laureate, Harvard University
- Henry Kendall, Nobel laureate, Massachusetts Institute of Technology (deceased)
- Alan Guth, physicist, Massachusetts Institute of Technology
- Sir Martin Rees, Astronomer Royal of Great Britain, Cambridge University
- Freeman Dyson, physicist, Institute for Advanced Study, Princeton University
- John Schwarz, physicist, California Institute of Technology
- Lisa Randall, physicist, Harvard University
- J. Richard Gott III, physicist, Princeton University
- Neil de Grasse Tyson, astronomer, Princeton University and Hayden Planetarium
- Paul Davies, physicist, University of Adelaide
- Ken Croswell, astronomer, University of California, Berkeley
- Don Goldsmith, astronomer, University of California, Berkeley
- Brian Greene, physicist, Columbia University

- Cumrun Vafa, physicist, Harvard University
- Stuart Samuel, physicist, University of California, Berkeley
- Carl Sagan, astronomer, Cornell University (deceased)
- Daniel Greenberger, physicist, City College of New York
- V. P. Nair, physicist, City College of New York
- Robert P. Kirshner, astronomer, Harvard University
- Peter D. Ward, geologist, University of Washington
- John Barrow, astronomer, University of Sussex
- Marcia Bartusiak, science journalist, Massachusetts Institute of Technology
- John Casti, physicist, Santa Fe Institute
- Timothy Ferris, science journalist
- Michael Lemonick, science writer, *Time* magazine
- Fulvio Melia, astronomer, University of Arizona
- John Horgan, science journalist
- Richard Muller, physicist, University of California, Berkeley
- Lawrence Krauss, physicist, Case Western Reserve University
- Ted Taylor, atomic bomb designer
- Philip Morrison, physicist, Massachusetts Institute of Technology
- Hans Moravec, computer scientist, Carnegie Mellon University
- Rodney Brooks, computer scientist, Artificial Intelligence Laboratory, Massachusetts Institute of Technology
- Donna Shirley, astrophysicist, Jet Propulsion Laboratory
- Dan Wertheimer, astronomer, SETI@home, University of California, Berkeley
- Paul Hoffman, science journalist, *Discover* magazine
- Francis Everitt, physicist, Gravity Probe B, Stanford University
- Sidney Perkowitz, physicist, Emory University

I would also like to thank the following scientists for stimulating discussions about physics over the years that have greatly helped to sharpen the content of this book:

- T. D. Lee, Nobel laureate, Columbia University
- Sheldon Glashow, Nobel laureate, Harvard University

- Richard Feynman, Nobel laureate, California Institute of Technology (deceased)
- Edward Witten, physicist, Institute for Advanced Study, Princeton University
- Joseph Lykken, physicist, Fermilab
- David Gross, physicist, Kavli Institute, Santa Barbara
- Frank Wilczek, physicist, University of California, Santa Barbara
- Paul Townsend, physicist, Cambridge University
- Peter Van Nieuwenhuizen, physicist, State University of New York, Stony Brook
- Miguel Virasoro, physicist, University of Rome
- Bunji Sakita, physicist, City College of New York (deceased)
- Ashok Das, physicist, University of Rochester
- Robert Marshak, physicist, City College of New York (deceased)
- Frank Tipler, physicist, Tulane University
- Edward Tryon, physicist, Hunter College
- Mitchell Begelman, astronomer, University of Colorado

I would like to thank Ken Croswell for numerous comments on the book.

I would also like to thank my editor, Roger Scholl, who has masterfully edited two of my books. His sure hand has greatly enhanced the books, and his comments have always helped to clarify and deepen the content and presentation of my books. Last, I would like to thank my agent, Stuart Krichevsky, who has ushered in my books for all these years.

PREFACE

Cosmology is the study of the universe as a whole, including its birth and perhaps its ultimate fate. Not surprisingly, it has undergone many transformations in its slow, painful evolution, an evolution often overshadowed by religious dogma and superstition.

The first revolution in cosmology was ushered in by the introduction of the telescope in the 1600s. With the aid of the telescope, Galileo Galilei, building on the work of the great astronomers Nicolaus Copernicus and Johannes Kepler, was able to open up the splendor of the heavens for the first time to serious scientific investigation. The advancement of this first stage of cosmology culminated in the work of Isaac Newton, who finally laid down the fundamental laws governing the motion of the celestial bodies. Instead of magic and mysticism, the laws of heavenly bodies were now seen to be subject to forces that were computable and reproducible.

A second revolution in cosmology was initiated by the introduction of the great telescopes of the twentieth century, such as the one at Mount Wilson with its huge 100-inch reflecting mirror. In the 1920s, astronomer Edwin Hubble used this giant telescope to overturn centuries of dogma, which stated that the universe was static and eternal, by demonstrating that the galaxies in the heavens are moving away from the earth at tremendous velocities—that is, the universe is expanding. This confirmed the results of Einstein's theory of general relativity, in which the architecture of space-time, instead of being flat and linear, is dynamic and curved. This gave the first plausible explanation of the origin of the universe, that the universe began with a cataclysmic explosion called the "big bang," which sent the stars and galaxies hurtling outward in space. With the pioneering work of George Gamow and his colleagues on the big bang theory and Fred Hoyle on the origin of the elements, a scaffolding was emerging giving the broad outlines of the evolution of the universe.

A third revolution is now under way. It is only about five years old. It has been ushered in by a battery of new, high-tech instruments, such as space satellites, lasers, gravity wave detectors, X-ray telescopes, and high-speed supercomputers. We now have the most authoritative data yet on the nature of the universe, including its age, its composition, and perhaps even its future and eventual death.

Astronomers now realize that the universe is expanding in a runaway mode, accelerating without limit, becoming colder and colder with time. If this continues, we face the prospect of the "big freeze," when the universe is plunged into darkness and cold, and all intelligent life dies out.

This book is about this third great revolution. It differs from my earlier books on physics, *Beyond Einstein* and *Hyperspace*, which helped to introduce to the public the new concepts of higher dimensions and superstring theory. In *Parallel Worlds*, instead of focusing on space-time, I concentrate on the revolutionary developments in cosmology unfolding within the last several years, based on new evidence from the world's laboratories and the outermost reaches of space, and new breakthroughs in theoretical physics. It is my intention that it can be read and grasped without any previous introduction to physics or cosmology.

In part 1 of the book, I focus on the study of the universe, summarizing the advances made in the early stages of cosmology, culminating in the theory called "inflation," which gives us the most advanced formulation to date of the big bang theory. In part 2, I focus specifically on the emerging theory of the multiverse—a world made up of multiple universes, of which ours is but one—and discuss the possibility of wormholes, space and time warps, and how higher dimensions might connect them. Superstring theory and M-theory have given us the first major step beyond Einstein's original theory; they give further evidence that our universe may be but one of many. Finally, in part 3, I discuss the big freeze and what scientists now see as the end of our universe. I also give a serious, though speculative, discussion of how an advanced civilization in the distant future might use the laws of physics to leave our universe trillions of years from now and enter another, more hospitable universe to begin the process of rebirth, or to go back in time when the universe was warmer.

With the flood of new data we are receiving today, with new tools such as space satellites which can scan the heavens, with new gravity wave detectors, and with new city-size atom smashers nearing completion, physicists feel that we are entering what may be the golden age of cosmology. It is, in short, a great time to be a physicist and a voyager on this quest to understand our origins and the fate of the universe.

PART
ONE

THE UNIVERSE

CHAPTER ONE

Baby Pictures of the Universe

The poet only asks to get his head into the heavens. It is
the logician who seeks to get the heavens into his head.
And it is his head that splits.

—G. K. Chesterson

WHEN I WAS A CHILD, I had a personal conflict over my beliefs. My
parents were raised in the Buddhist tradition. But I attended
Sunday school every week, where I loved hearing the biblical stories
about whales, arks, pillars of salt, ribs, and apples. I was fascinated
by these Old Testament parables, which were my favorite part of
Sunday school. It seemed to me that the parables about great floods,
burning bushes, and parting waters were so much more exciting
than Buddhist chanting and meditation. In fact, these ancient tales
of heroism and tragedy vividly illustrated deep moral and ethical
lessons which have stayed with me all my life.

One day in Sunday school we studied Genesis. To read about God
thundering from the heavens, "Let there be Light!" sounded so much
more dramatic than silently meditating about Nirvana. Out of naïve
curiosity, I asked my Sunday school teacher, "Did God have a
mother?" She usually had a snappy answer, as well as a deep moral
lesson to offer. This time, however, she was taken aback. No, she
replied hesitantly, God probably did not have a mother. "But then

where did God come from?" I asked. She mumbled that she would have to consult with the minister about that question.

I didn't realize that I had accidentally stumbled on one of the great questions of theology. I was puzzled, because in Buddhism, there is no God at all, but a timeless universe with no beginning or end. Later, when I began to study the great mythologies of the world, I learned that there were two types of cosmologies in religion, the first based on a single moment when God created the universe, the second based on the idea that the universe always was and always will be.

They couldn't both be right, I thought.

Later, I began to find that these common themes cut across many other cultures. In Chinese mythology, for example, in the beginning there was the cosmic egg. The infant god P'an Ku resided for almost an eternity inside the egg, which floated on a formless sea of Chaos. When it finally hatched, P'an Ku grew enormously, over ten feet per day, so the top half of the eggshell became the sky and the bottom half the earth. After 18,000 years, he died to give birth to our world: his blood became the rivers, his eyes the sun and moon, and his voice the thunder.

In many ways, the P'an Ku myth mirrors a theme found in many other religions and ancient mythologies, that the universe sprang into existence *creatio ex nihilo* (created from nothing). In Greek mythology, the universe started off in a state of Chaos (in fact, the word "chaos" comes from the Greek word meaning "abyss"). This featureless void is often described as an ocean, as in Babylonian and Japanese mythology. This theme is found in ancient Egyptian mythology, where the sun god Ra emerged from a floating egg. In Polynesian mythology, the cosmic egg is replaced by a coconut shell. The Mayans believed in a variation of this story, in which the universe is born but eventually dies after five thousand years, only to be resurrected again and again to repeat the unending cycle of birth and destruction.

These *creatio ex nihilo* myths stand in marked contrast to the cosmology according to Buddhism and certain forms of Hinduism. In these mythologies, the universe is timeless, with no beginning or end. There are many levels of existence, but the highest is Nirvana, which is eternal and can be attained only by the purest meditation. In the Hindu *Mahapurana*, it is written, "If God created the world, where was He before Creation? . . . Know that the world is uncreated, as time itself is, without beginning and end."

These mythologies stand in marked contradiction to each other, with no apparent resolution between them. They are mutually exclusive: either the universe had a beginning or it didn't. There is, apparently, no middle ground.

Today, however, a resolution seems to be emerging from an entirely new direction—the world of science—as the result of a new generation of powerful scientific instruments soaring through outer space. Ancient mythology relied upon the wisdom of storytellers to expound on the origins of our world. Today, scientists are unleashing a battery of space satellites, lasers, gravity wave detectors, interferometers, high-speed supercomputers, and the Internet, in the process revolutionizing our understanding of the universe, and giving us the most compelling description yet of its creation.

What is gradually emerging from the data is a grand synthesis of these two opposing mythologies. Perhaps, scientists speculate, Genesis occurs repeatedly in a timeless ocean of Nirvana. In this new picture, our universe may be compared to a bubble floating in a much larger "ocean," with new bubbles forming all the time. According to this theory, universes, like bubbles forming in boiling water, are in continual creation, floating in a much larger arena, the Nirvana of eleven-dimensional hyperspace. A growing number of physicists suggest that our universe did indeed spring forth from a fiery cataclysm, the big bang, but that it also coexists in an eternal ocean of other universes. If we are right, big bangs are taking place even as you read this sentence.

Physicists and astronomers around the world are now speculating about what these parallel worlds may look like, what laws they may obey, how they are born, and how they may eventually die. Perhaps these parallel worlds are barren, without the basic ingredients of life. Or perhaps they look just like our universe, separated by a single quantum event that made these universes diverge from ours. And a few physicists are speculating that perhaps one day, if life becomes untenable in our present universe as it ages and grows cold, we may be forced to leave it and escape to another universe.

The engine driving these new theories is the massive flood of data that is pouring from our space satellites as they photograph remnants of creation itself. Remarkably, scientists are now zeroing in on what happened a mere 380,000 years after the big bang, when the "afterglow" of creation first filled the universe. Perhaps the most compelling picture of this radiation from creation is coming from a new instrument called the WMAP satellite.

THE WMAP SATELLITE

"Incredible!" "A milestone!" were among the words uttered in February 2003 by normally reserved astrophysicists as they described the precious data harvested from their latest satellite. The WMAP (Wilkinson microwave anisotropy probe), named after pioneering cosmologist David Wilkinson and launched in 2001, has given scientists, with unprecedented precision, a detailed picture of the early universe when it was a mere 380,000 years old. The colossal energy left over from the original fireball that gave birth to stars and galaxies has been circulating around our universe for billions of years. Today, it has finally been captured on film in exquisite detail by the WMAP satellite, yielding a map never seen before, a photo of the sky showing with breathtaking detail the microwave radiation created by the big bang itself, what has been called the "echo of creation" by *Time* magazine. Never again will astronomers look at the sky in the same way again.

The findings of the WMAP satellite represent "a rite of passage for cosmology from speculation to precision science," declared John Bahcall of the Institute for Advanced Study at Princeton. For the first time, this deluge of data from this early period in the history of the universe has allowed cosmologists to answer precisely the most ancient of all questions, questions that have puzzled and intrigued humanity since we first gazed at the blazing celestial beauty of the night sky. How old is the universe? What is it made of? What is the fate of the universe?

(In 1992, a previous satellite, the COBE [Cosmic Background Explorer satellite] gave us the first blurry pictures of this background radiation filling the sky. Although this result was revolutionary, it was also disappointing because it gave such an out-of-focus picture of the early universe. This did not prevent the press from excitedly dubbing this photograph "the face of God." But a more accurate description of the blurry pictures from COBE would be that they represented a "baby picture" of the infant universe. If the universe today is an eighty-year-old man, the COBE, and later the WMAP, pictures showed him as a newborn, less than a day old.)

The reason the WMAP satellite can give us unprecedented pictures of the infant universe is that the night sky is like a time machine. Because light travels at a finite speed, the stars we see at night are seen as they once were, not as they are today. It takes a little over a second for light from the Moon to reach Earth, so when we gaze at the Moon we actually see it as it was a second earlier. It takes about eight minutes for light to travel from the Sun to Earth. Likewise, many of the familiar stars we see in the heavens are so distant that it takes from 10 to 100 years for their light to reach our eyes. (In other words, they lie 10 to 100 light-years from Earth. A light-year is roughly 6 trillion miles, or the distance light travels in a year.) Light from the distant galaxies may be hundreds of millions to billions of light-years away. As a result, they represent "fossil" light, some emitted even before the rise of the dinosaurs. Some of the farthest objects we can see with our telescopes are called quasars, huge galactic engines generating unbelievable amounts of power near the edge of the visible universe, which can lie up to 12 to 13 billion light-years from Earth. And now, the WMAP satellite has detected radiation emitted even before that, from the original fireball that created the universe.

To describe the universe, cosmologists sometimes use the example of looking down from the top of the Empire State Building, which soars more than a hundred floors above Manhattan. As you look down from the top, you can barely see the street level. If the base of the

Empire State Building represents the big bang, then, looking down from the top, the distant galaxies would be located on the tenth floor. The distant quasars seen by Earth telescopes would be on the seventh floor. The cosmic background measured by the WMAP satellite would be just half an inch above the street. And now the WMAP satellite has given us the precise measurement of the age of the universe to an astonishing 1 percent accuracy: 13.7 billion years.

The WMAP mission is the culmination of over a decade of hard work by astrophysicists. The concept of the WMAP satellite was first proposed to NASA in 1995 and was approved two years later. On June 30, 2001, NASA sent the WMAP satellite aboard a Delta II rocket into a solar orbit perched between Earth and the Sun. The destination was carefully chosen to be the Lagrange point 2 (or L2, a special point of relative stability near Earth). From this vantage point, the satellite always points away from the Sun, Earth, and Moon and hence has a totally unobstructed view of the universe. It completely scans the entire sky every six months.

Its instrumentation is state-of-the-art. With its powerful sensors, it can detect the faint microwave radiation left over from the big bang that bathes the universe, but is largely absorbed by our atmosphere. The aluminum-composite satellite measures 3.8 meters by 5 meters (about 11.4 feet by 15 feet) and weighs 840 kilograms (1,850 pounds). It has two back-to-back telescopes that focus the microwave radiation from the surrounding sky, and eventually it radios the data back to Earth. It is powered by just 419 watts of electricity (the power of five ordinary lightbulbs). Sitting a million miles from Earth, the WMAP satellite is well above Earth's atmospheric disturbances, which can mask the faint microwave background, and it is able to get continuous readings of the entire sky.

The satellite completed its first observation of the full sky in April 2002. Six months later, the second full sky observation was made. Today, the WMAP satellite has given us the most comprehensive, detailed map of this radiation ever produced. The background microwave radiation the WMAP detected was first predicted by George Gamow and his group in 1948, who also noted that this radiation has a temperature associated with it. The WMAP measured this temperature to be just above absolute zero, or between 2.7249 to 2.7251 degrees Kelvin.

To the unaided eye, the WMAP map of the sky looks rather uninteresting; it is just a collection of random dots. However, this collection of dots has driven some astronomers almost to tears, for they represent fluctuations or irregularities in the original, fiery cataclysm of the big bang shortly after the universe was created. These tiny fluctuations are like "seeds" that have since expanded enormously as the universe itself exploded outward. Today, these tiny seeds have blossomed into the galactic clusters and galaxies we see lighting up the heavens. In other words, our own Milky Way galaxy and all the galactic clusters we see around us were once one of these tiny fluctuations. By measuring the distribution of these fluctuations, we see the origin of the galactic clusters, like dots painted on the cosmic tapestry that hangs over the night sky.

Today, the volume of astronomical data is outpacing scientists' theories. In fact, I would argue that we are entering a golden age of cosmology. (As impressive as the WMAP satellite is, it will likely be

This is a "baby picture" of the universe, as it was when it was only 380,000 years old, taken by the WMAP satellite. Each dot most likely represents a tiny quantum fluctuation in the afterglow of creation that has expanded to create the galaxies and galactic clusters we see today.

dwarfed by the Planck satellite, which the Europeans are launching in 2007; the Planck will give astronomers even more detailed pictures of this microwave background radiation.) Cosmology today is finally coming of age, emerging from the shadows of science after languishing for years in a morass of speculation and wild conjecture. Historically, cosmologists have suffered from a slightly unsavory reputation. The passion with which they proposed grandiose theories of the universe was matched only by the stunning poverty of their data. As Nobel laureate Lev Landau used to quip, "cosmologists are often in error but never in doubt." The sciences have an old adage: "There's speculation, then there's more speculation, and then there's cosmology."

As a physics major at Harvard in the late 1960s, I briefly toyed with the possibility of studying cosmology. Since childhood, I've always had a fascination with the origin of the universe. However, a quick glance at the field showed that it was embarrassingly primitive. It was not an experimental science at all, where one can test hypotheses with precise instruments, but rather a collection of loose, highly speculative theories. Cosmologists engaged in heated debates about whether the universe was born in a cosmic explosion or whether it has always existed in a steady state. But with so little data, the theories quickly outpaced the data. In fact, the less the data, the fiercer the debate.

Throughout the history of cosmology, this paucity of reliable data also led to bitter, long-standing feuds between astronomers, which often raged for decades. (For example, just before astronomer Allan Sandage of the Mount Wilson Observatory was supposed to give a talk about the age of the universe, the previous speaker announced sarcastically, "What you will hear next is all wrong." And Sandage, hearing of how a rival group had generated a great deal of publicity, would roar, "That's a bunch of hooey. It's war—it's war!")

THE AGE OF THE UNIVERSE

Astronomers have been especially keen to know the age of the universe. For centuries, scholars, priests, and theologians have tried to estimate the age of the universe using the only method at their disposal: the genealogy of humanity since Adam and Eve. In the last century, geologists have used the residual radiation stored in rocks to give the best estimate of the age of Earth. In comparison, the WMAP satellite today has measured the echo of the big bang itself to give us the most authoritative age of the universe. The WMAP data reveals that the universe was born in a fiery explosion that took place 13.7 billion years ago.

(Over the years, one of the most embarrassing facts plaguing cosmology has been that the age of the universe was often computed to be younger than the age of the planets and stars, due to faulty data. Previous estimates for the age of the universe were as low as 1 to 2 billion years, which contradicted the age of Earth [4.5 billion years] and the oldest stars [12 billion years]. These contradictions have now been eliminated.)

The WMAP has added a new, bizarre twist to the debate over what the universe is made of, a question that the Greeks asked over two thousand years ago. For the past century, scientists believed that they knew the answer to this question. After thousands of painstaking experiments, scientists had concluded that the universe was basically made of about a hundred different types of atoms, arranged in an orderly periodic chart, beginning with elemental hydrogen. This forms the basis of modern chemistry and is, in fact, taught in every high school science class. The WMAP has now demolished that belief.

Confirming previous experiments, the WMAP satellite showed that the visible matter we see around us (including the mountains, planets, stars, and galaxies) makes up a paltry 4 percent of the total matter and energy content of the universe. (Of that 4 percent, most of it is in the form of hydrogen and helium, and probably only 0.03 percent takes the form of the heavy elements.) Most of the universe is actually made of mysterious, invisible material of totally unknown origin. The familiar elements that make up our world constitute only 0.03 percent of the universe. In some sense, science is being thrown back centuries into the past, before the rise of the atomic hypothesis, as physicists grapple with the fact that the universe is dominated by entirely new, unknown forms of matter and energy.

According to the WMAP, 23 percent of the universe is made of a strange, undetermined substance called dark matter, which has weight, surrounds the galaxies in a gigantic halo, but is totally invisible. Dark matter is so pervasive and abundant that, in our own Milky Way galaxy, it outweighs all the stars by a factor of 10. Although invisible, this strange dark matter can be observed indirectly by scientists because it bends starlight, just like glass, and hence can be located by the amount of optical distortion it creates.

Referring to the strange results obtained from the WMAP satellite, Princeton astronomer John Bahcall said, "We live in an implausible, crazy universe, but one whose defining characteristics we now know."

But perhaps the greatest surprise from the WMAP data, data that sent the scientific community reeling, was that 73 percent of the universe, by far the largest amount, is made of a totally unknown form of energy called dark energy, or the invisible energy hidden in the vacuum of space. Introduced by Einstein himself in 1917 and then later discarded (he called it his "greatest blunder"), dark energy, or the energy of nothing or empty space, is now re-emerging as the driving force in the entire universe. This dark energy is now believed to create a new antigravity field which is driving the galaxies apart. The ultimate fate of the universe itself will be determined by dark energy.

No one at the present time has any understanding of where this "energy of nothing" comes from. "Frankly, we just don't understand it. We know what its effects are [but] we're completely clueless . . . everybody's clueless about it," admits Craig Hogan, an astronomer at the University of Washington at Seattle.

If we take the latest theory of subatomic particles and try to compute the value of this dark energy, we find a number that is off by 10^{120} (that's the number 1 followed by 120 zeros). This discrepancy between theory and experiment is far and away the largest gap ever found in the history of science. It is one of our greatest embarrassments—our best theory cannot calculate the value of the largest source of energy in the entire universe. Surely, there is a shelf full of Nobel Prizes waiting for the enterprising individuals who can unravel the mystery of dark matter and dark energy.

INFLATION

Astronomers are still trying to wade through this avalanche of data from the WMAP. As it sweeps away older conceptions of the universe, a new cosmological picture is emerging. "We have laid the cornerstone of a unified coherent theory of the cosmos," declares Charles L. Bennett, who led an international team that helped to build and analyze the WMAP satellite. So far, the leading theory is the "inflationary universe theory," a major refinement of the big bang theory, first proposed by physicist Alan Guth of MIT. In the inflationary scenario, in the first trillionth of a trillionth of a second, a mysterious antigravity force caused the universe to expand much faster than originally thought. The inflationary period was unimaginably explosive, with the universe expanding much faster than the speed of light. (This does not violate Einstein's dictum that nothing can travel faster than light, because it is empty space that is expanding. For material objects, the light barrier cannot be broken.) Within a fraction of a second, the universe expanded by an unimaginable factor of 10^{50}.

To visualize the power of this inflationary period, imagine a balloon that is being rapidly inflated, with the galaxies painted on the surface. The universe that we see populated by the stars and galaxies all lies on the surface of this balloon, rather than in the interior. Now draw a microscopic circle on the balloon. This tiny circle represents the visible universe, everything we can see with our telescopes. (By comparison, if the entire visible universe were as small as a subatomic particle, then the actual universe would be much larger than the visible universe that we see around us.) In other words, the inflationary expansion was so intense that there are whole regions of the universe beyond our visible universe that will forever be beyond our reach.

The inflation was so enormous, in fact, that the balloon seems flat in our vicinity, a fact that has been experimentally verified by the WMAP satellite. In the same way that the earth appears flat to us because we are so small compared to the radius of Earth, the universe appears flat only because it is curved on a much larger scale.

By assuming that the early universe underwent this process of inflation, one can almost effortlessly explain many of the puzzles concerning the universe, such as why it appears to be flat and uniform. Commenting on the inflation theory, physicist Joel Primack has said, "No theory as beautiful as this has ever been wrong before."

THE MULTIVERSE

The inflationary universe, although it is consistent with the data from the WMAP satellite, still does not answer the question: what caused inflation? What set off this antigravity force that inflated the universe? There are over fifty proposals explaining what turned on inflation and what eventually terminated it, creating the universe we see around us. But there is no universal consensus. Most physicists rally around the core idea of a rapid inflationary period, but there is no definitive proposal to answer what the engine behind inflation is.

Because no one knows precisely how inflation started, there is always the possibility that the same mechanism can take place again—that inflationary explosions can happen repeatedly. This is the idea proposed by Russian physicist Andrei Linde of Stanford University—that whatever mechanism caused part of the universe to suddenly inflate is still at work, perhaps randomly causing other distant regions of the universe to inflate as well.

According to this theory, a tiny patch of a universe may suddenly inflate and "bud," sprouting a "daughter" universe or "baby" universe, which may in turn bud another baby universe, with this budding process continuing forever. Imagine blowing soap bubbles into the air. If we blow hard enough, we see that some of the soap bubbles split in half and generate new soap bubbles. In the same way, universes may be continually giving birth to new universes. In this scenario, big bangs have been happening continually. If true, we may live in a sea of such universes, like a bubble floating in an ocean of other bubbles. In fact, a better word than "universe" would be "multiverse" or "megaverse."

Linde calls this theory eternal, self-reproducing inflation, or "chaotic inflation," because he envisions a never-ending process of continual inflation of parallel universes. "Inflation pretty much forces the idea of multiple universes upon us," declares Alan Guth, who first proposed the inflation theory.

This theory also means that our universe may, at some time, bud a baby universe of its own. Perhaps our own universe may have gotten its start by budding off from a more ancient, earlier universe.

As the Astronomer Royal of Great Britain, Sir Martin Rees, has said, "What's conventionally called 'the universe' could be just one member of an ensemble. Countless other ways may exist in which the laws are different. The universe in which we've emerged belongs to the unusual subset that permits complexity and consciousness to develop."

All this research activity on the subject of the multiverse has given rise to speculation about what these other universes may look like, whether they harbor life, and even whether it's possible to eventually make contact with them. Calculations have been made by

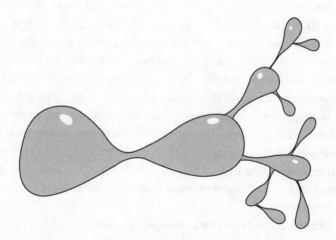

Theoretical evidence is mounting to support the existence of the multiverse, in which entire universes continually sprout or "bud" off other universes. If true, it would unify two of the great religious mythologies, Genesis and Nirvana. Genesis would take place continually within the fabric of timeless Nirvana.

scientists at Cal Tech, MIT, Princeton, and other centers of learning to determine whether entering a parallel universe is consistent with the laws of physics.

M-THEORY AND THE ELEVENTH DIMENSION

The very idea of parallel universes was once viewed with suspicion by scientists as being the province of mystics, charlatans, and cranks. Any scientist daring to work on parallel universes was subject to ridicule and was jeopardizing his or her career, since even today there is no experimental evidence proving their existence.

But recently, the tide has turned dramatically, with the finest minds on the planet working furiously on the subject. The reason for this sudden change is the arrival of a new theory, string theory, and its latest version, M-theory, which promise not only to unravel the nature of the multiverse but also to allow us to "read the Mind of God," as Einstein once eloquently put it. If proved correct, it would represent the crowning achievement of the last two thousand years of research in physics, ever since the Greeks first began the search for a single coherent and comprehensive theory of the universe.

The number of papers published in string theory and M-theory is staggering, amounting to tens of thousands. Hundreds of international conferences have been held on the subject. Every single major university in the world either has a group working on string theory or is desperately trying to learn it. Although the theory is not testable with our feeble present-day instruments, it has sparked enormous interest among physicists, mathematicians, and even experimentalists who hope to test the periphery of the theory in the future with powerful gravity wave detectors in outer space and huge atom smashers.

Ultimately, this theory may answer the question that has dogged cosmologists ever since the big bang theory was first proposed: what happened before the big bang?

This requires us to bring to bear the full force of our physical knowledge, of every physical discovery accumulated over the centuries. In other words, we need a "theory of everything," a theory of every physical force that drives the universe. Einstein spent the last thirty years of his life chasing after this theory, but he ultimately failed.

At present, the leading (and only) theory that can explain the diversity of forces we see guiding the universe is string theory or, in its latest incarnation, M-theory. (M stands for "membrane" but can also mean "mystery," "magic," even "mother." Although string theory and M-theory are essentially identical, M-theory is a more mysterious and more sophisticated framework which unifies various string theories.)

Ever since the Greeks, philosophers have speculated that the ultimate building blocks of matter might be made of tiny particles called atoms. Today, with our powerful atom smashers and particle accelerators, we can break apart the atom itself into electrons and nuclei, which in turn can be broken into even smaller subatomic particles. But instead of finding an elegant and simple framework, it was distressing to find that there were hundreds of subatomic particles streaming from our accelerators, with strange names like neutrinos, quarks, mesons, leptons, hadrons, gluons, W-bosons, and so forth. It is hard to believe that nature, at its most fundamental level, could create a confusing jungle of bizarre subatomic particles.

String theory and M-theory are based on the simple and elegant idea that the bewildering variety of subatomic particles making up the universe are similar to the notes that one can play on a violin string, or on a membrane such as a drum head. (These are no ordinary strings and membranes; they exist in ten- and eleven-dimensional hyperspace.)

Traditionally, physicists viewed electrons as being point particles, which were infinitesimally small. This meant physicists had to introduce a different point particle for each of the hundreds of subatomic particles they found, which was very confusing. But according to string theory, if we had a supermicroscope that could peer into the heart of an electron, we would see that it was not a point particle at all but a tiny vibrating string. It only appeared to be a point particle because our instruments were too crude.

This tiny string, in turn, vibrates at different frequencies and resonances. If we were to pluck this vibrating string, it would change mode and become another subatomic particle, such as a quark. Pluck it again, and it turns into a neutrino. In this way, we can explain the blizzard of subatomic particles as nothing but different musical notes of the string. We can now replace the hundreds of subatomic particles seen in the laboratory with a single object, the string.

In this new vocabulary, the laws of physics, carefully constructed after thousands of years of experimentation, are nothing but the laws of harmony one can write down for strings and membranes. The laws of chemistry are the melodies that one can play on these strings. The universe is a symphony of strings. And the "Mind of God," which Einstein wrote eloquently about, is cosmic music resonating throughout hyperspace. (Which raises another question: If the universe is a symphony of strings, then is there a composer? I address this question in chapter 12.)

MUSICAL ANALOGY	STRING COUNTERPART
Musical notation	Mathematics
Violin strings	Superstrings
Notes	Subatomic particles
Laws of harmony	Physics
Melodies	Chemistry
Universe	Symphony of strings
"Mind of God"	Music resonating through hyperspace
Composer	?

THE END OF THE UNIVERSE

The WMAP not only gives the most accurate glimpse of the early universe, it also gives the most detailed picture of how our universe will die. Just as the mysterious antigravity force pushed the galaxies apart at the beginning of time, this same antigravity force is now pushing the universe to its final fate. Previously, astronomers thought that the expansion of the universe was gradually winding down. Now, we realize that the universe is actually accelerating, with the galaxies hurtling away from us at increasing speed. The same dark energy that makes up 73 percent of the matter and energy in the universe is accelerating the expansion of the universe, pushing the galaxies apart at ever increasing speeds. "The universe is behaving like a driver who slows down approaching a red stoplight and then hits the accelerator when the light turns green," says Adam Riess of the Space Telescope Institute.

Unless something happens to reverse this expansion, within 150 billion years our Milky Way galaxy will become quite lonely, with 99.99999 percent of all the nearby galaxies speeding past the edge of the visible universe. The familiar galaxies in the night sky will be rushing so fast away from us that their light will never reach us. The galaxies themselves will not disappear, but they will be too far for our telescopes to observe them anymore. Although the visible universe contains approximately 100 billion galaxies, in 150 billion years only a few thousand galaxies in the local supercluster of galaxies will be visible. Even further in time, only our local group, consisting of about thirty-six galaxies, will comprise the entire visible universe, with billions of galaxies drifting past the edge of the horizon. (This is because the gravity within the local group is sufficient to overcome this expansion. Ironically, as the distant galaxies slip away from view, any astronomer living in this dark era may fail to detect an expansion in the universe at all, since the local group of galaxies itself does not expand internally. In the far future, astronomers analyzing the night sky for the first time might not realize that there is any expansion and conclude that the universe is static and consists of only thirty-six galaxies.)

If this antigravity force continues, the universe will ultimately die in a big freeze. All intelligent life in the universe will eventually freeze in an agonizing death, as the temperature of deep space plunges toward absolute zero, where the molecules themselves can hardly move. At some point trillions upon trillions of years from now, the stars will cease to shine, their nuclear fires extinguished as they exhaust their fuels, forever darkening the night sky. The cosmic expansion will leave only a cold, dead universe of black dwarf stars, neutron stars, and black holes. And even further into the future, the black holes themselves will evaporate their energy away, leaving a lifeless, cold mist of drifting elementary particles. In such a bleak, cold universe, intelligent life by any conceivable definition is physically impossible. The iron laws of thermodynamics forbid the transfer of any information in such a freezing environment, and all life will necessarily cease.

The first realization that the universe may eventually die in ice was made in the eighteenth century. Commenting on the depressing concept that the laws of physics seemingly doom all intelligent life, Charles Darwin wrote, "Believing as I do that man in the distant future will be a far more perfect creature than he now is, it is an intolerable thought that he and all other sentient beings are doomed to complete annihilation after such long-continued slow progress." Unfortunately, the latest data from the WMAP satellite seem to confirm Darwin's worst fears.

ESCAPE INTO HYPERSPACE

It is a law of physics that intelligent life within the universe will necessarily face this ultimate death. But it is also a law of evolution that when the environment changes, life must either leave, adapt, or die. Because it is impossible to adapt to a universe that is freezing to death, the only options are to die—or to leave the universe itself. When facing the ultimate death of the universe, is it possible that civilizations trillions of years ahead of us will assemble the necessary technology to leave our universe in a dimensional "lifeboat" and drift toward another, much younger and hotter universe? Or will they use their superior technology to build a "time warp" and travel back into their own past, when temperatures were much warmer?

Some physicists have proposed a number of plausible, although extremely speculative schemes, using the most advanced physics available, to provide the most realistic look at dimensional portals or gateways to another universe. The blackboards of physics laboratories around the world are full of abstract equations, as physicists compute whether or not one might use "exotic energy" and black holes to find a passageway to another universe. Can an advanced civilization, perhaps millions to billions of years ahead of ours in technology, exploit the known laws of physics to enter other universes?

Cosmologist Stephen Hawking of Cambridge University once quipped, "Wormholes, if they exist, would be ideal for rapid space travel. You might go through a wormhole to the other side of the galaxy and be back in time for dinner."

And if wormholes and dimensional portals are simply too small to permit the final exodus from the universe, then there is another final option: to reduce the total information content of an advanced, intelligent civilization to the molecular level and inject this through the gateway, where it will then self-assemble on the other side. In this way, an entire civilization may inject its seed through a dimensional gateway and reestablish itself, in its full glory. Hyperspace, instead of being a plaything for theoretical physicists, could potentially become the ultimate salvation for intelligent life in a dying universe.

But to fully understand the implications of this event, we must first understand how cosmologists and physicists have painstakingly arrived at these astounding conclusions. In the course of *Parallel Worlds*, we review the history of cosmology, stressing the paradoxes that have infested the field for centuries, culminating in the theory of inflation, which, while consistent with all the experimental data, forces us to entertain the concept of multiple universes.

CHAPTER TWO

The Paradoxical Universe

Had I been present at the creation, I would have given
some useful hints for the better ordering of the universe.
—Alphonse the Wise

Damn the solar system. Bad light; planets too distant;
pestered with comets; feeble contrivance; could make a
better [universe] myself.

—Lord Jeffrey

IN THE PLAY *As You Like It*, Shakespeare wrote the immortal
words

All the world's a stage,
And all the men and women merely players.
They have their exits and their entrances.

During the Middle Ages, the world was indeed a stage, but it was a
small, static one, consisting of a tiny, flat Earth around which the
heavenly bodies moved mysteriously in their perfect celestial orbs.
Comets were seen as omens foretelling the death of kings. When the
great comet of 1066 sailed over England, it terrified the Saxon sol-
diers of King Harold, who quickly lost to the advancing, victorious

troops of William the Conqueror, setting the stage for the formation of modern England.

That same comet sailed over England once again in 1682, again instilling awe and fear throughout Europe. Everyone, it seemed, from peasants to kings, was mesmerized by this unexpected celestial visitor which swept across the heavens. Where did the comet come from? Where was it going, and what did it mean?

One wealthy gentleman, Edmund Halley, an amateur astronomer, was so intrigued by the comet that he sought out the opinions of one of the greatest scientists of the day, Isaac Newton. When he asked Newton what force might possibly control the motion of the comet, Newton calmly replied that the comet was moving in an ellipse as a consequence of an inverse square force law (that is, the force on the comet diminished with the square of its distance from the sun). In fact, said Newton, he had been tracking the comet with a telescope that he had invented (the reflecting telescope used today by astronomers around the world) and its path was following his law of gravitation that he had developed twenty years earlier.

Halley was shocked beyond belief. "How do you know?" demanded Halley. "Why, I have calculated it," replied Newton. Never in his wildest dreams did Halley expect to hear that the secret of the celestial bodies, which had mystified humanity since the first humans gazed at the heavens, could be explained by a new law of gravity.

Staggered by the significance of this monumental breakthrough, Halley generously offered to pay for the publication of this new theory. In 1687, with Halley's encouragement and funding, Newton published his epic work *Philosophiae Naturalis Principia Mathematica* (*Mathematical Principles of Natural Philosophy*). It has been hailed as one of the most important works ever published. In a single stroke, scientists who were ignorant of the larger laws of the solar system were suddenly able to predict, with pinpoint precision, the motion of heavenly bodies.

So great was the impact of *Principia* in the salons and courts of Europe that the poet Alexander Pope wrote:

Nature and nature's laws lay hid in the night,
God said, Let Newton Be! and all was light.

(Halley realized that if the comet's orbit was an ellipse, one might be able to calculate when it might sail over London again. Searching old records, he found that the comets of 1531, 1607, and 1682 were indeed the same comet. The comet that was so pivotal to the creation of modern England in 1066 was seen by people throughout recorded history, including Julius Caesar. Halley predicted that the comet would return in 1758, long after Newton and Halley had passed away. When the comet did indeed return on Christmas Day that year, on schedule, it was christened Halley's comet.)

Newton had discovered the universal law of gravity twenty years earlier, when the black plague shut down Cambridge University and he was forced to retreat to his country estate at Woolsthorpe. He fondly recalled that while walking around his estate, he saw an apple fall. Then he asked himself a question that would eventually change the course of human history: if an apple falls, does the moon also fall? In a brilliant stroke of genius, Newton realized that apples, the moon, and the planets all obeyed the same law of gravitation, that they were all falling under an inverse square law. When Newton found that the mathematics of the seventeenth century were too primitive to solve this force law, he invented a new branch of mathematics, the calculus, to determine the motion of falling apples and moons.

In *Principia*, Newton had also written down the laws of mechanics, the laws of motion that determine the trajectories of all terrestrial and celestial bodies. These laws laid the basis for designing machines, harnessing steam power, and creating locomotives, which in turn helped pave the way for the Industrial Revolution and modern civilization. Today, every skyscraper, every bridge, and every rocket is constructed using Newton's laws of motion.

Newton not only gave us the eternal laws of motion; he also overturned our worldview, giving us a radically new picture of the universe in which the mysterious laws governing celestial bodies were identical to the laws governing Earth. The stage of life was no longer surrounded by terrifying celestial omens; the same laws that applied to the actors also applied to the set.

BENTLEY'S PARADOX

Because *Principia* was such an ambitious work, it raised the first disturbing paradoxes about the construction of the universe. If the world is a stage, then how big is it? Is it infinite or finite? This is an age-old question; even the Roman philosopher Lucretius was fascinated by it. "The Universe is not bounded in any direction," he wrote. "If it were, it would necessarily have a limit somewhere. But clearly a thing cannot have a limit unless there is something outside to limit it . . . In all dimensions alike, on this side or that, upward or downward throughout the universe, there is no end."

But Newton's theory also revealed the paradoxes inherent in any theory of a finite or infinite universe. The simplest questions lead to a morass of contradictions. Even as Newton was basking in the fame brought to him by the publication of *Principia*, he discovered that his theory of gravity was necessarily riddled with paradoxes. In 1692, a clergyman, Rev. Richard Bentley, wrote a disarmingly simple but distressing letter to Newton. Since gravity was always attractive and never repulsive, wrote Bentley, this meant that any collection of stars would naturally collapse into themselves. If the universe was finite, then the night sky, instead of being eternal and static, should be a scene of incredible carnage, as stars plowed into each other and coalesced into a fiery superstar. But Bentley also pointed out that if the universe were infinite, then the force on any object, tugging it to the left or right, would also be infinite, and therefore the stars should be ripped to shreds in fiery cataclysms.

At first, it seemed as if Bentley had Newton checkmated. Either the universe was finite (and it collapsed into a fireball), or it was infinite (in which case all the stars would be blown apart). Either possibility was a disaster for the young theory being proposed by

Newton. This problem, for the first time in history, revealed the subtle but inherent paradoxes that riddle any theory of gravity when applied to the entire universe.

After careful thought, Newton wrote back that he found a loophole in the argument. He preferred an infinite universe, but one that was totally uniform. Thus, if a star is tugged to the right by an infinite number of stars, this is canceled exactly by an equal tug of another infinite sequence of stars in the other direction. All forces are balanced in each direction, creating a static universe. Thus, if gravity is always attractive, the only solution to Bentley's paradox is to have a uniform, infinite universe.

Newton had indeed found a loophole in Bentley's argument. But Newton was clever enough to realize the weakness of his own response. He admitted in a letter that his solution, although technically correct, was inherently unstable. Newton's uniform but infinite universe was like a house of cards: seemingly stable, but liable to collapse at the slightest disturbance. One could calculate that if even a single star is jiggled by a tiny amount, it would set off a chain reaction, and star clusters would immediately begin to collapse. Newton's feeble response was to appeal to "a divine power" that prevented his house of cards from collapsing. "A continual miracle is needed to prevent the Sun and the fixt stars from rushing together through gravity," he wrote.

To Newton, the universe was like a gigantic clock wound up at the beginning of time by God which has been ticking away ever since, according to his three laws of motion, without Divine interference. But at times, even God himself had to intervene and tweak the universe a bit, to keep it from collapsing. (In other words, occasionally God has to intervene to prevent the sets on the stage of life from collapsing on top of the actors.)

OLBERS' PARADOX

In addition to Bentley's paradox, there was an even deeper paradox inherent in any infinite universe. Olbers' paradox begins by asking why the night sky is black. Astronomers as early as Johannes Kepler realized that if the universe were uniform and infinite, then wherever you looked, you would see the light from an infinite number of stars. Gazing at any point in the night sky, our line of sight will eventually cross an uncountable number of stars and thus receive an infinite amount of starlight. Thus, the night sky should be on fire! The fact that the night sky is black, not white, has been a subtle but profound cosmic paradox for centuries.

Olbers' paradox, like Bentley's paradox, is deceptively simple but has bedeviled many generations of philosophers and astronomers. Both Bentley's and Olbers' paradoxes depend on the observation that, in an infinite universe, gravitational forces and light beams can add to give infinite, meaningless results. Over the centuries, scores of incorrect answers have been proposed. Kepler was so disturbed by this paradox that he simply postulated that the universe was finite, enclosed within a shell, and hence only a finite amount of starlight could ever reach our eyes.

The confusion over this paradox is so great that a 1987 study showed that fully 70 percent of astronomy textbooks gave the incorrect answer.

At first, one might try to solve Olbers' paradox by stating that starlight is absorbed by dust clouds. This was the answer given by Heinrich Wilhelm Olbers himself in 1823 when he first clearly stated the paradox. Olbers wrote, "How fortunate that the Earth does not receive starlight from every point of the celestial vault! Yet, with such unimaginable brightness and heat, amounting to 90,000 times more than what we now experience, the Almighty could easily have designed organisms capable of adapting to such extreme conditions." In order that the earth not be bathed "against a background as brilliant as the Sun's disk," Olbers suggested that dust clouds must absorb the intense heat to make life on earth possible. For example, the fiery center of our own Milky Way galaxy, which should by rights dominate the night sky, is actually hidden behind dust clouds. If we look in the direction of the constellation Sagittarius, where the center of the Milky Way is located, we see not a blazing ball of fire but a patch of darkness.

But dust clouds cannot genuinely explain Olbers' paradox. Over an infinite period of time, the dust clouds will absorb sunlight from an infinite number of stars and eventually will glow like the surface of a star. Thus, even the dust clouds should be blazing in the night sky.

Similarly, one might suppose that the farther a star is, the fainter it is. This is true, but this also cannot be the answer. If we look at a portion of the night sky, the very distant stars are indeed faint, but there are also more stars the farther you look. These two effects would exactly cancel in a uniform universe, leaving the night sky white. (This is because the intensity of starlight decreases as the square of the distance, which is canceled by the fact that the number of stars goes up as the square of the distance.)

Oddly enough, the first person in history to solve the paradox was the American mystery writer Edgar Allan Poe, who had a long-term interest in astronomy. Just before he died, he published many of his observations in a rambling, philosophical poem called *Eureka: A Prose Poem*. In a remarkable passage, he wrote:

> Were the succession of stars endless, then the background of the sky would present us an uniform luminosity, like that displayed by the Galaxy—*since there could be absolutely no point, in all that background, at which would not exist a star*. The only mode, therefore, in which, under such a state of affairs, we could comprehend the *voids* which our telescopes find in innumerable directions, would be by supposing that the distance of the invisible background [is] so immense that no ray from it has yet been able to reach us at all.

He concluded by noting that the idea "is by far too beautiful *not* to possess Truth as its essentiality."

This is the key to the correct answer. The universe is not infinitely old. There was a Genesis. There is a finite cutoff to the light that reaches our eye. Light from the most distant stars has not yet had time to reach us. Cosmologist Edward Harrison, who was the first to discover that Poe had solved Olbers' paradox, has written, "When I first read Poe's words I was astounded: How could a poet, at best an amateur scientist, have perceived the right explanation 140 years ago when in our colleges the wrong explanation . . . is still being taught?"

In 1901, Scottish physicist Lord Kelvin also discovered the correct answer. He realized that when you look at the night sky, you are looking at it as it was in the past, not as it is now, because the speed of light, although enormous by earth standards (186,282 miles per second), is still finite, and it takes time for light to reach Earth from the distant stars. Kelvin calculated that for the night sky to be white, the universe would have to extend hundreds of trillions of light-years. But because the universe is not trillions of years old, the sky is necessarily black. (There is also a second, contributing reason why the night sky is black, and that is the finite lifespan of the stars, which is measured in billions of years.)

Recently, it has become possible to experimentally verify the correctness of Poe's solution, using satellites like the Hubble space telescope. These powerful telescopes, in turn, allow us to answer a question even children ask: Where is the farthest star? And what lies beyond the farthest star? To answer these questions, astronomers programmed the Hubble space telescope to perform a historic task: to take a snapshot of the farthest point in the universe. To capture extremely faint emissions from the deepest corners of space, the telescope had to perform an unprecedented task: to aim at precisely the same point in the sky near the constellation Orion for a total of several hundred hours, which required the telescope to be aligned perfectly for four hundred orbits of Earth. The project was so difficult that it had to be spread out over four months.

In 2004, a stunning photograph was released which made front-page headlines around the world. It showed a collection of ten thousand infant galaxies as they condensed out of the chaos of the big bang itself. "We might have seen the end of the beginning," declared Anton Koekemoer of the Space Telescope Science Institute. The photograph showed a jumble of faint galaxies over 13 billion light-years from Earth—that is, it took over 13 billion years for their light to reach Earth. Since the universe itself is only 13.7 billion years old, this means these galaxies were formed roughly half a billion years after creation, when the first stars and galaxies were condensing out of the "soup" of gases left over from the big bang. "Hubble takes us to within a stone's throw of the big bang itself," said astronomer Massimo Stivavelli of the Institute.

But this raises the question: What lies beyond the farthest galaxies? When peering at this remarkable photograph, what is quite apparent is that there is only blackness between these galaxies. This blackness is what causes the night sky to be black. It is the ultimate cutoff for light from the distant stars. However, this blackness in turn is actually the background microwave radiation. So the final answer to the question of why the night sky is black is that the night sky is not really black at all. (If our eyes could somehow see microwave radiation, and not just visible light, we would see radiation from the big bang itself flooding the night sky. In some sense, radiation from the big bang comes out every night. If we had eyes able to see microwaves, we could see that beyond the farthest star lies creation itself.)

EINSTEIN THE REBEL

Newton's laws were so successful that it took over two hundred years for science to take the next fateful step, with the work of Albert Einstein. Einstein started his career as a most unlikely candidate for such a revolutionary. After he graduated with a bachelor's degree from the Polytechnic Institute in Zurich, Switzerland, in 1900, he found himself hopelessly unemployable. His career was sabotaged by his professors, who disliked this impudent, cocky student who often cut classes. His pleading, depressing letters show the depths to which he descended. He considered himself to be a failure and a painful financial burden on his parents. In one poignant letter, he confessed that he even considered ending his life: "The misfortune of my poor parents, who for so many years have not had a happy moment, weighs most heavily on me . . . I am nothing but a burden to my relatives . . . It would surely be better if I did not live at all," he wrote dejectedly.

In desperation, he thought of switching careers and joining an insurance company. He even took a job tutoring children but got into an argument with his employer and was fired. When his girlfriend, Mileva Maric, unexpectedly became pregnant, he realized sadly that their child would be born illegitimate because he did not have the resources to marry her. (No one knows what eventually happened to his illegitimate daughter, Lieseral.) And the deep, personal shock he felt when his father suddenly died left an emotional scar from which he never fully recovered. His father died thinking his son was a failure.

Although 1901–02 was perhaps the worst period in Einstein's life, what saved his career from oblivion was the recommendation of a classmate, Marcel Grossman, who was able to pull some strings and secure a job for him as a lowly clerk at the Swiss Patent Office in Bern.

PARADOXES OF RELATIVITY

On the surface, the Patent Office was an unlikely place from which to launch the greatest revolution in physics since Newton. But it had its advantages. After quickly disposing of the patent applications piling up on his desk, Einstein would sit back and return to a dream he had when he was a child. In his youth, Einstein had read a book, Aaron Bernstein's *People's Book on Natural Science,* "a work which I read with breathless attention," he recalled. Bernstein asked the reader to imagine riding alongside electricity as it raced down a telegraph wire. When he was sixteen, Einstein asked himself a similar question: what would a light beam look like if you could catch up to it? Einstein would recall, "Such a principle resulted from a paradox upon which I had already hit at the age of sixteen: If I pursue a beam of light with the velocity c (velocity of light in a vacuum), I should observe such a beam of light as a spatially oscillatory electromagnetic field at rest. However, there seems to be no such thing, whether on the basis of experience or according to Maxwell's equations." As a child, Einstein thought that if you could race alongside a light beam, it should appear frozen, like a motionless wave. However, no one had ever seen frozen light, so something was terribly wrong.

At the turn of the century, there were two great pillars of physics upon which everything rested: Newton's theory of mechanics and gravity, and Maxwell's theory of light. In the 1860s, Scottish physicist James Clerk Maxwell had shown that light consists of vibrating electric and magnetic fields constantly changing into each other. What Einstein discovered, much to his shock, was that these two pillars were in contradiction to each other, and that one of them had to fall.

Within Maxwell's equations, he found the solution to the puzzle that had haunted him for ten years. Einstein found something that Maxwell himself had missed: Maxwell's equations showed that light traveled at a constant velocity, no matter how fast you tried to catch up to it. The speed of light c was the same in all inertial frames (that is, frames traveling at constant velocity). Whether you were standing still, riding on a train, or sitting on a speeding comet, you would see a light beam racing ahead of you at the same speed. No matter how fast you moved, you could never outrace light.

This immediately led to a thicket of paradoxes. Imagine, for the moment, an astronaut trying to catch up to a speeding light beam. The astronaut blasts off in his rocket ship until he is racing neck-and-neck with the light beam. A bystander on Earth witnessing this hypothetical chase would claim that the astronaut and the light beam were moving side by side to each other. However, the astronaut would say something completely different, that the light beam sped away from him, just as if his rocket ship were at rest.

The question confronting Einstein was: how can two people have such different interpretations of the same event? In Newton's theory, one could always catch up to a light beam; in Einstein's world, this was impossible. There was, he suddenly realized, a fundamental flaw in the very foundation of physics. In the spring of 1905, Einstein recalled, "a storm broke out in my mind." In one stroke, he finally found the solution: *time beats at different rates, depending on how fast you move.* In fact, the faster you move, the slower time progresses.

Time is not an absolute, as Newton once thought. According to Newton, time beat uniformly throughout the universe, so that the passage of one second on Earth was identical to one second on Jupiter or Mars. Clocks beat in absolute synchronization throughout the universe. To Einstein, however, different clocks beat at different rates throughout the universe.

If time could change depending on your velocity, Einstein realized, then other quantities, such as length, matter, and energy, should also change. He found that the faster you moved, the more distances contracted (which is sometimes called the Lorentz-FitzGerald contraction). Similarly, the faster you moved, the heavier you became. (In fact, as you approached the speed of light, time would slow down to a stop, distances would contract to nothing, and your mass would become infinite, which are all absurd. This is the reason why you cannot break the light barrier, which is the ultimate speed limit in the universe.)

This strange distortion of space-time led one poet to write:

There was a young fellow named Fisk
Whose fencing was exceedingly brisk.
So fast was his action,
The FitzGerald contraction
Reduced his rapier to a disk.

In the same way that Newton's breakthrough unified Earth-bound physics with heavenly physics, Einstein unified space with time. But he also showed that matter and energy are unified and hence can change into each other. If an object becomes heavier the faster it moves, then it means that the energy of motion is being transformed into matter. The reverse is also true—matter can be converted into energy. Einstein computed how much energy would be converted into matter, and he came up with the formula $E = mc^2$, that is, even a tiny amount of matter m is multiplied by a huge number (the square of the speed of light) when it turns into energy E. Thus, the secret energy source of the stars themselves was revealed to be the conversion of matter into energy via this equation, which lights up the universe. The secret of the stars could be derived from the simple statement that the speed of light is the same in all inertial frames.

Like Newton before him, Einstein changed our view of the stage of life. In Newton's world, all the actors knew precisely what time it was and how distances were measured. The beating of time and the dimensions of the stage never changed. But relativity gave us a bizarre way of understanding space and time. In Einstein's universe, all the actors have wristwatches that read different times. This means that it is impossible to synchronize all the watches on the stage. Setting rehearsal time for noon means different things to different actors. In fact, strange things happen when actors race across the stage. The faster they move, the slower their watches beat and the heavier and flatter their bodies become.

It would take years before Einstein's insight would be recognized by the larger scientific community. But Einstein did not stand still; he wanted to apply his new theory of relativity to gravity itself. He realized how difficult this would be; he would be tampering with the most successful theory of his time. Max Planck, founder of the quantum theory, warned him, "As an older friend, I must advise you against it for in the first place you will not succeed, and even if you succeed, no one will believe you."

Einstein realized that his new theory of relativity violated the Newtonian theory of gravity. According to Newton, gravity traveled instantaneously throughout the universe. But this raised a question that even children sometimes ask: "What happens if the Sun disappears?" To Newton, the entire universe would witness the disappearance of the Sun instantly, at the same time. But according to special relativity, this is impossible, since the disappearance of a star was limited by the speed of light. According to relativity, the sudden disappearance of the Sun should set off a spherical shock wave of gravity that spreads outward at the speed of light. Outside the shock wave, observers would say that the Sun is still shining, since gravity has not had time to reach them. But inside the wave, an observer would say that the Sun has disappeared. To resolve this problem, Einstein introduced an entirely different picture of space and time.

FORCE AS THE BENDING OF SPACE

Newton embraced space and time as a vast, empty arena in which events could occur, according to his laws of motion. The stage was full of wonder and mystery, but it was essentially inert and motionless, a passive witness to the dance of nature. Einstein, however, turned this idea upside down. To Einstein, the stage itself would become an important part of life. In Einstein's universe, space and time were not a static arena as Newton had assumed, but were dynamic, bending and curving in strange ways. Assume the stage of life is replaced by a trampoline net, such that the actors gently sink under their own weight. On such an arena, we see that the stage becomes just as important as the actors themselves.

Think of a bowling ball placed on a bed, gently sinking into the mattress. Now shoot a marble along the warped surface of the mattress. It will travel in a curved path, orbiting around the bowling ball. A Newtonian, witnessing the marble circling the bowling ball from a distance, might conclude that there was a mysterious force that the bowling ball exerted on the marble. A Newtonian might say that the bowling ball exerted an instantaneous pull which forced the marble toward the center.

To a relativist, who can watch the motion of the marble on the bed from close up, it is obvious that there is no force at all. There is just the bending of the bed, which forces the marble to move in a curved line. To the relativist, there is no pull, there is only a push, exerted by the curved bed on the marble. Replace the marble with Earth, the bowling ball with the Sun, and the bed with empty spacetime, and we see that Earth moves around the Sun not because of the pull of gravity but because the Sun warps the space around Earth, creating a push that forces Earth to move in a circle.

Einstein was thus led to believe that gravity was more like a fabric than an invisible force that acted instantaneously throughout the universe. If one rapidly shakes this fabric, waves are formed which travel along the surface at a definite speed. This resolves the paradox of the disappearing sun. If gravity is a by-product of the bending of the fabric of space-time itself, then the disappearance of the Sun can be compared to suddenly lifting the bowling ball from the bed. As the bed bounces back to its original shape, waves are sent down the bed sheet traveling at a definite speed. Thus, by reducing gravity to the bending of space and time, Einstein was able to reconcile gravity and relativity.

Imagine an ant trying to walk across a crumpled sheet of paper. He will walk like a drunken sailor, swaying to the left and right, as he tries to walk across the wrinkled terrain. The ant would protest that he is not drunk, but that a mysterious force is tugging on him, yanking him to the left and to the right. To the ant, empty space is full of mysterious forces that prevent him from walking in a straight path. Looking at the ant from a close distance, however, we see that there is no force at all pulling him. He is being pushed by the folds in the crumpled sheet of paper. The forces acting on the ant are an illusion caused by the bending of space itself. The "pull" of the force is actually the "push" created when he walks over a fold in the paper. In other words, gravity does not pull; space pushes.

By 1915, Einstein was finally able to complete what he called the general theory of relativity, which has since become the architecture upon which all of cosmology is based. In this startling new picture, gravity was not an independent force filling the universe but the apparent effect of the bending of the fabric of space-time. His theory was so powerful that he could summarize it in an equation about an inch long. In this brilliant new theory, the amount of bending of space and time was determined by the amount of matter and energy it contained. Think of throwing a rock into a pond, which creates a series of ripples emanating from the impact. The larger the rock, the more the warping of the surface of the pond. Similarly, the larger the star, the more the bending of space-time surrounding the star.

THE BIRTH OF COSMOLOGY

Einstein tried to use this picture to describe the universe as a whole. Unknown to him, he would have to face Bentley's paradox, formulated centuries earlier. In the 1920s, most astronomers believed that the universe was uniform and static. So Einstein started by assuming that the universe was filled uniformly with dust and stars. In one model, the universe could be compared to a large balloon or bubble. We live on the skin of the bubble. The stars and galaxies that we see surrounding us can be compared to dots painted on the surface of the balloon.

To his surprise, whenever he tried to solve his equations, he found that the universe became dynamic. Einstein faced the same problem identified by Bentley over two hundred years earlier. Since gravity is always attractive, never repulsive, a finite collection of stars should collapse into a fiery cataclysm. This, however, contradicted the prevailing wisdom of the early twentieth century, which stated that the universe was static and uniform.

As revolutionary as Einstein was, he could not believe that the universe could be in motion. Like Newton and legions of others, Einstein believed in a static universe. So in 1917, Einstein was forced to introduce a new term into his equations, a "fudge factor" that produced a new force into his theory, an "antigravity" force that pushed the stars apart. Einstein called this the "cosmological constant," an ugly duckling that seemed like an afterthought to Einstein's theory. Einstein then arbitrarily chose this antigravity to cancel precisely the attraction of gravity, creating a static universe. In other words, the universe became static by fiat: the inward contraction of the universe due to gravity was canceled by the outward force of dark energy. (For seventy years, this antigravity force was considered to be something of an orphan, until the discoveries of the last few years.)

In 1917, the Dutch physicist Willem de Sitter produced another solution to Einstein's theory, one in which the universe was infinite but was completely devoid of any matter; in fact, it consisted only of energy contained in the vacuum, the cosmological constant. This pure antigravity force was sufficient to drive a rapid, exponential expansion of the universe. Even without matter, this dark energy could create an expanding universe.

Physicists were now faced with a dilemma. Einstein's universe had matter, but no motion. De Sitter's universe had motion, but no matter. In Einstein's universe, the cosmological constant was necessary to neutralize the attraction of gravity and create a static universe. In de Sitter's universe, the cosmological constant alone was sufficient to create an expanding universe.

Finally, in 1919, when Europe was trying to dig its way out of the rubble and carnage of World War I, teams of astronomers were sent around the world to test Einstein's new theory. Einstein had earlier proposed that the curvature of space-time by the Sun would be suf-

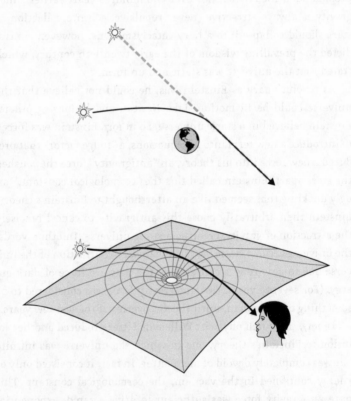

In 1919, two groups confirmed Einstein's prediction that light from a distant star would bend when passing by the Sun. Thus, the position of the star would appear to move from its normal position in the presence of the Sun. This is because the Sun has warped the space-time surrounding it. Thus, gravity does not "pull." Rather, space "pushes."

ficient to bend starlight that is passing in its vicinity. Starlight should bend around the Sun in a precise, calculable way, similar to the way glass bends light. But since the brilliance of Sun's light masks any stars during the day, scientists would have to wait for an eclipse of the Sun to make the decisive experiment.

A group led by British astrophysicist Arthur Eddington sailed to the island of Principe in the Gulf of Guinea off the coast of West Africa to record the bending of starlight around the Sun during the next solar eclipse. Another team, led by Andrew Crommelin, set sail to Sobral in northern Brazil. The data they gathered indicated an average deviation of starlight to be 1.79 arc seconds, which confirmed Einstein's prediction of 1.74 arc seconds (to within experimental error). In other words, light did bend near the Sun. Eddington later claimed that verifying Einstein's theory was the greatest moment in his life.

On November 6, 1919, at a joint meeting of the Royal Society and the Royal Astronomical Society in London, Nobel laureate and Royal Society president J. J. Thompson said solemnly that this was "one of the greatest achievements in the history of human thought. It is not the discovery of an outlying island but of a whole continent of new scientific ideas. It is the greatest discovery in connection with gravitation since Newton enunciated his principles."

(According to legend, Eddington was later asked by a reporter, "There's a rumor that only three people in the entire world understand Einstein's theory. You must be one of them." Eddington stood in silence, so the reporter said, "Don't be modest, Eddington." Eddington shrugged, and said, "Not at all. I was wondering who the third might be.")

The next day, the London *Times* splashed the headline: "Revolution in Science—New Theory of the Universe—Newton's Ideas Overthrown." The headline marked the moment when Einstein became a world-renowned figure, a messenger from the stars.

So great was this announcement, and so radical was Einstein's departure from Newton, that it also caused a backlash, as distinguished physicists and astronomers denounced the theory. At Columbia University, Charles Lane Poor, a professor of celestial mechanics, led the criticism of relativity, saying, "I feel as if I had been wandering with Alice in Wonderland and had tea with the Mad Hatter."

The reason that relativity violates our common sense is not that relativity is wrong, but that our common sense does not represent reality. *We* are the oddballs of the universe. We inhabit an unusual piece of real estate, where temperatures, densities, and velocities are quite mild. However, in the "real universe," temperatures can be blisteringly hot in the center of stars, or numbingly cold in outer space, and subatomic particles zipping through space regularly travel near light-speed. In other words, our common sense evolved in a highly unusual, obscure part of the universe, Earth; it is not surprising that our common sense fails to grasp the true universe. The problem lies not in relativity but in assuming that our common sense represents reality.

THE FUTURE OF THE UNIVERSE

Although Einstein's theory was successful in explaining astronomical phenomena such as the bending of starlight around the Sun and the slight wobbling of the orbit of the planet Mercury, its cosmological predictions were still confusing. Matters were greatly clarified by the Russian physicist Aleksandr Friedmann, who found the most general and realistic solutions of Einstein's equations. Even today, they are taught in every graduate course in general relativity. (He discovered them in 1922, but he died in 1925, and his work was largely forgotten until years later.)

Normally, Einstein's theory consists of a series of extraordinarily difficult equations which often require a computer to solve. However, Friedmann assumed that the universe was dynamic and then made two simplifying assumptions (called the cosmological principle): that the universe is isotropic (it looks the same no matter where we look from a given point), and that the universe is homogeneous (it is uniform no matter where you go in the universe).

Under these two simplifying assumptions, we find that these

equations collapse. (In fact, both Einstein's and de Sitter's solutions were special cases of Friedmann's more general solution.) Remarkably, his solutions depend on just three parameters:

1. *H*, which determines the rate of expansion of the universe. (Today, this is called Hubble's constant, named after the astronomer who actually measured the expansion of the universe.)
2. *Omega*, which measures the average density of matter in the universe.
3. *Lambda*, the energy associated with empty space, or dark energy.

Many cosmologists have spent their entire professional careers trying to nail down the precise value of these three numbers. The subtle interplay between these three constants determines the future evolution of the entire universe. For example, since gravity attracts, the density of the universe Omega acts as a kind of brake, to slow the expansion of the universe, reversing some of the effects of the big bang's rate of expansion. Think of throwing a rock into the air. Normally, gravity is strong enough to reverse the direction of the rock, which then tumbles back to Earth. However, if one throws the rock fast enough, then it can escape Earth's gravity and soar into outer space forever. Like a rock, the universe originally expanded because of the big bang, but matter, or Omega, acts as a brake on the expansion of the universe, in the same way that Earth's gravity acts as a brake on the rock.

For the moment, let's assume that Lambda, the energy associated with empty space, equals zero. Let Omega be the density of the universe divided by the critical density. (The critical density of the universe is approximately 10 hydrogen atoms per cubic meter. To appreciate how empty the universe is, the critical density of the universe corresponds to finding a single hydrogen atom within the volume of three basketballs, on average.)

If Omega is less than 1, scientists conclude that there is not enough matter in the universe to reverse the original expansion from the big bang. (Like throwing the rock in the air, if Earth's mass is not great enough, the rock will eventually leave Earth.) As a result, the universe will expand forever, eventually plunging the universe into a big freeze until temperatures approach absolute zero. (This is the principle behind a refrigerator or air conditioner. When gas expands, it cools down. In your air conditioner, for example, gas circulating in a pipe expands, cooling the pipe and your room.)

If Omega is greater than 1, then there is sufficient matter and gravity in the universe to ultimately reverse the cosmic expansion. As a result, the expansion of the universe will come to a halt, and the universe will begin to contract. (Like the rock thrown in the air, if Earth's mass is great enough, the rock will eventually reach a maximum height and then come tumbling back to Earth.) Temperatures will begin to soar, as the stars and galaxies rush toward each other. (Anyone who has ever inflated a bicycle tire knows that the compression of gas creates heat. The mechanical work of pumping air is converted into heat energy. In the same way, the compression of the universe converts gravitational energy into heat energy.) Eventually, temperatures would become so hot that all life would be extinguished,

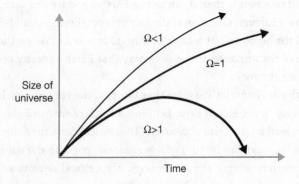

The evolution of the universe has three possible histories. If Omega is less than 1 (and Lambda is 0), the universe will expand forever into the big freeze. If Omega is greater than 1, the universe will recollapse into the big crunch. If Omega is equal to 1, then the universe is flat and will expand forever. (The WMAP satellite data shows that Omega plus Lambda is equal to 1, meaning that the universe is flat. This is consistent with the inflationary theory.)

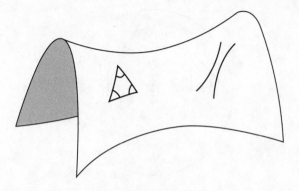

If the Omega is less than 1 (and Lambda is 0), then the universe is open and its curvature is negative, as in a saddle. Parallel lines never meet, and the interior angles of triangles sum to less than 180 degrees.

as the universe heads toward a fiery "big crunch." (Astronomer Ken Croswell labels this process "from Creation to Cremation.")

A third possibility is that Omega is perched precisely at 1; in other words, the density of the universe equals the critical density, in which case the universe hovers between the two extremes but will still expand forever. (This scenario, we will see, is favored by the inflationary picture.)

And last, there is the possibility that the universe, in the aftermath of a big crunch, can reemerge into a new big bang. This theory is referred to as the oscillating universe.

Friedmann showed that each of these scenarios, in turn, determines the curvature of space-time. If Omega is less than 1 and the universe expands forever, Friedmann showed that not only is time infinite, but space is infinite as well. The universe is said to be "open," that is, infinite in both space and time. When Friedmann computed the curvature of this universe, he found it to be negative. (This is like the surface of a saddle or a trumpet. If a bug lived on the surface of this surface, it would find that parallel lines never meet, and the interior angles of a triangle sum up to less than 180 degrees.)

If Omega is larger than 1, then the universe will eventually con-

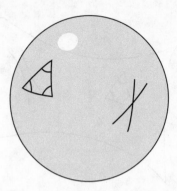

If Omega is greater than 1, then the universe is closed and its curvature is positive, like in a sphere. Parallel lines always meet, and the angles of a triangle sum to greater than 180 degrees.

tract into a big crunch. Time and space are finite. Friedmann found that the curvature of this universe is positive (like a sphere). Finally, if Omega equals 1, then space is flat and both time and space are unbounded.

Not only did Friedmann provide the first comprehensive approach to Einstein's cosmological equations, he also gave the most realistic conjecture about Doomsday, the ultimate fate of the universe—whether it will perish in a big freeze, fry in a big crunch, or oscillate forever. The answer depends upon the crucial parameters: the density of the universe and the energy of the vacuum.

But Friedmann's picture left a gaping hole. If the universe is expanding, then it means that it might have had a beginning. Einstein's theory said nothing about the instant of this beginning. What was missing was the moment of creation, the big bang. And three scientists would eventually give us a most compelling picture of the big bang.

CHAPTER THREE

The Big Bang

The universe is not only queerer than we suppose, it is queerer than we can suppose.

—J. B. S. Haldane

What we humans are looking for in a creation story is a way of experiencing the world that will open to us the transcendent, that informs us and at the same time forms ourselves within it. That is what people want. This is what the soul asks for.

—Joseph Campbell

THE COVER OF *Time* magazine on March 6, 1995, showing the great spiral galaxy M100, claimed "Cosmology is in chaos." Cosmology was being thrown into turmoil because the latest data from the Hubble space telescope seemed to indicate that the universe was younger than its oldest star, a scientific impossibility. The data indicated that the universe was between 8 billion and 12 billion years old, while some believed the oldest star to be as much as 14 billion years old. "You can't be older than your ma," quipped Christopher Impey of the University of Arizona.

But once you read the fine print, you realized that the theory of the big bang is quite healthy. The evidence disproving the big bang theory was based on a single galaxy, M100, which is a dubious way of conducting science. The loopholes were, as the article acknowledged, "big enough to drive the Starship Enterprise through." Based on the Hubble space telescope's rough data, the age of the universe could not be calculated to better than 10 to 20 percent accuracy.

My point is that the big bang theory is not based on speculation but on hundreds of data points taken from several different sources, each of which converge to support a single, self-consistent theory. (In science, not all theories are created equal. While anyone is free to propose their own version of the creation of the universe, it should be required that it explain the hundreds of data points we have collected that are consistent with the big bang theory.)

The three great "proofs" of the big bang theory are based on the work of three larger-than-life scientists who dominated their respective fields: Edwin Hubble, George Gamow, and Fred Hoyle.

EDWIN HUBBLE, PATRICIAN ASTRONOMER

While the theoretical foundation of cosmology was laid by Einstein, modern observational cosmology was almost single-handedly created by Edwin Hubble, who was perhaps the most important astronomer of the twentieth century.

Born in 1889 in the backwoods of Marshfield, Missouri, Hubble was a modest country boy with high ambitions. His father, a lawyer and insurance agent, urged him to pursue a career in law. Hubble, however, was enthralled by the books of Jules Verne and enchanted by the stars. He devoured science fiction classics like *Twenty Thousand Leagues Under the Sea* and *From the Earth to the Moon*. He was also an accomplished boxer; promoters wanted him to turn professional and fight the world heavyweight champion, Jack Johnson.

He won a prestigious Rhodes scholarship to study law at Oxford, where he began to adopt the mannerisms of British upper-crust society. (He started wearing tweed suits, smoking a pipe, adopting a distinguished British accent, and speaking of his dueling scars, which were rumored to be self-inflicted.)

Hubble, however, was unhappy. What really motivated him was

not torts and lawsuits; his romance was with the stars, one that had started when he was a child. He bravely switched careers and headed for the University of Chicago and the observatory at Mount Wilson, California, which then housed the largest telescope on Earth, with a 100-inch mirror. Starting so late in his career, Hubble was a man in a hurry. To make up for lost time, he rapidly set out to answer some of the deepest, most enduring mysteries in astronomy.

In the 1920s, the universe was a comfortable place; it was widely believed that the entire universe consisted of just the Milky Way galaxy, the hazy swath of light that cuts across the night sky resembling spilt milk. (The word "galaxy," in fact, comes from the Greek word for milk.) In 1920, the "Great Debate" took place between astronomers Harlow Shapley of Harvard and Heber Curtis of Lick Observatory. Entitled "The Scale of the Universe," it concerned the size of the Milky Way galaxy and the universe itself. Shapley took the position that the Milky Way made up the entire visible universe. Curtis believed that beyond the Milky Way lay the "spiral nebulae," strange but beautiful wisps of swirling haze. (As early as the 1700s, the philosopher Immanuel Kant had speculated that these nebulae were "island universes.")

Hubble was intrigued by the debate. The key problem was that determining the distance to the stars is (and still remains) one of the most fiendishly difficult tasks in astronomy. A bright star that is very distant can look identical to a dim star that is close by. This confusion was the source of many great feuds and controversies in astronomy. Hubble needed a "standard candle," an object that emits the same amount of light anywhere in the universe, to resolve the problem. (A large part, in fact, of the effort in cosmology to this day consists of attempting to find and calibrate such standard candles. Many of the great debates in astronomy center around how reliable these standard candles really are.) If one had a standard candle that burned uniformly with the same intensity throughout the universe, then a star that was four times dimmer than normal would simply be twice as far from Earth.

One night, when analyzing a photograph of the spiral nebula Andromeda, Hubble had a eureka moment. What he found within

Andromeda was a type of variable star (called a Cepheid) which had been studied by Henrietta Leavitt. It was known that this star regularly grew and dimmed with time, and the time for one complete cycle was correlated with its brightness. The brighter the star, the longer its cycle of pulsation. Thus, by simply measuring the length of this cycle, one could calibrate its brightness and hence determine its distance. Hubble found that it had a period of 31.4 days, which, much to his surprise, translated to a distance of a million light-years, far outside the Milky Way galaxy. (The Milky Way's luminous disk is only 100,000 light-years across. Later calculations would show that Hubble in fact underestimated the true distance to Andromeda, which is closer to 2 million light-years away.)

When he performed the same experiment on other spiral nebulae, Hubble found that they too were well outside the Milky Way galaxy. In other words, it was clear to him that these spiral nebulae were entire island universes in their own right—that the Milky Way galaxy was just one galaxy in a firmament of galaxies.

In one stroke, the size of the universe became vastly larger. From a single galaxy, the universe was suddenly populated with millions, perhaps billions, of sister galaxies. From a universe just 100,000 light-years across, the universe suddenly was perhaps billions of light-years across.

That discovery alone would have guaranteed Hubble a place in the pantheon of astronomers. But he topped even that discovery. Not only was he determined to find the distance to the galaxies, he wanted to calculate how fast they moved, as well.

DOPPLER EFFECT AND THE EXPANDING UNIVERSE

Hubble knew that the simplest way of calculating the speed of distant objects is to analyze the change in sound or light they emit, otherwise known as the Doppler effect. Cars make this sound as they pass us on the highway. Police use the Doppler effect to calculate your speed; they flash a laser beam onto your car, which reflects back to the police car. By analyzing the shift in frequency of the laser light, the police can calculate your velocity.

If a star, for example, is moving toward you, the light waves it emits are squeezed like an accordion. As a result, its wavelength gets shorter. A yellow star will appear slightly bluish (because the color blue has a shorter wavelength than yellow). Similarly, if a star is moving away from you, its light waves are stretched, giving it a longer wavelength, so that a yellow star appears slightly reddish. The greater the distortion, the greater the velocity of the star. Thus, if we know the shift in frequency of starlight, we can determine the star's speed.

In 1912, astronomer Vesto Slipher had found that the galaxies were moving away from Earth at great velocity. Not only was the universe much larger than previously expected, it was also expanding and at great speed. Outside of small fluctuations, he found that the galaxies exhibited a redshift, caused by galaxies moving away from us, rather than a blue one. Slipher's discovery showed that the universe was indeed dynamic and not static, as Newton and Einstein had assumed.

In all the centuries that scientists had studied the paradoxes of Bentley and Olbers, no one had seriously considered the possibility that the universe was expanding. In 1928, Hubble made a fateful trip to Holland to meet with Willem de Sitter. What intrigued Hubble was de Sitter's prediction that the farther away a galaxy is, the faster it should be moving. Think of an expanding balloon with galaxies marked on its surface. As the balloon expands, the galaxies that are close to each other move apart relatively slowly. The closer they are to each other, the slower they move apart. But galaxies that are farther apart on the balloon move apart much faster.

De Sitter urged Hubble to look for this effect in his data, which could be verified by analyzing the redshift of the galaxies. The greater the redshift of a galaxy, the faster it was moving away, and hence the farther it should be. (According to Einstein's theory, the redshift of a galaxy was not, technically speaking, caused by the galaxy speeding away from Earth; instead, it was caused by the expansion of space itself between the galaxy and Earth. The origin of the redshift is that light emanating from a distant galaxy is stretched or lengthened by the expansion of space, and hence it appears reddened.)

HUBBLE'S LAW

When Hubble went back to California, he heeded de Sitter's advice and looked for evidence of this effect. By analyzing twenty-four galaxies, he found that the farther the galaxy was, the faster it was moving away from Earth, just as Einstein's equations had predicted. The ratio between the two (speed divided by distance) was roughly a constant. It quickly became known as Hubble's constant, or H. It is perhaps the single most important constant in all of cosmology, because Hubble's constant tells you the rate at which the universe is expanding.

If the universe is expanding, scientists pondered, then perhaps it had a beginning, as well. The inverse of the Hubble constant, in fact, gives a rough calculation of the age of the universe. Imagine a videotape of an explosion. In the videotape, we see the debris leaving the site of the explosion and can calculate the velocity of expansion. But this also means that we can run the videotape backward, until all the debris collects into a single point. Since we know the velocity of expansion, we can roughly work backward and calculate the time at which the explosion took place.

(Hubble's original estimate put the age of the universe at about 1.8 billion years, which gave generations of cosmologists headaches because that was younger than the reputed age of Earth and the stars. Years later, astronomers realized that errors in measuring the light from the Cepheid variables in Andromeda had given an incorrect value of Hubble's constant. In fact, the "Hubble wars" concerning the precise value of the Hubble constant have raged for the past seventy years. The most definitive figure today comes from the WMAP satellite.)

In 1931, on Einstein's triumphant visit to the Mount Wilson

Observatory, he first met Hubble. Realizing that the universe was indeed expanding, he called the cosmological constant his "biggest blunder." (However, even a blunder by Einstein is enough to shake the foundations of cosmology, as we will see in discussing the WMAP satellite data in later chapters.) When Einstein's wife was shown around the mammoth observatory, she was told that the gigantic telescope was determining the ultimate shape of the universe. Mrs. Einstein replied nonchalantly, "My husband does that on the back of an old envelope."

THE BIG BANG

A Belgian priest, Georges Lemaître, who learned of Einstein's theory, was fascinated by the idea that the theory logically led to a universe that was expanding and therefore had a beginning. Because gases heat up as they are compressed, he realized that the universe at the beginning of time must have been fantastically hot. In 1927, he stated that the universe must have started out as a "superatom" of incredible temperature and density, which suddenly exploded outward, giving rise to Hubble's expanding universe. He wrote, "The evolution of the world can be compared to a display of fireworks that has just ended: some few red wisps, ashes and smoke. Standing on a well-chilled cinder, we see the slow fading of the suns, and we try to recall the vanished brilliance of the origin of worlds."

(The first person to propose this idea of a "superatom" at the beginning of time was, once again, Edgar Allan Poe. He argued that matter attracts other forms of matter, therefore at the beginning of time there must have been a cosmic concentration of atoms.)

Lemaître would attend physics conferences and pester other scientists with his idea. They would listen to him with good humor and then quietly dismiss his idea. Arthur Eddington, one of the leading physicists of his time, said, "As a scientist, I simply do not believe that the present order of things started off with a bang . . . The notion of an abrupt beginning to this present order of Nature is repugnant to me."

But, over the years, his persistence gradually wore down the resistance of the physics community. The scientist who would become the most important spokesman and popularizer of the big bang theory would eventually provide the most convincing proof of the theory.

GEORGE GAMOW, COSMIC JESTER

While Hubble was the sophisticated patrician of astronomy, his work was continued by yet another larger-than-life figure, George Gamow. Gamow was in many respects his opposite: a jester, a cartoonist, famous for his practical jokes and his twenty books on science, many of them for young adults. Several generations of physicists (myself included) were raised on his entertaining and informative books about physics and cosmology. In a time when relativity and the quantum theory were revolutionizing science and society, his books stood alone: they were the only credible books on advanced science available to teenagers.

While lesser scientists are often barren of ideas, content to merely grind through mountains of dry data, Gamow was one of the creative geniuses of his time, a polymath who rapidly spun off ideas that would change the course of nuclear physics, cosmology, and even DNA research. It was perhaps no accident that the autobiography of James Watson, who with Francis Crick unraveled the secret of the DNA molecule, was titled *Genes, Gamow, and Girls*. As his colleague Edward Teller recalled, "Ninety percent of Gamow's theories were wrong, and it was easy to recognize that they were wrong. But he didn't mind. He was one of those people who had no particular pride in any of his inventions. He would throw out his latest idea and then treat it as a joke." But the remaining 10 percent of his ideas would go on to change the entire scientific landscape.

Gamow was born in Odessa, Russia, in 1904, during that country's early social upheavals. Gamow recalled that "classes were often suspended when Odessa was bombarded by some enemy warship, or when Greek, French, or British expeditionary forces staged a bayonet attack along the main streets of the city against entrenched, White, Red, or even green Russian forces, or when Russian forces of different colors fought one another."

The turning point in his early life came when he went to church and secretly took home some communion bread after the service. Looking through a microscope, he could see no difference between the communion bread, representing the flesh of Jesus Christ, and ordinary bread. He concluded, "I think this was the experiment which made me a scientist."

He was educated at the University of Leningrad and studied under physicist Aleksandr Friedmann. Later, at the University of Copenhagen, he met many of the giants of physics, like Niels Bohr. (In 1932, he and his wife tried unsuccessfully to defect from the Soviet Union by sailing on a raft from the Crimean to Turkey. Later, he succeeded in defecting while attending a physics conference in Brussels, which earned him a death sentence from the Soviets.)

Gamow was famous for sending limericks to his friends. Most are unprintable, but one limerick captures the anxieties cosmologists feel when they face the enormity of astronomical numbers and stare infinity in the face:

> There was a young fellow from Trinity
> Who took the square root of infinity
> But the number of digits
> Gave him the fidgits;
> He dropped Math and took up Divinity.

In the 1920s in Russia, Gamow scored his first big success when he solved the mystery of why radioactive decay was possible. Thanks to the work of Madame Curie and others, scientists knew that the uranium atom was unstable and emitted radiation in the form of an alpha ray (the nucleus of a helium atom). But according to Newtonian mechanics, the mysterious nuclear force that held the nucleus together should have been a barrier that prevented this leakage. How was this possible?

Gamow (and R. W. Gurney and E. U. Condon) realized that radioactive decay was possible because in the quantum theory, the uncertainty principle meant that one never knew precisely the location and velocity of a particle; hence there was a small probability that it might "tunnel" or penetrate right through a barrier. (Today, this idea of tunneling is central to all of physics and is used to explain the properties of electronic devices, black holes, and the big bang. The universe itself might have been created via tunneling.)

By analogy, Gamow envisioned a prisoner sealed in a jail, surrounded by huge prison walls. In a classical Newtonian world, escape is impossible. But in the strange world of the quantum theory, you don't know precisely where the prisoner is at any point or his velocity. If the prisoner bangs against the prison walls often enough, you can calculate the chances that one day he will pass right through them, in direct violation of common sense and Newtonian mechanics. There is a finite, calculable probability that he will be found outside the gates of the prison walls. For large objects like prisoners, you would have to wait longer than the lifetime of the universe for this miraculous event to happen. But for alpha particles and subatomic particles, it happens all the time, because these particles hit against the walls of the nucleus repeatedly with vast amounts of energy. Many feel that Gamow should have been given the Nobel Prize for this vitally important work.

In the 1940s, Gamow's interests began to shift from relativity to cosmology, which he viewed as a rich, undiscovered country. All that was known about the universe at that time was that the sky was black and that the universe was expanding. Gamow was guided by a single idea: to find any evidence or "fossils" proving that there was a big bang billions of years ago. This was frustrating, because cosmology is not an experimental science in the true sense of the word. There are no experiments one can conduct on the big bang. Cosmology is more like a detective story, an observational science where you look for "relics" or evidence at the scene of the crime, rather than an experimental science where you can perform precise experiments.

NUCLEAR KITCHEN OF THE UNIVERSE

Gamow's next great contribution to science was his discovery of the nuclear reactions that gave birth to the lightest elements that we see in the universe. He liked to call it the "prehistoric kitchen of the universe," where all the elements of the universe were originally cooked by the intense heat of the big bang. Today, this process is called "nucleosynthesis," or calculating the relative abundances of the elements in the universe. Gamow's idea was that there was an unbroken chain, starting with hydrogen, that could be built by simply adding successively more particles to the hydrogen atom. The entire Mendeleev periodic chart of the chemical elements, he believed, could be created from the heat of the big bang.

Gamow and his students reasoned that because the universe was an incredibly hot collection of protons and neutrons at the instant of creation, then perhaps fusion took place, with hydrogen atoms being fused together to produce helium atoms. As in a hydrogen bomb or a star, the temperatures are so hot that the protons of a hydrogen atom are smashed into each other until they merge, creating helium nuclei. Subsequent collisions between hydrogen and helium would, according to this scenario, produce the next set of elements, including lithium and beryllium. Gamow assumed that the higher elements could be sequentially built up by adding more and more subatomic particles to the nucleus—in other words, that all of the hundred or so elements that make up the visible universe were "cooked" in the fiery heat of the original fireball.

In typical fashion, Gamow laid out the broad outlines of this ambitious program and let his Ph.D. student Ralph Alpher fill in the details. When the paper was finished, he couldn't resist a practical joke. He put physicist Hans Bethe's name on the paper without his permission, and it became the celebrated alpha-beta-gamma paper.

What Gamow had found was that the big bang indeed was hot enough to create helium, which makes up about 25 percent of the universe, by mass. Working in reverse, one "proof" of the big bang can be found by simply looking at many of the stars and galaxies of today and realizing that they are made of approximately 75 percent hydrogen, 25 percent helium, and a few trace elements. (As David Spergel, an astrophysicist at Princeton, has said, "Every time you buy a balloon, you are getting atoms [some of which] were made in the first few minutes of the big bang.")

However, Gamow also found problems with the calculation. His theory worked well for the very light elements. But elements with 5 and 8 neutrons and protons are extremely unstable and hence cannot act as a "bridge" to create elements that have a greater number of protons and neutrons. The bridge was washed out at 5 and 8 particles. Since the universe is composed of heavy elements with a great many more than 5 and 8 neutrons and protons, this left a cosmic mystery. The failure of Gamow's program to extend beyond the 5-particle and 8-particle gap remained a stubborn problem for years, dooming his vision of showing that all the elements of the universe were created at the moment of the big bang.

MICROWAVE BACKGROUND RADIATION

At the same time, another idea intrigued him: if the big bang was so incredibly hot, perhaps some of its residual heat is still circulating around the universe today. If so, it would give a "fossil record" of the big bang itself. Perhaps the big bang was so colossal that its aftershocks are still filling up the universe with a uniform haze of radiation.

In 1946, Gamow assumed that the big bang began with a superhot core of neutrons. This was a reasonable assumption, since very little was known about subatomic particles other than the electron, proton, and neutron. If he could estimate the temperature of this ball of neutrons, he realized he could calculate the amount and nature of radiation that it emitted. Two years later, Gamow showed that radiation given off by this superhot core would act like "black body radiation." This is a very specific type of radiation given off by a hot object; it absorbs all light hitting it, emitting radiation back in a characteristic way. For example, the Sun, molten lava, hot coals in a fire, and hot ceramics in an oven all glow yellow-red and emit black body radiation. (Black body radiation was first discovered by the famed maker of porcelain, Thomas Wedgwood, in 1792. He noticed that when raw materials were baked in his ovens, they changed in color from red to yellow to white, as he raised the temperature.)

This is important because once one knows the color of a hot object, one also knows roughly its temperature, and vice versa; the precise formula relating the temperature of a hot object and the radiation it emits was first obtained by Max Planck in 1900, which led to the birth of the quantum theory. (This is, in fact, one way in which scientists determine the temperature of the Sun. The Sun radiates mainly yellow light, which in turn corresponds to a black body temperature of roughly 6,000 K. Thus we know the temperature of the Sun's outer atmosphere. Similarly, the red giant star Betelgeuse has a surface temperature of 3,000 K, the black body temperature corresponding to the color red, which is also emitted by a red-hot piece of coal.)

Gamow's 1948 paper was the first time anyone had suggested that the radiation of the big bang might have a specific characteristic—black body radiation. The most important characteristic of black body radiation is its temperature. Next, Gamow had to compute the current temperature of black body radiation.

Gamow's Ph.D. student Ralph Alpher and another student, Robert Herman, tried to complete Gamow's calculation by computing its temperature. Gamow wrote, "Extrapolating from the early days of the universe to the present time, we found that during the eons which had passed, the universe must have cooled to about 5 degrees above the absolute temperature."

In 1948, Alpher and Herman published a paper giving detailed arguments why the temperature of the afterglow of the big bang today should be 5 degrees above absolute zero (their estimate was remarkably close to what we now know is the correct temperature of 2.7 degrees above zero). This radiation, which they identified as being in the microwave range, should still be circulating around the universe today, they postulated, filling up the cosmos with a uniform afterglow.

(The reasoning is as follows. For years after the big bang, the temperature of the universe was so hot that anytime an atom formed, it would be ripped apart; hence there were many free electrons that could scatter light. Thus, the universe was opaque, not transparent. Any light beam moving in this super-hot universe would be absorbed after traveling a short distance, so the universe looked cloudy. After 380,000 years, however, the temperature dropped to 3,000 degrees. Below that temperature, atoms were no longer ripped apart by collisions. As a result, stable atoms could form, and light beams could now travel for light-years without being absorbed. Thus, for the first time, empty space became transparent. This radiation, which was no longer instantly absorbed as soon as it was created, is circulating around the universe today.)

When Alpher and Herman showed Gamow their final calculation of the temperature of the universe, Gamow was disappointed. The temperature was so cold that it would be extremely difficult to measure. It took Gamow a year to finally agree that the details of their calculation were correct. But he despaired of ever being able to measure such a faint radiation field. Instruments available in the 1940s were hopelessly inadequate to measure this faint echo. (In a later calculation, using an incorrect assumption, Gamow pushed the temperature of the radiation up to 50 degrees.)

They gave a series of talks to publicize their work. But unfortunately, their prophetic result was ignored. Alpher has said, "We expended a hell of a lot of energy giving talks about the work. Nobody bit; nobody said it could be measured . . . And so over the period 1948 to 1955, we sort of gave up."

Undaunted, Gamow, via his books and lectures, became the leading personality pushing the big bang theory. But he met his match in a fierce adversary very much his equal. While Gamow could charm his audience with his impish jokes and witticisms, Fred Hoyle could overpower audiences with his sheer brilliance and aggressive audacity.

FRED HOYLE, CONTRARIAN

The microwave background radiation gives us the "second proof" of the big bang. But the man least likely to provide the third great proof of the big bang via nucleosynthesis was Fred Hoyle, a man who ironically spent almost his entire professional life trying to disprove the big bang theory.

Hoyle was the personification of an academic misfit, a brilliant contrarian who dared to defy conventional wisdom with his some-times pugnacious style. While Hubble was the ultimate patrician, emulating the mannerisms of an Oxford don, and Gamow was the entertaining jester and polymath who could dazzle audiences with his quips, limericks, and pranks, Hoyle's style resembled that of a rough-hewn bulldog; he seemed strangely out of place in the ancient halls of Cambridge University, the old haunt of Isaac Newton.

Hoyle was born in 1915 in northern England, the son of a textile merchant, in an area dominated by the wool industry. As a child, he was excited by science; radio was just coming to the village, and, he recalled, twenty to thirty people eagerly wired up their homes with radio receivers. But the turning point in his life came when his parents gave him a telescope for a present.

Hoyle's combative style started when he was a child. He had mas-tered the multiplication tables at age three, and then his teacher asked him to learn Roman numerals. "How could anybody be so daft as to write VIII for 8?" he recalled scornfully. But when he was told that the law required him to attend school, he wrote, "I concluded that, unhappily, I'd been born into a world dominated by a rampag-ing monster called 'law' that was both all-powerful and all-stupid."

His disdain for authority was also cemented by a run-in with an-other teacher, who told the class that a particular flower had five petals. Proving her wrong, he brought the flower with six petals into class. For that impudent act of insubordination, she whacked him hard in his left ear. (Hoyle later became deaf in that ear.)

STEADY STATE THEORY

In the 1940s, Hoyle was not enamored of the big bang theory. One de-fect in the theory was that Hubble, because of errors in measuring light from distant galaxies, had miscalculated the age of the universe to be 1.8 billion years. Geologists claimed that Earth and the solar system were probably many billions of years old. How could the universe be younger than its planets?

With colleagues Thomas Gold and Hermann Bondi, Hoyle set out to construct a rival to the theory. Legend has it that their theory, the steady state theory, was inspired by a 1945 ghost movie called *Dead of Night*, starring Michael Redgrave. The movie consists of a series of ghost stories, but in the final scene there is a memorable twist: the movie ends just as it began. Thus the movie is circular, with no beginning or end. This allegedly inspired the three to propose a theory of the universe that also had no beginning or end. (Gold later clarified this story. He recalled, "I think we saw that movie several months before, and after I proposed the steady state, I said to them, 'Isn't that a bit like *Dead of Night*?'")

In this model, portions of the universe were in fact expanding, but new matter was constantly being created out of nothing, so that the density of the universe remained the same. Although he could give no details of how matter mysteriously emerged out of nowhere, the theory immediately attracted a band of loyalists who battled the big bang theorists. To Hoyle, it seemed illogical that a fiery cataclysm could appear out of nowhere to send the galaxies hurtling in all directions; he preferred the smooth creation of mass out of nothing. In other words, the universe was timeless. It had no end, nor a beginning. It just was.

(The steady state–big bang controversy was similar to the controversy affecting geology and other sciences. In geology, there was the enduring debate between uniformitarianism [the belief that Earth has been shaped by gradual changes in the past] and catastrophism [which postulated that change took place via violent events]. Although uniformitarianism still explains much of the geologic and ecological features of Earth, no one can now deny the impact of comets and asteroids, which have generated mass extinctions, or the breakup and movements of the continents via tectonic drift.)

BBC LECTURES

Hoyle never shied away from a good fight. In 1949, both Hoyle and
Gamow were invited by the British Broadcasting Corporation to debate the origin of the universe. During the broadcast, Hoyle made
history when he took a swipe at the rival theory. He said fatefully,
"These theories were based on the hypothesis that all the matter in
the universe was created in one big bang at a particular time in the
remote past." The name stuck. The rival theory was now officially
christened "the big bang" by its greatest enemy. (He later claimed
that he did not mean it to be derogatory. He confessed, "There is no
way in which I coined the phrase to be derogatory. I coined it to be
striking.")

(Over the years, proponents of the big bang have tried heroically
to change the name. They are dissatisfied with the common, almost
vulgar connotation of the name and the fact that it was coined by its
greatest adversary. Purists are especially irked that it was also fac-
tually incorrect. First, the big bang was not big (since it originated
from a tiny singularity of some sort much smaller than an atom) and
second, there was no bang (since there is no air in outer space). In
August 1993, *Sky and Telescope* magazine sponsored a contest to rename
the big bang theory. The contest garnered thirteen thousand entries,
but the judges could not find any that was better than the original.)

What sealed Hoyle's fame to a whole generation was his celebrated BBC radio series on science. In the 1950s, the BBC planned to
air lectures on science every Saturday evening. However, when the
original guest canceled, the producers were pressed to find a substi-
tute. They contacted Hoyle, who agreed to come on. Then they
checked his file, where there was a note that said, "DO NOT USE THIS
MAN."

Fortuitously, they ignored this dire warning from a previous pro-
ducer, and he gave five spell-binding lectures to the world. These
classic BBC broadcasts mesmerized the nation and in part inspired
the next generation of astronomers. Astronomer Wallace Sargent recalls the impact that these broadcasts had on him: "When I was fif-

teen, I heard Fred Hoyle give lectures on the BBC called 'The Nature of the Universe.' The idea that you knew what the temperature and density were at the center of the Sun came as a hell of a shock. At the age of fifteen, that sort of thing seemed beyond knowledge. It was not just the amazing numbers, but the fact that you could know them at all."

NUCLEOSYNTHESIS IN THE STARS

Hoyle, who disdained idle armchair speculation, set out to test his steady state theory. He relished the idea that the elements of the universe were cooked not in the big bang, as Gamow believed, but in the center of stars. If the hundred or so chemical elements were all created by the intense heat of the stars, then there would be no need for a big bang at all.

In a series of seminal papers published in the 1940s and 1950s, Hoyle and his colleagues laid out in vivid detail how the nuclear reactions inside the core of a star, not the big bang, could add more and more protons and neutrons to the nuclei of hydrogen and helium, until they could create all the heavier elements, at least up to iron. (They solved the mystery of how to create elements beyond mass number 5, which had stumped Gamow. In a stroke of genius, Hoyle realized that if there were a previously unnoticed unstable form of carbon, created out of three helium nuclei, it might last just long enough to act as a "bridge," allowing for the creation of higher elements. In the core of stars, this new unstable form of carbon might last just long enough so that, by successively adding more neutrons and protons, one could create elements beyond mass number 5 and 8. When this unstable form of carbon was actually found, it brilliantly demonstrated that nucleosynthesis could take place in stars, rather than the big bang. Hoyle even created a large computer program that could determine, almost from first principles, the relative abundances of elements we see in nature.)

But even the intense heat of the stars is not sufficient to "cook" elements beyond iron, such as copper, nickel, zinc, and uranium. (It is extremely difficult to extract energy by fusing elements beyond iron, for a variety of reasons, including the repulsion of the protons in the nucleus and the lack of binding energy.) For those heavy elements, one needs an even larger oven—the explosion of massive stars, or supernovae. Since trillions of degrees can be attained in the final death throes of a supergiant star when it violently collapses, there is enough energy there to "cook" the elements beyond iron. This means that most of the elements beyond iron were, in fact, blasted out of the atmospheres of exploding stars, or supernovae.

In 1957, Hoyle, as well as Margaret and Geoffrey Burbidge and William Fowler, published perhaps the most definitive work detailing the precise steps necessary to build up the elements of the universe and predict their known abundances. Their arguments were so precise, powerful, and persuasive that even Gamow had to concede that Hoyle had given the most compelling picture of nucleosynthesis. Gamow, in typical fashion, even coined the following passage, written in biblical style. In the beginning, when God was creating the elements,

In the excitement of counting, He missed calling for mass five and so, naturally no heavier elements could have been formed. God was very much disappointed, and wanted first to contract the Universe again, and to start all over from the beginning. But it would be much too simple. Thus, being almighty, God decided to correct His mistake in a most impossible way. And God said, "Let there be Hoyle." And there was Hoyle. And God looked at Hoyle . . . And told him to make heavy elements in any way he pleased. And Hoyle decided to make heavy elements in stars, and to spread them around by supernova explosions.

EVIDENCE AGAINST THE STEADY STATE

Over the decades, however, evidence began to slowly mount against the steady state universe on a number of fronts. Hoyle found himself fighting a losing battle. In his theory, since the universe did not evolve but was continually creating new matter, the early universe should look very much like the present-day universe. Galaxies seen today should look very similar to galaxies billions of years ago. The steady state theory could then be disproved if there were signs of dramatic evolutionary changes during the course of billions of years.

In the 1960s, mysterious sources of immense power were found in outer space, dubbed "quasars," or quasi-stellar objects. (The name was so catchy that a TV set was later named after it.) Quasars generated enormous amounts of power and had huge redshifts, meaning that they were billions of light-years away, and they also lit up the heavens when the universe was very young. (Today, astronomers believe that these are gigantic young galaxies, driven by the power of huge black holes.) We do not see evidence of any quasars today, though according to the steady state theory they should exist. Over billions of years, they have disappeared.

There was another problem with Hoyle's theory. Scientists realized that there was simply too much helium in the universe to fit the predictions of the steady state universe. Helium, familiar as the gas found in children's balloons and blimps, is actually quite rare on Earth, but it's the second most plentiful element in the universe after hydrogen. It's so rare, in fact, that it was first found in the Sun, rather than the Earth. (In 1868, scientists analyzed light from the Sun that was sent through a prism. The deflected sunlight broke up into the usual rainbow of colors and spectral lines, but the scientists also detected faint spectral lines caused by a mysterious element never seen before. They mistakenly thought it was a metal, whose names usually end in "ium," like lithium and uranium. They named this mystery metal after the Greek word for sun, "helios." Finally in 1895, helium was found on Earth in uranium deposits, and scientists embarrassingly discovered that it was a gas, not a metal. Thus, helium, first discovered in the Sun, was born as a misnomer.)

If primordial helium was mainly created in the stars, as Hoyle believed, then it should be quite rare and found near the cores of stars. But all the astronomical data showed that helium was actually quite plentiful, making up about 25 percent of the mass of the atoms in the universe. It was found to be uniformly distributed around the universe (as Gamow believed).

Today, we know that both Gamow and Hoyle had pieces of the truth concerning nucleosynthesis. Gamow originally thought that all the chemical elements were fallout or ashes of the big bang. But his theory fell victim to the 5-particle and 8-particle gap. Hoyle thought he could sweep away the big bang theory altogether by showing that stars "cook" all the elements, without any need to resort to a big bang at all. But his theory failed to account for the huge abundance of helium we now know exists in the universe.

In essence, Gamow and Hoyle have given us a complementary picture of nucleosynthesis. The very light elements up to mass 5 and 8 were indeed created by the big bang, as Gamow believed. Today, as the result of discoveries in physics, we know that the big bang did produce most of the deuterium, helium-3, helium-4, and lithium-7 we see in nature. But the heavier elements up to iron were mostly cooked in the cores of the stars, as Hoyle believed. If we add the elements beyond iron (such as copper, zinc, and gold) that were blasted out by the blistering heat of a supernova, then we have a complete picture explaining the relative abundances of all the elements in the universe. (Any rival theory to modern-day cosmology would have a formidable task: to explain the relative abundances of the hundred-odd elements in the universe and their myriad isotopes.)

HOW STARS ARE BORN

One by-product of this intense debate over nucleosynthesis is that it has given us a rather complete description of the life cycle of stars. A typical star like our Sun begins its life as a large ball of diffuse hydrogen gas called a protostar and gradually contracts under the force of gravity. As it begins to collapse, it begins to spin rapidly (which often leads to the formation of a double-star system, where two stars chase each other in elliptical orbits, or the formation of planets in the plane of rotation of the star). The core of the star also heats up tremendously until it hits approximately 10 million degrees or more, when the fusion of hydrogen to helium takes place.

After the star ignites, it is called a main sequence star and it may burn for about 10 billion years, slowly turning its core from hydrogen to waste helium. Our Sun is currently midway through this process. After the era of hydrogen burning ends, the star begins to burn helium, whereupon it expands enormously to the size of the orbit of Mars and becomes a "red giant." After the helium fuel in the core is exhausted, the outer layers of the star dissipate, leaving the core itself, a "white dwarf" star about the size of Earth. Smaller stars like our Sun will die in space as hunks of dead nuclear material in white dwarf stars.

But in stars, perhaps ten to forty times the mass of our Sun, the fusion process proceeds much more rapidly. When the star becomes a red supergiant, its core rapidly fuses the lighter elements, so it resembles a hybrid star, a white dwarf inside a red giant. In this white dwarf star, the lighter elements up to iron on the periodic table of elements may be created. When the fusion process reaches the stage where the element iron is created, no more energy can be extracted from the fusion process, so the nuclear furnace, after billions of years, finally shuts down. At this point, the star abruptly collapses, creating huge pressures that actually push the electrons into the nuclei. (The density can exceed 400 billion times the density of water.) This causes temperatures to soar to trillions of degrees. The gravitational energy compressed into this tiny object explodes outward into a supernova. The intense heat of this process causes fusion to start once again, and the elements beyond iron on the periodic table are synthesized.

The red supergiant Betelgeuse, for example, which can be easily seen in the constellation Orion, is unstable; it can explode at any time as a supernova, spewing large quantities of gamma rays and X rays into the surrounding neighborhood. When that happens, this supernova will be visible in daytime and might outshine the Moon at night. (It was once thought that the titanic energy released by a supernova destroyed the dinosaurs 65 million years ago. A supernova about ten light-years away could, in fact, end all life on Earth. Fortunately, the giant stars Spica and Betelgeuse are 260 and 430 light-years away, respectively, too far to cause much serious damage to Earth when they finally explode. But some scientists believe that a minor extinction of sea creatures 2 million years ago was caused by a supernova explosion of a star 120 light-years away.)

This also means that our Sun is not Earth's true "mother." Although many peoples of Earth have worshipped the Sun as a god that gave birth to Earth, this is only partially correct. Although Earth was originally created from the Sun (as part of the ecliptic plane of debris and dust that circulated around the Sun 4.5 billion years ago), our Sun is barely hot enough to fuse hydrogen to helium. This means that our true "mother" sun was actually an unnamed star or collection of stars that died billions of years ago in a supernova, which then seeded nearby nebulae with the higher elements beyond iron that make up our body. Literally, our bodies are made of stardust, from stars that died billions of years ago.

In the aftermath of a supernova explosion, there is a tiny remnant called a neutron star, which is made of solid nuclear matter compressed to the size of Manhattan, almost 20 miles in size. (Neutron stars were first predicted by Swiss astronomer Fritz Zwicky in 1933, but they seemed so fantastic that they were ignored by scientists for decades.) Because the neutron star is emitting radiation irregularly and is also spinning rapidly, it resembles a spinning lighthouse, spewing radiation as it rotates. As seen from Earth, the neutron star appears to pulsate and is hence called a pulsar.

Extremely large stars, perhaps larger than 40 solar masses, when they eventually undergo a supernova explosion, might leave behind a neutron star that is larger than 3 solar masses. The gravity of this neutron star is so large that it can counteract the repulsive force between neutrons, and the star will make its final collapse into perhaps the most exotic object in the universe, a black hole, which I discuss in chapter 5.

BIRD DROPPINGS AND THE BIG BANG

The final stake in the heart of the steady state theory was the discovery of Arno Penzias and Robert Wilson in 1965. Working on the 20-foot Bell Laboratory Holmdell Horn Radio Telescope in New Jersey, they were looking for radio signals from the heavens when they picked up an unwanted static. They thought it was probably an aberration, because it seemed to be coming uniformly from all directions, rather than from a single star or galaxy. Thinking the static might have come from dirt and debris, they carefully cleaned off what Penzias described as "a white coating of dieletric material" (commonly known as bird droppings) that had covered the opening of the radio telescope. The static seemed even larger. Although they did not yet know it, they had accidentally stumbled upon the microwave background predicted by Gamow's group back in 1948.

Now the cosmological history reads a little bit like the Keystone cops, with three groups groping for an answer without any knowledge of the others. On one hand, Gamow, Alpher, and Hermann had laid out the theory behind the microwave background back in 1948; they had predicted the temperature of the microwave radiation to be 5 degrees above absolute zero. They gave up trying to measure the background radiation of space, however, because the instruments back then were not sensitive enough to detect it. In 1965, Penzias and Wilson found this black body radiation but didn't know it. Meanwhile, a third group, led by Robert Dicke of Princeton University, had independently rediscovered the theory of Gamow and his colleagues and were actively looking for the background radiation, but their equipment was too woefully primitive to find it.

This comical situation ended when a mutual friend, astronomer Bernard Burke, informed Penzias of the work of Robert Dicke. When the two groups finally connected, it became clear that Penzias and Wilson had detected signals from the big bang itself. For this momentous discovery, Penzias and Wilson won the Nobel Prize in 1978.

In hindsight, Hoyle and Gamow, the two most visible proponents of the opposite theories, had a fateful encounter in a Cadillac in 1956 that could have changed the course of cosmology. "I recall George driving me around in a white Cadillac," recalled Hoyle. Gamow repeated his conviction to Hoyle that the big bang left an afterglow that should be seen even today. However, Gamow's latest figures placed the temperature of that afterglow at 50 degrees. Then Hoyle made an astounding revelation to Gamow. Hoyle was aware of an obscure paper, written in 1941 by Andrew McKellar, that showed that the temperature of outer space cannot exceed 3 degrees. At higher temperatures, new reactions can occur which would create excited carbon-hydrogen (CH) and carbon-nitrogen (CN) radicals in outer space. By measuring the spectra of these chemicals, one could then determine the temperature of outer space. In fact, he found that the density of CN molecules he detected in space indicated a temperature of about 2.3 degrees K. In other words, unknown to Gamow, the 2.7 K background radiation had already been indirectly detected in 1941.

Hoyle recalled, "Whether it was the too-great comfort of the Cadillac, or because George wanted a temperature higher than 3 K, whereas I wanted a temperature of zero degrees, we missed the chance of spotting the discovery made nine years later by Arno Penzias and Bob Wilson." If Gamow's group had not made a numerical error and had come up with a lower temperature, or if Hoyle had not been so hostile to the big bang theory, perhaps history might have been written differently.

PERSONAL AFTERSHOCKS OF THE BIG BANG

The discovery of the microwave background by Penzias and Wilson had a decided effect on the careers of Gamow and Hoyle. To Hoyle, the work of Penzias and Wilson was a near-death experience. Finally, in *Nature* magazine in 1965, Hoyle officially conceded defeat, citing the microwave background and helium abundance as reasons for abandoning his steady state theory. But what really disturbed him was that the steady state theory had lost its predictive power: "It is widely believed that the existence of the microwave background killed the 'steady state' cosmology, but what really killed the steady-state theory was psychology . . . Here, in the microwave background, was an important phenomenon which it had not predicted . . . For many years, this knocked the stuffing out of me." (Hoyle later reversed himself, trying to tinker with newer variations of the steady state theory of the universe, but each variation became less and less plausible.)

Unfortunately, the question of priority left a bad taste in Gamow's mouth. Gamow, if one reads between the lines, was not pleased that his work and the work of Alpher and Hermann were rarely mentioned, if at all. Ever polite, he kept mum about his feelings, but in private letters he wrote that it was unfair that physicists and historians would completely ignore their work.

Although the work of Penzias and Wilson was a huge blow to the steady state theory and helped put the big bang on firm experimental footing, there were huge gaps in our understanding of the structure of the expanding universe. In a Friedmann universe, for example, one must know the value of Omega, the average distribution of matter in the universe, to understand its evolution. However, the determination of Omega became quite problematic when it was realized that most of the universe was not made of familiar atoms and molecules but a strange new substance called "dark matter," which outweighed ordinary matter by a factor of 10. Once again, the leaders in this field were not taken seriously by the rest of the astronomical community.

OMEGA AND DARK MATTER

The story of dark matter is perhaps one of the strangest chapters in cosmology. Back in the 1930s, maverick Swiss astronomer Fritz Zwicky of Cal Tech noticed that the galaxies in the Coma cluster of galaxies were not moving correctly under Newtonian gravity. These galaxies, he found, moved so fast that they should fly apart and the cluster should dissolve, according to Newton's laws of motion. The only way, he thought, that the Coma cluster can be kept together, rather than flying apart, was if the cluster had hundreds of times more matter than could be seen by telescope. Either Newton's laws were somehow incorrect at galactic distances or else there was a huge amount of missing, invisible matter in the Coma cluster that was holding it together.

This was the first indication in history that there was something terribly amiss with regard to the distribution of matter in the universe. Astronomers universally rejected or ignored the pioneering work of Zwicky, unfortunately, for several reasons.

First, astronomers were reluctant to believe that Newtonian gravity, which had dominated physics for several centuries, could be incorrect. There was a precedent for handling crises like this in astronomy. When the orbit of Uranus was analyzed in the nineteenth century, it was found that it wobbled—it deviated by a tiny amount from the equations of Isaac Newton. So either Newton was wrong, or there must be a new planet whose gravity was tugging on Uranus. The latter was correct, and Neptune was found on the first attempt in 1846 by analyzing the location predicted by Newton's laws.

Second, there was the question of Zwicky's personality and how astronomers treated "outsiders." Zwicky was a visionary who was often ridiculed or ignored in his lifetime. In 1933, with Walter Baade, he coined the word "supernova" and correctly predicted that a tiny neutron star, about 14 miles across, would be the ultimate remnant of an exploding star. The idea was so utterly outlandish that it was lampooned in a *Los Angeles Times* cartoon on January 19, 1934. Zwicky was furious at a small, elite group of astronomers whom, he thought, tried to exclude him from recognition, stole his ideas, and denied him time on the 100- and 200-inch telescopes. (Shortly before he died in 1974, Zwicky self-published a catalog of the galaxies. The catalog opened with the heading, "A Reminder to the High Priests of American Astronomy and to their Sycophants." The essay gave a blistering criticism of the clubby, ingrown nature of the astronomy elite, which tended to shut out mavericks like him. "Today's sycophants and plain thieves seem to be free, in American Astronomy in particular, to appropriate discoveries and inventions made by lone wolves and non-conformists," he wrote. He called these individuals "spherical bastards," because "they are bastards any way you look at them." He was incensed that he was passed over when the Nobel Prize was awarded to someone else for the discovery of the neutron star.)

In 1962, the curious problem with galactic motion was rediscovered by astronomer Vera Rubin. She studied the rotation of the Milky Way galaxy and found the same problem; she, too, received a cold shoulder from the astronomy community. Normally, the farther a planet is from the Sun, the slower it travels. The closer it is, the faster it moves. That's why Mercury is named after the god of speed, because it is so close to the Sun, and why Pluto's velocity is ten times slower than Mercury's, because it is the farthest from the Sun. However, when Vera Rubin analyzed the blue stars in our galaxy, she found that the stars rotated around the galaxy at the same rate, independent of their distance from the galactic center (which is called a flat rotation curve), thereby violating the precepts of Newtonian mechanics. In fact, she found that the Milky Way galaxy was rotating so fast that, by rights, it should fly apart. But the galaxy has been quite stable for about 10 billion years; it was a mystery why the rotation curve was flat. To keep the galaxy from disintegrating, it had to be ten times heavier than scientists currently imagined. Apparently, 90 percent of the mass of the Milky Way galaxy was missing!

Vera Rubin was ignored, in part because she was a woman. With a certain amount of pain, she recalls that, when she applied to Swarthmore College as a science major and casually told the admissions officer that she liked to paint, the interviewer said, "Have you ever considered a career in which you paint pictures of astronomical objects?" She recalled, "That became a tag line in my family: for many years, whenever anything went wrong for anyone, we said, 'Have you ever considered a career in which you paint pictures of astronomical objects?' " When she told her high school physics teacher that she got accepted to Vassar, he replied, "You should do okay as long as you stay away from science." She would later recall, "It takes an enormous amount of self-esteem to listen to things like that and not be demolished."

After she graduated, she applied and was accepted to Harvard, but she declined because she got married and followed her husband, a chemist, to Cornell. (She got a letter back from Harvard, with the handwritten words written on the bottom, "Damn you women. Every time I get a good one ready, she goes off and gets married.") Recently, she attended an astronomy conference in Japan, and she was the only woman there. "I really couldn't tell that story for a long time without weeping, because certainly in one generation . . . not an awful lot has changed," she confessed.

Nevertheless, the sheer weight of her careful work, and the work of others, slowly began to convince the astronomical community of the missing mass problem. By 1978, Rubin and her colleagues had examined eleven spiral galaxies; all of them were spinning too fast to stay together, according to the laws of Newton. That same year, Dutch radio astronomer Albert Bosma published the most complete analysis of dozens of spiral galaxies yet; almost all of them exhibited the same anomalous behavior. This finally seemed to convince the astronomical community that dark matter did indeed exist.

The simplest solution to this distressing problem was to assume that the galaxies were surrounded by an invisible halo that contained ten times more matter than the stars themselves. Since that time other, more sophisticated means have been developed to measure the presence of this invisible matter. One of the most impressive is to measure the distortion of starlight as it travels through invisible matter. Like the lens of your glasses, dark matter can bend light (because of its enormous mass and hence gravitational pull). Recently, by carefully analyzing the photographs of the Hubble space telescope with a computer, scientists were able to construct maps of the distribution of dark matter throughout the universe.

A fierce scramble has been going on to find out what dark matter is made of. Some scientists think it might consist of ordinary matter, except that it is very dim (that is, made of brown dwarf stars, neutron stars, black holes, and so on, which are nearly invisible). Such objects are lumped together as "baryonic matter," that is, matter made of familiar baryons (like neutrons and protons). Collectively, they are called MACHOs (short for Massive Compact Halo Objects).

Others think dark matter may consist of very hot nonbaryonic matter, such as neutrinos (called hot dark matter). However, neutrinos move so fast that they cannot account for most of the clumping of dark matter and galaxies that we see in nature. Still others throw up their hands and think that dark matter was made of an entirely new type of matter, called "cold dark matter," or WIMPS (weakly interacting massive particles), which are the leading candidate to explain most of dark matter.

COBE SATELLITE

Using an ordinary telescope, the workhorse of astronomy since the time of Galileo, one cannot possibly solve the mystery of dark matter. Astronomy has progressed remarkably far by using standard Earthbound optics. However, in the 1990s a new generation of astronomical instruments was coming of age that used the latest in satellite technology, lasers, and computers and completely changed the face of cosmology.

One of the first fruits of this harvest was the COBE (Cosmic Background Explorer) satellite, launched in November 1989. While the original work of Penzias and Wilson confirmed just a few data points consistent with the big bang, the COBE satellite was able to measure scores of data points that matched precisely the prediction of black body radiation made by Gamow and his colleagues in 1948.

In 1990, at a meeting of the American Astronomical Society, 1,500 scientists in the audience burst into a sudden thunderous standing ovation when they saw the COBE results placed on a viewgraph, showing a near-perfect agreement with a microwave background with a temperature of 2.728 K.

The Princeton astronomer Jeremiah P. Ostriker remarked, "When fossils were found in the rocks, it made the origin of species absolutely clear-cut. Well, COBE found [the universe's] fossils."

However, the viewgraphs from COBE were quite fuzzy. For example, scientists wanted to analyze "hot spots" or fluctuations within the cosmic background radiation, fluctuations that should be about a degree across in the sky. But COBE's instruments could only detect fluctuations that were 7 or more degrees across; they weren't sensitive enough to detect these small hot spots. Scientists were forced to wait for the results of the WMAP satellite, due to be launched after the turn of the century, which they hoped would settle a host of such questions and mysteries.

CHAPTER FOUR

Inflation and Parallel Universes

Nothing cannot come from nothing.

—Lucretius

I assume that our Universe did indeed appear from nowhere about 10^{10} years ago . . . I offer the modest proposal that our Universe is simply one of those things which happens from time to time.

—Edward Tryon

The universe is the ultimate free lunch.

—Alan Guth

IN THE CLASSIC science fiction novel *Tau Zero*, written by Poul Anderson, a starship named Leonora Christine is launched on a mission to reach the nearby stars. Carrying fifty people, the ship can attain velocities near the speed of light as it travels to a new star system. More important, the ship uses a principle of special relativity, which says that time slows down inside the starship the faster it moves. Hence, a trip to the nearby stars that takes decades, as viewed from Earth, appears to last only a few years to the astronauts. To an observer on Earth watching the astronauts by telescope, it would appear as if they were frozen in time, so that they are in a kind of suspended animation. But to the astronauts on board, time progresses normally. When the starship decelerates and the astronauts disembark on a new world, they will find that they have traveled thirty light-years in just a few years.

The ship is an engineering marvel; it is powered by ramjet fusion engines that scoop the hydrogen of deep space and then burn it for unlimited energy. It travels so fast that the crew can even see the Doppler shifting of starlight; stars in front of them appear bluish, while stars behind them appear reddish.

Then disaster strikes. About ten light-years from Earth, the ship experiences turbulence as it passes through an interstellar dust cloud, and its deceleration mechanism becomes permanently disabled. The horrified crew find themselves trapped on a runaway starship, speeding faster and faster as it approaches the speed of light. They watch helplessly as the out-of-control ship passes entire star systems in a matter of minutes. Within a year, the starship zips through half the Milky Way galaxy. As it accelerates beyond control, it speeds past galaxies in a matter of months, even as millions of years have passed on Earth. Soon, they are traveling so close to the speed of light, tau zero, that they witness cosmic events, as the universe itself begins to age before their eyes.

Eventually, they see that the original expansion of the universe is reversing, and that the universe is contracting on itself. Temperatures begin to rise dramatically, as they realize that they are headed for the big crunch. Crew members silently say their prayers as temperatures skyrocket, galaxies begin to coalesce, and a cosmic primordial atom forms before them. Death by incineration, it appears, is inevitable.

Their only hope is that matter will collapse into a finite area of finite density, and that, traveling at their great speed, they might slip rapidly through it. Miraculously, their shielding protects them as they fly through the primordial atom, and they find themselves witnessing the creation of a new universe. As the universe re-expands, they are awed to witness the creation of new stars and galaxies before their eyes. They fix their spacecraft and carefully chart their course for a galaxy old enough to have the higher elements that will make life possible. Eventually, they locate a planet that can harbor life and create a colony on that planet to start humanity all over again.

This story was written in 1967, when a vigorous debate raged among astronomers as to the ultimate fate of the universe: whether it would die in a big crunch or a big freeze, would oscillate indefinitely, or would live forever in a steady state. Since then, the debate seems to be settled, and a new theory called inflation has emerged.

BIRTH OF INFLATION

"SPECTACULAR REALIZATION," Alan Guth wrote in his diary in 1979. He felt exhilarated, realizing that he might have stumbled across one of the great ideas of cosmology. Guth had made the first major revision of the big bang theory in fifty years by making a seminal observation: he could solve some of the deepest riddles of cosmology if he assumed that the universe underwent a turbocharged hyperinflation at the instant of its birth, astronomically faster than the one believed by most physicists. With this hyperexpansion, he found he could effortlessly solve a host of deep cosmological questions that had defied explanation. It was an idea that would come to revolutionize cosmology. (Recent cosmological data, including the results of the WMAP satellite, are consistent with its predictions.) It is not the only cosmological theory, but is by far the simplest and most credible.

It is remarkable that such a simple idea could solve so many thorny cosmological questions. One of several problems that inflation elegantly solved was the flatness problem. Astronomical data has shown that the curvature of the universe is remarkably close to zero, in fact much closer to zero than previously believed by most astronomers. This could be explained if the universe, like a balloon that is rapidly being inflated, was flattened out during the inflation period. We, like ants walking on the surface of a balloon, are simply too small to observe the tiny curvature of the balloon. Inflation has stretched space-time so much that it appears flat.

What was also historic about Guth's discovery was that it represented the application of elementary particle physics, which involves analyzing the tiniest particles found in nature, to cosmology, the study of the universe in its entirety, including its origin. We now realize that the deepest mysteries of the universe cannot be solved without the physics of the extremely small: the world of the quantum theory and elementary particle physics.

SEARCH FOR UNIFICATION

Guth was born in 1947 in New Brunswick, New Jersey. Unlike Einstein, Gamow, or Hoyle, there was no instrument or seminal moment that propelled him into the world of physics. Neither of his parents graduated from college or showed much interest in science. But by his own admission he was always fascinated by the relationship between math and the laws of nature.

At MIT in the 1960s, he seriously considered a career in elementary particle physics. In particular, he was fascinated by the excitement generated by a new revolution sweeping through physics, the search for the unification of all fundamental forces. For ages, the holy grail of physics has been to search for unifying themes that can explain the complexities of the universe in the simplest, most coherent fashion. Since the time of the Greeks, scientists have thought that the universe we see today represents the broken, shattered remnants of a greater simplicity, and our goal is to reveal this unification.

After two thousand years of investigation into the nature of matter and energy, physicists have determined that just four fundamental forces drive the universe. (Scientists have tried to look for a possible fifth force, but so far all results in this direction have been negative or inconclusive.)

The first force is gravity, which holds the Sun together and guides planets in their celestial orbits in the solar system. If gravity were suddenly "turned off," the stars in the heavens would explode, Earth would disintegrate, and we would all be flung into outer space at about a thousand miles an hour.

The second great force is electromagnetism, the force that lights up our cities, fills our world with TV, cell phones, radio, laser beams, and the Internet. If the electromagnetic force were suddenly shut down, civilization would be instantly hurled a century or two into the past into darkness and silence. This was graphically illustrated by the great blackout of 2003, which paralyzed the entire Northeast. If we examine the electromagnetic force microscopically, we see that it is actually made of tiny particles, or quanta, called photons.

The third force is the weak nuclear force, which is responsible for radioactive decay. Because the weak force is not strong enough to hold the nucleus of the atom together, it allows the nucleus to break up or decay. Nuclear medicine in hospitals relies heavily on the nuclear force. The weak force also helps to heat up the center of Earth via radioactive materials, which drive the immense power of volcanoes. The weak force, in turn, is based on the interactions of electrons and neutrinos (ghost-like particles that are nearly massless and can pass through trillions of miles of solid lead without interacting with anything). These electrons and neutrinos interact by exchanging other particles, called W- and Z-bosons.

The strong nuclear force holds the nuclei of the atoms together. Without the nuclear force, the nuclei would all disintegrate, atoms would fall apart, and reality as we know it would dissolve. The strong nuclear force is responsible for the approximately one hundred elements we see filling up the universe. Together, the weak and strong nuclear forces are responsible for the light emanating from stars via Einstein's equation $E = mc^2$. Without the nuclear force, the entire universe would be darkened, plunging the temperature on Earth and freezing the oceans solid.

The astonishing feature of these four forces is that they are entirely different from each other, with different strengths and properties. For example, gravity is by far the weakest of the four forces, 10^{36} times weaker than the electromagnetic force. The earth weighs 6 trillion trillion kilograms, yet its massive weight and its gravity can easily be canceled by the electromagnetic force. Your comb, for example, can pick up tiny pieces of paper via static electricity, thereby canceling the gravity of the entire earth. Also, gravity is strictly attractive. The electromagnetic force can be both attractive or repulsive, depending on the charge of a particle.

UNIFICATION AT THE BIG BANG

One of the fundamental questions facing physics is: why should the universe be ruled by four distinct forces? And why should these four forces look so dissimilar, with different strengths, different interactions, and different physics?

Einstein was the first to embark upon a campaign to unify these forces into a single, comprehensive theory, starting by uniting gravity with the electromagnetic force. He failed because he was too far ahead of his time; too little was known about the strong force to make a realistic unified field theory. But Einstein's pioneering work opened the eyes of the physics world to the possibility of a "theory of everything."

The goal of a unified field theory seemed utterly hopeless in the 1950s, especially when elementary particle physics was in total chaos, with atom smashers blasting nuclei apart to find the "elementary constituents" of matter, only to find hundreds more particles streaming out of the experiments. "Elementary particle physics" became a contradiction in terms, a cosmic joke. The Greeks thought that, as we broke down a substance to its basic building blocks, things would get simpler. The opposite happened: physicists struggled to find enough letters in the Greek alphabet to label these particles. J. Robert Oppenheimer joked that the Nobel Prize in physics should go to the physicist who did not discover a new particle that year. Nobel laureate Steven Weinberg began to wonder whether the human mind was even capable of solving the secret of the nuclear force.

This bedlam of confusion, however, was somewhat tamed in the early 1960s when Murray Gell-Mann and George Zweig of Cal Tech proposed the idea of quarks, the constituents that make up the protons and neutrons. According to quark theory, three quarks make up a proton or a neutron, and a quark and antiquark make up a meson (a particle that holds the nucleus together). This was only a partial solution (since today we are flooded with different types of quarks), but it did serve to inject new energy into a once dormant field.

In 1967, a stunning breakthrough was made by physicists Steven Weinberg and Abdus Salam, who showed that it was possible to unify the weak and electromagnetic forces. They created a new theory whereby electrons and neutrinos (which are called leptons) interacted with each other by exchanging new particles called the W- and Z-bosons as well as photons. By treating the W- and Z-bosons and photons on the very same footing, they created a theory which unified the two forces. In 1979, Steven Weinberg, Sheldon Glashow, and Abdus Salam were awarded the Nobel Prize for their collective work in unifying two of the four forces, the electromagnetic force with the weak force, and providing insight into the strong nuclear force.

In the 1970s, physicists analyzed the data coming from the particle accelerator at the Stanford Linear Accelerator Center (SLAC), which fired intense beams of electrons at a target in order to probe deep into the interior of the proton. They found that the strong nuclear force that held the quarks together inside the proton could be explained by introducing new particles called gluons, which are the quanta of the strong nuclear force. The binding force holding the proton together could be explained by the exchange of gluons between the constituent quarks. This led to a new theory of the strong nuclear force called Quantum Chromodynamics.

So by the mid 1970s, it was possible to splice three of the four forces together (excluding gravity) to get what is called the Standard Model, a theory of quarks, electrons, and neutrinos, which interact by exchanging gluons, W- and Z-bosons, and photons. It is the culmination of decades of painfully slow research in particle physics. At present, the Standard Model fits all the experimental data concerning particle physics, without exception.

Although the Standard Model is one of the most successful physical theories of all time, it is remarkably ugly. It is hard to believe that nature at a fundamental level can operate on a theory that seems to be so cobbled together. For example, there are nineteen arbitrary parameters in the theory that are simply put in by hand, without any rhyme or reason (that is, the various masses and inter-action strengths are not determined by the theory but have to be de-termined by experiment; ideally, in a true unified theory, these constants would be determined by the theory itself, without relying on outside experiments).

Furthermore, there are three exact copies of elementary parti-cles, called generations. It is hard to believe that nature, at its most fundamental level, would include three exact copies of subatomic particles. Except for the masses of these particles, these generations are duplicates of each other. (For example, the carbon copies of the electron include the muon, which weighs 200 times more than the

	Quarks		Gluons	
First Generation	up	down	electron	neutrino
Second Generation	charm	strange	muon	muon-neutrino
Third Generation	top	bottom	tau	tau-neutrino

| W-Boson | Z-Boson | Gluons | Higgs |

These are the subatomic particles contained within the Standard Model, the most successful theory of elementary particles. It is built out of quarks, which make up the protons and neutrons, leptons like the electron and neutrino, and many other particles. Notice that the model results in three identical copies of subatomic particles. Since the Standard Model fails to account for gravity (and seems so awkward), theoretical physicists feel it cannot be the final theory.

electron, and the tau particle, which weighs 3,500 times more.) And last, the Standard Model makes no mention of gravity, although gravity is perhaps the most pervasive force in the universe.

Because the Standard Model, notwithstanding its stunning experimental successes, seems so contrived, physicists tried to develop yet another theory, or the grand unified theory (GUT), which put the quarks and leptons on the same footing. It also treated the gluon, the W- and Z-boson, and the photon on the same level. (It could not be the "final theory," however, because gravity was still conspicuously left out; it was considered too difficult to merge with the other forces, as we shall see.)

This program of unification, in turn, introduced a new paradigm to cosmology. The idea was simple and elegant: at the instant of the big bang, all four fundamental forces were unified into a single, coherent force, a mysterious "superforce." All four forces had the same strength and were part of a larger, coherent whole. The universe started out in a state of perfection. However, as the universe began to expand and cool rapidly, the original superforce began to "crack," with different forces breaking off one after the other.

According to this theory, the cooling of the universe after the big bang is analogous to the freezing of water. When water is in liquid form, it is quite uniform and smooth. However, when it freezes, millions of tiny ice crystals form inside. When liquid water is totally frozen, its original uniformity is quite broken, with the ice containing cracks, bubbles, and crystals.

In other words, today we see that the universe is horribly broken. It is not uniform or symmetrical at all but consists of jagged mountain ranges, volcanoes, hurricanes, rocky asteroids, and exploding stars, without any coherent unity; moreover, we also see the four fundamental forces without any relationship to each other. But the reason why the universe is so broken is that it is quite old and cold.

Although the universe started in a state of perfect unity, today it has gone through many phase transitions, or changes of state, with the forces of the universe breaking free of the others one by one as it cooled. It is the job of physicists to go backward, to reconstruct the steps by which the universe originally started (in a state of perfection) and which led to the broken universe we see around us.

The key, therefore, is to understand precisely how these phase transitions occurred at the beginning of the universe, which physicists call "spontaneous breaking." Whether it is the melting of ice, the boiling of water, the creation of rain clouds, or the cooling of the big bang, phase transitions can connect two entirely different phases of matter. (To illustrate how powerful these phase transitions can be, the artist Bob Miller has asked the riddle: "How would you suspend 500,000 pounds of water in the air with no visible means of support? The answer: build a cloud.")

FALSE VACUUM

When one force breaks off from the other forces, the process can be compared to the breaking of a dam. Rivers flow downhill because water flows in the direction of the lowest energy, which is sea level. The lowest energy state is called a vacuum. However, there is an unusual state called the false vacuum. If we dam a river, for example, the dam appears to be stable, but it is actually under tremendous pressure. If a tiny crack occurs in the dam, the pressure can suddenly burst the dam and release a torrent of energy from the false vacuum (the dammed river) and cause a catastrophic flood toward the true vacuum (sea level). Entire villages can be flooded if we have spontaneous breaking of the dam and a sudden transition to the true vacuum.

Similarly, in GUT theory, the universe originally started out in the state of the false vacuum, with the three forces unified into a single force. However, the theory was unstable, and the theory spontaneously broke and made the transition from the false vacuum, where the forces were unified, to the true vacuum, where the forces are broken.

This was already known before Guth began to analyze GUT theory. But Guth noticed something that had been overlooked by others. In the state of the false vacuum, the universe expands exponentially, just the way de Sitter predicted back in 1917. It is the cosmological constant, the energy of the false vacuum, that drives the universe to expand at such an enormous rate. Guth asked himself a fateful question: can this exponential de Sitter expansion solve some of the problems of cosmology?

MONOPOLE PROBLEM

One prediction of many GUT theories was the production of copious numbers of monopoles at the beginning of time. A monopole is a single magnetic north or south pole. In nature, these poles are always found in pairs. If you take a magnet, you invariably find both a north pole and a south pole bound together. If you take a hammer and split a magnet in half, then you do not find two monopoles; instead, you find two smaller magnets, each with its own pair of north and south poles.

The problem, however, was that scientists, after centuries of experiments, had found no conclusive evidence for monopoles. Since no one had ever seen a monopole, Guth was puzzled why GUT theories predicted so many of them. "Like the unicorn, the monopole has continued to fascinate the human mind despite the absence of confirmed observations," Guth remarked.

Then it suddenly hit him. In a flash, all the pieces fit together. He realized that if the universe started in a state of false vacuum, it could expand exponentially, as de Sitter had proposed decades earlier. In this false vacuum state, the universe could suddenly inflate by an incredible amount, thereby diluting the density of monopoles. If scientists had never seen a monopole before, it was only because monopoles were spread out over a universe that was much larger than previously thought.

To Guth, this revelation was a source of amazement and joy. Such a simple observation could explain the monopole problem in a single stroke. But Guth realized that this prediction would have cosmological implications far beyond his original idea.

FLATNESS PROBLEM

Guth realized that his theory solved another problem, the flatness problem, discussed earlier. The standard picture of the big bang could not explain why the universe was so flat. In the 1970s, it was believed that the matter density in the universe, called Omega, was around 0.1. The fact that this was relatively close to the critical density of 1.0 so many billions of years after the big bang was deeply disturbing. As the universe expanded, Omega should have changed with time. This number was uncomfortably close to the value of 1.0, which describes a perfectly flat space.

For any reasonable value of Omega at the beginning of time, Einstein's equations show that it should almost be zero today. For Omega to be so close to 1 so many billions of years after the big bang would require a miracle. This is what is called in cosmology the fine-tuning problem. God, or some creator, had to "choose" the value of Omega to within fantastic accuracy for Omega to be about 0.1 today. For Omega to be between 0.1 and 10 today, it means that Omega had to be 1.00000000000000 one second after the big bang. In other words, at the beginning of time the value of Omega had to be "chosen" to equal the number 1 to within one part in a hundred trillion, which is difficult to comprehend.

Think of trying to balance a pencil vertically on its tip. No matter how we try to balance the pencil, it usually falls down. In fact, it requires a fine-tuning of great precision to start the pencil balanced just right so it doesn't fall over. Now try to balance the pencil on its tip so that it stays vertical not just for one second but for years! You see the enormous fine-tuning that is involved to get Omega to be 0.1 today. The slightest error in fine-tuning Omega would have created Omega vastly different from 1. So why is Omega so close to 1 day, when by rights it should be astronomically different?

To Guth, the answer was obvious. The universe simply inflated by such a remarkable degree that it flattened the universe. Like a person concluding that Earth is flat because he cannot see the horizon, astronomers concluded that Omega is around 1 because inflation flattened the universe.

HORIZON PROBLEM

Not only did inflation explain the data supporting the flatness of the universe, it also solved the horizon problem. This problem is based on the simple realization that the night sky seems to be relatively uniform, no matter where you look. If you turn your head 180 degrees, you observe that the universe is uniform, even though you have just seen parts of the universe separated by tens of billions of light-years. Powerful telescopes scanning the heavens can find no appreciable deviation from this uniformity either. Our space satellites have shown that the cosmic microwave radiation is also extremely uniform. No matter where we look in space, the temperature of the background radiation deviates no more than a thousandth of a degree.

But this is a problem, because the speed of light is the ultimate speed limit in the universe. There is no way, in the lifetime of the universe, that light or information could have traveled from one part of the night sky to the other side. For example, if we look at the microwave radiation in one direction, it has traveled over 13 billion years since the big bang. If we turn our heads around and look in the opposite direction, we see microwave radiation that is identical that has also traveled over 13 billion years. Since they are at the same temperature, they must have been in thermal contact at the beginning of time. But there is no way that information could have traveled from opposite points in the night sky (separated by over 26 billion light-years) since the big bang.

The situation is even worse if we look at the sky 380,000 years after the big bang, when the background radiation was first formed. If we look in opposite points in the sky, we see that the background radiation is nearly uniform. But according to calculations from the big bang theory, these opposite points are separated by 90 million light-years (because of the expansion of space since the explosion). But there is no way that light could have traveled by 90 million light-years in just 380,000 years. Information would have had to travel much faster than the speed of light, which is impossible.

By rights, the universe should appear quite lumpy, with one part too distant to have made contact with another distant part. How can the universe appear so uniform, when light simply did not have enough time to mix and spread information from one distant part of the universe to the other? (Princeton physicist Robert Dicke called this the horizon problem, since the horizon is the farthest point you can see, the farthest point that light can travel.)

But Guth realized that inflation was the key to explain this problem, as well. He reasoned that our visible universe was probably a tiny patch in the original fireball. The patch itself was uniform in density and temperature. But inflation suddenly expanded this tiny patch of uniform matter by a factor of 10^{50}, much faster than the speed of light, so that the visible universe today is remarkably uniform. So the reason why the night sky and the microwave radiation is so uniform is that the visible universe was once a tiny but uniform patch of the original fireball that suddenly inflated to become the universe.

REACTION TO INFLATION

Although Guth was confident the inflationary idea was correct, he was a bit nervous when he first began to give talks publicly. When he presented his theory in 1980, "I was still worried that some consequence of theory might be spectacularly wrong. There was also the fear that I would reveal my status as a greenhorn cosmologist," he confessed. But his theory was so elegant and powerful that physicists around the world immediately saw its importance. Nobel laureate Murray Gell-Mann exclaimed, "You've solved the most important problem in cosmology!" Nobel laureate Sheldon Glashow confided to Guth that Steven Weinberg was "furious" when he heard about inflation. Anxiously, Guth asked, "Did Steve have any objections to it?" Glashow replied, "No, he just didn't think of it himself."

How could they have missed such a simple solution, scientists asked themselves. The reception to Guth's theory was enthusiastic among theoretical physicists, who were amazed at its scope.

It also had an impact on Guth's job prospects. One day, because of the tight job market, he was staring unemployment in the face. "I was in a marginal situation on the job market," he confessed. Suddenly, job offers began to pour in from top universities, but not from his first choice, MIT. But then he read a fortune cookie that said, "An exciting opportunity lies just ahead of you if you are not too timid." This gave him the nerve to boldly phone MIT and inquire about a job. He was stunned when MIT called a few days later and offered him a professorship. The next fortune cookie he read said, "You should not act on the impulse of the moment." Ignoring its advice, he decided to accept the MIT position. "What would a Chinese fortune cookie know, anyhow?" he asked himself.

However, there were still serious problems. The astronomers were less than impressed by Guth's theory, since it was glaringly deficient in one area: it gave the wrong prediction for Omega. The fact that Omega was roughly close to 1 could be explained by inflation. However, inflation went much further and predicted that Omega (or Omega plus Lambda) should be precisely 1.0, corresponding to a flat universe. In the following years, as more and more experimental data were collected locating vast amounts of dark matter in the universe, Omega budged slightly, rising to 0.3. But this was still potentially fatal for inflation. Although inflation would generate over three thousand papers in the next decade among physicists, it continued to be a curiosity for astronomers. To them, the data seemed to rule out inflation.

Some astronomers complained privately that particle physicists were so obsessed with the beauty of inflation that they were willing to ignore experimental fact. (Astronomer Robert Kirshner of Harvard wrote, "This 'inflation' idea sounds crazy. The fact that it is taken seriously by people who sit firmly in endowed chairs doesn't automatically make it right." Roger Penrose of Oxford called inflation "a fashion the high-energy physicists have visited on the cosmologists . . . Even aardvarks think their offspring are beautiful.")

Guth believed that sooner or later the data would show that the universe was flat. But what did bother him was that his original picture suffered from a small but crucial defect, one that is still not completely understood today. Inflation was ideally suited to solving a series of deep cosmological problems. The problem was he didn't know how to turn inflation off.

Think of heating up a pot of water to its boiling point. Just before it boils, it is momentarily in the state of high energy. It wants to boil, but it can't because it needs some impurity to start a bubble. But once a bubble starts, it quickly enters a lower energy state of the true vacuum, and the pot becomes full of bubbles. Eventually, the bubbles become so large that they coalesce, until the pot is uniformly full of steam. When all the bubbles merge, the phase of transition from water to steam is complete.

In Guth's original picture, each bubble represented a piece of our universe that was inflating out of the vacuum. But when Guth did this calculation, he found that the bubbles did not coalesce properly, leaving the universe incredibly lumpy. In other words, his theory left the pot full of steam bubbles that never quite merged to become a uniform pot of steam. Guth's vat of boiling water never seemed to settle down to the universe of today.

In 1981, Andrei Linde of the P. N. Lebedev Institute in Russia and Paul J. Steinhardt and Andreas Albrecht, then at the University of Pennsylvania, found a way around this puzzle, realizing that if a single bubble of false vacuum inflated long enough, it would eventually fill up the entire pot and create a uniform universe. In other words, our entire world could be the by-product of a single bubble that inflated to fill up the universe. You did not need a large number of bubbles to coalesce in order to create a uniform pot of steam. Just a single bubble would do, if it inflated long enough.

Think back to the analogy of the dam and the false vacuum. The thicker the dam, the longer it takes for water to tunnel through the dam. If the wall of the dam is thick enough, then the tunneling will be delayed arbitrarily long. If the universe is allowed to inflate by a factor of 10^{50}, then a single bubble has enough time to solve the horizon, flatness, and monopole problem. In other words, if tunneling is sufficiently delayed, the universe inflates long enough to flatten the universe and dilute the monopoles. But this still leaves the question: what mechanism can prolong inflation that huge amount?

Eventually, this sticky problem became known as the "graceful exit problem," that is, how to inflate the universe long enough so that a single bubble can create the entire universe. Over the years, at least fifty different mechanisms have been proposed to solve the graceful exit problem. (This is a deceptively difficult problem. I've tried several solutions myself. It was relatively easy to generate a modest amount of inflation in the early universe. But what is extremely difficult is getting the universe to inflate by a factor of 10^{50}. Of course, one might simply put in this 10^{50} factor by hand, but this is artificial and contrived.) In other words, the process of inflation was widely believed to have solved the monopole, horizon, and flatness problems, but no one knew precisely what drove inflation and what shut it off.

CHAOTIC INFLATION AND PARALLEL UNIVERSES

Physicist Andrei Linde, for one, was unfazed by the fact that no one agreed on a solution to the graceful exit problem. Linde confessed, "I just had the feeling that it was impossible for God not to use such a good possibility to simplify his work."

Eventually, Linde proposed a new version of inflation that seemed to eliminate some of the defects of the early versions. He envisioned a universe in which, at random points in space and time, spontaneous breaking occurs. At each point where breaking occurs, a universe is created which inflates a little. Most of the time, the amount of inflation is minor. But because this process is random, eventually there will be a bubble where the inflation lasts long enough to create our universe. Taken to its logical conclusion, this means that inflation is continuous and eternal, with big bangs happening all the time, with universes sprouting from other universes. In this picture, universes can "bud" off into other universes, creating a "multiverse."

In this theory, spontaneous breaking may occur anywhere within our universe, allowing an entire universe to bud off our universe. It also means that our own universe might have budded from a previous universe. In the chaotic inflationary model, the multiverse is eternal, even if individual universes are not. Some universes may have a very large Omega, in which case they immediately vanish into a big crunch after their big bang. Some universes only have a tiny Omega and expand forever. Eventually, the multiverse becomes dominated by those universes that inflate by a huge amount.

In retrospect, the idea of parallel universes is forced upon us. Inflation represents the merger of traditional cosmology with advances in particle physics. Being a quantum theory, particle physics states that there is a finite probability for unlikely events to occur, such as the creation of parallel universes. Thus, as soon as we admit the possibility of one universe being created, we open the door to the probability of an endless number of parallel universes being created. Think, for example, of how the electron is described in the quantum theory. Because of uncertainty, the electron does not exist at any single point, but exists in all possible points around the nucleus. This electron "cloud" surrounding the nucleus represents the electron being many places at the same time. This is the fundamental basis of all of chemistry which allows electrons to bind molecules together. The reason why our molecules do not dissolve is that parallel electrons dance around them and hold them together. Likewise, the universe was once smaller than an electron. When we apply the quantum theory to the universe, we are then forced to admit the possibility that the universe exists simultaneously in many states. In other words, once we open the door to applying quantum fluctuations to the universe, we are almost forced to admit the possibility of parallel universes. It seems we have little choice.

THE UNIVERSE FROM NOTHING

At first, one might object to the notion of a multiverse, because it seems to violate known laws, such as the conservation of matter and energy. However, the total matter/energy content of a universe may actually be very small. The matter content of the universe, including all the stars, planets, and galaxies, is huge and positive. However, the energy stored within gravity may be negative. If you add the positive energy due to matter to the negative energy due to gravity, the sum may be close to zero! In some sense, *such universes are free.* They can spring out of the vacuum almost effortlessly. (If the universe is closed, then the total energy content of the universe must be precisely zero.)

(To grasp this, think of a donkey that falls into a large hole in the ground. We have to add energy to the donkey in order to pull him out of the hole. Once he is out and he is standing on the ground, he is considered to have zero energy. Thus, because we had to add energy to the donkey to get him to a state of zero energy, he must have had negative energy while in the hole. Similarly, it takes energy to pull a planet out of a solar system. Once it is out in free space, the planet has zero energy. Since we have to add energy to extract a planet out of a solar system to attain a state of zero energy, the planet has negative gravitational energy while inside the solar system.)

In fact, to create a universe like ours may require a ridiculously small net amount of matter, perhaps as little as an ounce. As Guth likes to say, "the universe may be a free lunch." This idea of creating a universe from nothing was first introduced by physicist Edward Tryon of Hunter College of the City University of New York, in a paper published in *Nature* magazine in 1973. He speculated that the universe is something "which happens from time to time" due to a quantum fluctuation in the vacuum. (Although the net amount of matter necessary to create a universe may be close to zero, this matter must be compressed to incredible densities, as we see in chapter 12.)

Like the P'an Ku mythologies, this is an example of *creatio ex nihilo* cosmology. Although the universe-from-nothing theory cannot be proved with conventional means, it does help to answer very practical questions about the universe. For example, why doesn't the universe spin? Everything we see around us spins, from tops, hurricanes, planets, and galaxies, to quasars. It seems to be a universal characteristic of matter in the universe. But the universe itself does not spin. When we look at the galaxies in the heavens, their total spin cancels out to zero. (This is quite fortunate, because, as we see in chapter 5, if the universe did spin, then time travel would become commonplace and history would be impossible to write.) The reason why the universe does not spin may be that our universe came from nothing. Since the vacuum does not spin, we do not expect to see any net spin arising in our universe. In fact, all the bubble-universes within the multiverse may have zero net spin.

Why do positive and negative electrical charges balance out exactly? Normally, when we think of the cosmic forces governing the universe, we think more about gravity than the electromagnetic force, even though the gravitational force is infinitesimally small compared to the electromagnetic force. The reason for this is the perfect balance between positive and negative charges. As a result, the net charge of the universe appears to be zero, and gravity dominates the universe, not the electromagnetic force.

Although we take this for granted, the cancellation of positive and negative charges is quite remarkable, and has been experimentally checked to 1 part in 10^{21}. (Of course, there are local imbalances between the charges, and that's why we have lightning bolts. But the total number of charges, even for thunderstorms, adds up to zero.) If there were just 0.00001 percent difference in the net positive and negative electrical charges within your body, you would be ripped to shreds instantly, with your body parts thrown into outer space by the electrical force.

The answer to these enduring puzzles may be that the universe came from nothing. Since the vacuum has net zero spin and charge, any baby universe springing forth from nothing must also have net zero spin and charge.

There is one apparent exception to this rule. That exception is that the universe is made of matter rather than antimatter. Since matter and antimatter are opposites (with antimatter having exactly the opposite charge from matter), we might assume that the big bang must have created equal amount of matter and antimatter. The problem, however, is that matter and antimatter will annihilate each other on contact into a burst of gamma rays. Thus, we should not exist. The universe should be a random collection of gamma rays instead of teeming with ordinary matter. If the big bang were perfectly symmetrical (or if it came from nothing), then we should expect equal amounts of matter and antimatter to be formed. So why do we exist? The solution proposed by Russian physicist Andrei Sakharov is that the original big bang was not perfectly symmetrical at all. There was a tiny amount of symmetry breaking between matter and antimatter at the instant of creation, so that matter dominated over antimatter, which made possible the universe we see around us. (The symmetry that was broken at the big bang is called CP symmetry, the symmetry that reverses charges and the parity of matter and antimatter particles.) If the universe came from "nothing," then perhaps nothing was not perfectly empty but had a slight amount of symmetry breaking, which allows for the slight dominance of matter over antimatter today. The origin of this symmetry breaking is still not understood.

WHAT MIGHT OTHER UNIVERSES LOOK LIKE?

The multiverse idea is appealing, because all we have to do is assume that spontaneous breaking occurs randomly. No other assumptions have to be made. Each time a universe sprouts off another universe, the physical constants differ from the original, creating new laws of physics. If this is true, then an entirely new reality can emerge within each universe. But this raises the intriguing question: what do these other universes look like? The key to understanding the physics of parallel universes is to understand how universes are created, that is, to understand precisely how spontaneous breaking occurs.

When a universe is born and spontaneous breaking takes place, this also breaks the symmetry of the original theory. To a physicist, beauty means symmetry and simplicity. If a theory is beautiful, this means it has a powerful symmetry that can explain a large body of data in the most compact, economical manner. More precisely, an equation is considered to be beautiful if it remains the same when we interchange its components among themselves. One great advantage to finding the hidden symmetries of nature is that we can show that phenomena that are seemingly distinct are actually manifestations of the same thing, linked together by a symmetry. For example, we can show that electricity and magnetism are actually two aspects of the same object, because there is a symmetry that can interchange them within Maxwell's equations. Similarly, Einstein showed that relativity can turn space into time and vice versa, because they are part of the same object, the fabric of space-time.

Think of a snowflake, which has a beautiful six-fold symmetry, a source of endless fascination. The essence of its beauty is that it remains the same if we rotate the snowflake by 60 degrees. This also means that any equation we write down to describe the snowflake should reflect this fact, that it remains invariant under rotations of multiples of 60 degrees. Mathematically, we say that the snowflake has C_6 symmetry.

Symmetries then encode the hidden beauty of nature. But in reality, today these symmetries are horribly broken. The four great forces of the universe do not resemble each other at all. In fact, the universe is full of irregularities and defects; surrounding us are the fragments and shards of the original, primordial symmetry shattered by the big bang. Thus, the key to understanding possible parallel universes is to understand "symmetry breaking"—that is, how these symmetries might have broken after the big bang. As physicist David Gross has said, "The secret of nature is symmetry, but much of the texture of the world is due to mechanisms of symmetry breaking."

Think of the way a beautiful mirror shatters into a thousand pieces. The original mirror possessed great symmetry. You can rotate a mirror at any angle and it still reflects light in the same way. But after it is shattered, the original symmetry is broken. Determining precisely how the symmetry is broken determines how the mirror shatters.

SYMMETRY BREAKING

To see this, think of the development of an embryo. In its early stages, a few days after conception, an embryo consists of a perfect sphere of cells. Each cell is no different from the others. It looks the same no matter how we rotate it. Physicists say that the embryo at this stage has O(3) symmetry—that is, it remains the same no matter how you rotate it on any axis.

Although the embryo is beautiful and elegant, it is also rather useless. Being a perfect sphere, it cannot perform any useful functions or interact with the environment. In time, however, the embryo breaks this symmetry, developing a tiny head and torso, so it resembles a bowling pin. Although the original spherical symmetry is now broken, the embryo still has a residual symmetry; it remains the same if we spin it along its axis. Thus, it has cylindrical symmetry. Mathematically, we say that the original O(3) of the sphere has now been broken down to the O(2) symmetry of the cylinder.

The breaking of O(3) symmetry, however, could have proceeded in a different way. Starfish, for example, do not have cylindrical or bilateral symmetry; instead, when the spherical symmetry is broken, they have a C_5 symmetry (which remains the same under rotations by 72 degrees), giving it its five-pointed-star shape. Thus, the way in which the symmetry O(3) breaks determines the shape of the organism when it is born.

Similarly, scientists believe the universe started out in a state of perfect symmetry, with all the forces unified into a single force. The universe was beautiful, symmetrical, but rather useless. Life as we know it could not exist in this perfect state. In order for the possibility of life to exist, the symmetry of the universe had to break as it cooled.

SYMMETRY AND THE STANDARD MODEL

In the same way, to understand what parallel universes might look like, we must first understand the symmetries of the strong, weak, and electromagnetic interactions. The strong force, for example, is based on three quarks, which scientists label by giving them a fictitious "color" (for example, red, white, and blue). We want the equations to remain the same if we interchange these three colored quarks. We say that the equations have SU(3) symmetry, that is, when we reshuffle the three quarks, the equations remain the same. Scientists believe that a theory with SU(3) symmetry forms the most accurate description of the strong interactions (called Quantum Chromodynamics). If we had a gigantic supercomputer, starting with just the masses of the quarks and the strength of their interactions, we could, in theory, calculate all the properties of the proton and neutron and all the characteristics of nuclear physics.

Similarly, let's say we have two leptons, the electron and the neutrino. If we interchange them in an equation, we have SU(2) symmetry. We can also throw in light, which has the symmetry group U(1). (This symmetry group shuffles the various components or polarizations of light among each other.) Thus, the symmetry group of the weak and electromagnetic interactions is SU(2) × U(1).

If we simply glue these three theories together, not surprisingly we have the symmetry SU(3) × SU(2) × U(1), in other words, the symmetry that separately mixes three quarks among themselves and two leptons among themselves (but does not mix quarks with leptons). The resulting theory is the Standard Model, which, as we saw earlier, is perhaps one of the most successful theories of all time. As Gordon Kane of the University of Michigan says, "Everything that happens in our world (except for the effects of gravity) results from Standard Model particle interactions." Some of its predictions have been tested in the laboratory to hold within one part in a hundred million. (In fact, twenty Nobel Prizes have been awarded to physicists who have pieced together parts of the Standard Model.)

Finally, one might construct a theory that combines the strong, weak, and electromagnetic interaction into a single symmetry. The simplest GUT theory that can do this interchanges all five particles (three quarks and two leptons) into each other simultaneously. Unlike the Standard Model symmetry, the GUT symmetry can mix quarks and leptons together (which means that protons can decay into electrons). In other words, GUT theories contain SU(5) symmetry (reshuffling all five particles—three quarks and two leptons—among themselves). Over the years, many other symmetry groups have been analyzed, but SU(5) is perhaps the minimal group that fits the data.

When spontaneous breaking occurs, the original GUT symmetry can break in several ways. In one way, the GUT symmetry breaks down to SU(3) × SU(2) × U(1) with precisely 19 free parameters that we need to describe our universe. This gives us the known universe. However, there are actually many ways in which to break GUT symmetry. Other universes would most likely have a completely different residual symmetry. At the very minimum, these parallel universes might have different values of these 19 parameters. In other words, the strengths of the various forces would be different in different universes, leading to vast changes in the structure of the universe. By weakening the strength of the nuclear force, for example, one might prevent the formation of stars, leaving the universe in perpetual darkness, making life impossible. If the nuclear force is strengthened too much, stars could burn their nuclear fuel so fast that life would not have enough time to form.

The symmetry group may also be changed, creating an entirely different universe of particles. In some of these universes, the proton might not be stable and would rapidly decay into antielectrons. Such universes cannot have life as we know it, but would rapidly disintegrate into a lifeless mist of electrons and neutrinos. Other universes could break the GUT symmetry in yet another way, so there would be more stable particles, like protons. In such a universe, a huge variety of strange new chemical elements could exist. Life in those universes could be more complex than our own, with more chemical elements out of which to create DNA-like chemicals.

We can also break the original GUT symmetry so that we have more than one U(1) symmetry, so there is more than one form of light. This would be a strange universe, indeed, in which beings might "see" using not just one kind of force but several. In such a universe, the eyes of any living being could have a large variety of receptors to detect various forms of light-like radiation.

Not surprisingly, there are hundreds, perhaps even an infinite number of ways to break these symmetries. Each of these solutions, in turn, might correspond to an entirely separate universe.

TESTABLE PREDICTIONS

Unfortunately, the possibility of testing the multiverse theory, involving multiple universes with different sets of physical laws, is at present impossible. One would have to travel faster than light to reach these other universes. But one advantage of the inflation theory is that it makes predictions about the nature of our universe that *are* testable.

Since the inflationary theory is a quantum theory, it is based on the Heisenberg uncertainty principle, the cornerstone of the quantum theory. (The uncertainty principle states that you cannot make measurements with infinite accuracy, such as measuring the velocity and position of an electron. No matter how sensitive your instruments are, there will always be uncertainty in your measurements. If you know an electron's velocity, you cannot know its precise location; if you know its location, you cannot know its velocity.) Applied to the original fireball that set off the big bang, it means that the original cosmic explosion could not have been infinitely "smooth." (If it had been perfectly uniform, then we would know precisely the trajectories of the subatomic particles emanating from the big bang, which violates the uncertainty principle.) The quantum theory allows us to compute the size of these ripples or fluctuations in the original fireball. If we then inflate these tiny quantum ripples, we can calculate the minimum number of ripples we should see on the microwave background 380,000 years after the big bang. (And if we expand these ripples to the present day, we should find the current distribution of galactic clusters. Our galaxy itself started out in one of these tiny fluctuations.)

Initially, a superficial glance at the data from the COBE satellite found no deviations or fluctuations in the microwave background. This caused some anxiety among physicists, because a perfectly smooth microwave background would violate not just inflation but the entire quantum theory as well, violating the uncertainty principle. It would shake physics to its very core. The entire foundation of twentieth-century quantum physics might have to be thrown out.

Much to scientists' relief, a painstakingly detailed look at the computer-enhanced data from the COBE satellite found a blurry set of ripples, variations in temperature of 1 part in 100,000—the minimum amount of deviation tolerated by the quantum theory. These infinitesimal ripples were consistent with the inflationary theory. Guth confessed, "I'm completely snowed by the cosmic background radiation. The signal was so weak it wasn't even detected until 1965, and now they're measuring fluctuations of one part in 100,000."

Although the experimental evidence being gathered was slowly favoring inflation, scientists still had to resolve the nagging problem of the value of Omega—the fact that Omega was 0.3 rather than 1.0.

SUPERNOVAE—RETURN OF LAMBDA

While inflation turned out to be consistent with the COBE data scientists gathered, astronomers still grumbled in the 1990s that inflation was in flagrant violation of the experimental data on Omega. The tide first began to turn in 1998, as a result of data from a totally unexpected direction. Astronomers tried to recalculate the rate of expansion of the universe in the distant past. Instead of analyzing Cepheid variables, as Hubble did in the 1920s, they begin to examine supernovae in distant galaxies billions of light-years into the past. In particular, they examined type Ia supernovae, which are ideally suited for being used as standard candles.

Astronomers know that supernovae of this type have nearly the same brightness. (The brightness of type Ia supernovae is known so well that even small deviations can be calibrated: the brighter the supernova, the slower it declines in brightness.) Such supernovae are caused when a white dwarf star in a binary system slowly sucks matter from its companion star. By feeding off its sister star, this white dwarf gradually grows in mass until it weighs 1.4 solar masses, the maximum possible for a white dwarf. When they exceed this limit, they collapse and explode in a type Ia supernova. This trigger point is why type Ia supernovae are so uniform in brightness—it is the natural result of white dwarf stars reaching a precise mass and then collapsing under gravity. (As Subrahmanyan Chandrasekhar showed in 1935, in a white dwarf star the force of gravity crushing the star is balanced by a repulsive force between the electrons, called electron degeneracy pressure. If a white dwarf star weighs more than 1.4 solar masses, then gravity overcomes this force and the star is crushed, creating the supernova.) Since distant supernovae took place in the early universe, by analyzing them one can calculate the rate of expansion of the universe billions of years ago.

Two independent groups of astronomers (led by Saul Perlmutter of the Supernova Cosmology Project and Brian P. Schmidt of the High-Z Supernova Search Team) expected to find that the universe, although still expanding, was gradually slowing down. For several generations of astronomers, this was an article of faith, taught in every cosmology class—that the original expansion was gradually decelerating.

After analyzing about a dozen supernovae each, they found that the early universe was not expanding as fast as previously thought (that is, the redshifts of the supernovae and hence their velocity were smaller than originally suspected). When comparing the expansion rate of the early universe to today's expansion, they concluded that the expansion rate was relatively greater today. Much to their shock, these two groups came to the astounding conclusion that the universe is *accelerating*.

Much to their dismay, they found that it was impossible to fit the data with any value of Omega. The only way to make the data fit the theory was to reintroduce Lambda, the energy of the vacuum first introduced by Einstein. Moreover, they found that Omega was overwhelmed by an unusually large Lambda that was causing the universe to accelerate in a de Sitter–type expansion. The two groups independently came to this startling realization but were hesitant to publish their findings because of the strong historical prejudice that the value of Lambda was zero. As George Jacoby of the Kitt's

Peak Observatory has said, "The Lambda thing has always been a wild-eyed concept, and anybody crazy enough to say it's not zero was treated as kind of nuts."

Schmidt recalls, "I was still shaking my head, but we had checked everything . . . I was very reluctant about telling people, because I truly thought that we were going to get massacred." However, when both groups released their results in 1998, the sheer mountain of data they amassed could not be easily dismissed. Lambda, Einstein's "biggest blunder," which had been almost completely forgotten in modern cosmology, was now staging a remarkable comeback after ninety years of obscurity!

Physicists were dumbfounded. Edward Witten of the Institute for Advanced Study at Princeton said it was "the strangest experimental finding since I've been in physics." When the value of Omega, 0.3, was added to the value of Lambda, 0.7, the sum was (to within experimental error) equal to 1.0, the prediction of the inflationary theory. Like a jigsaw puzzle being assembled before our eyes, cosmologists were seeing the missing piece of inflation. It came from the vacuum itself.

This result was spectacularly reconfirmed by the WMAP satellite, which showed that the energy associated with Lambda, or dark energy, makes up 73 percent of all matter and energy in the universe, making it the dominant piece of the jigsaw puzzle.

PHASES OF THE UNIVERSE

Perhaps the greatest contribution of the WMAP satellite is that it gives scientists confidence that they are headed toward a "Standard Model" of cosmology. Although huge gaps still exist, astrophysicists are beginning to see outlines of a standard theory emerging from the data. According to the picture we are putting together now, the evolution of the universe proceeded in distinct stages as it cooled. The transition from these stages represents the breaking of a symmetry and the splitting off of a force of nature. Here are the phases and milestones as we know them today:

1. Before 10^{-43} seconds—Planck era

Almost nothing is certain about the Planck era. At the Planck energy (10^{19} billion electron volts), the gravitational force was as strong as the other quantum forces. As a consequence, the four forces of the universe were probably unified into a single "superforce." Perhaps the universe existed in a perfect phase of "nothingness," or empty higher-dimensional space. The mysterious symmetry that mixes all four forces, leaving the equations the same, is most likely "supersymmetry" (for a discussion of supersymmetry, see chapter 7). For reasons unknown, this mysterious symmetry that unified all four forces was broken, and a tiny bubble formed, our embryonic universe, perhaps as the result of a random, quantum fluctuation. This bubble was the size of the "Planck length," which is 10^{-33} centimeters.

2. 10^{-43} seconds—GUT era

Symmetry breaking occurred, creating a rapidly expanding bubble. As the bubble inflated, the four fundamental forces rapidly split off from each other. Gravity was the first force to be split off from the other three, releasing a shock wave throughout the universe. The original symmetry of the superforce was broken down to a smaller symmetry, perhaps containing the GUT symmetry SU(5). The remaining strong, weak, and electromagnetic interactions were still unified by this GUT symmetry. The universe inflated by an enormous factor, perhaps 10^{50}, during this phase, for reasons that are not understood, causing space to expand astronomically faster than the speed of light. The temperature was 10^{32} degrees.

3. 10^{-34} seconds—end of inflation

The temperature dropped to 10^{27} degrees as the strong force split off from the other two forces. (The GUT symmetry group broke down into SU(3) $\times$ SU(2) $\times$ U(1).) The inflationary period ended, allowing the universe to coast in a standard Friedmann expansion. The universe consisted of a hot plasma "soup" of free quarks, gluons, and leptons. Free quarks condensed into the protons and neutrons of today. Our universe was still quite small, only the size of the present solar system. Matter and antimatter were annihilated, but the tiny excess of matter over antimatter (one part in a billion) left behind the matter we see around us today. (This is the energy range that we hope will be duplicated in the next few years by the particle accelerator the Large Hadron Collider.)

4. 3 minutes—nuclei form

Temperatures dropped sufficiently for nuclei to form without being ripped apart from the intense heat. Hydrogen fused into helium (creating the current 75 percent hydrogen/25 percent helium ratio found today). Trace amounts of lithium were formed, but the fusion of higher elements stopped because nuclei with 5 particles were too unstable. The universe was opaque, with light being scattered by free electrons. This marks the end of the primeval fireball.

5. 380,000 years—atoms are born

The temperature dropped to 3,000 degrees Kelvin. Atoms formed as electrons settled around nuclei without being ripped apart by the heat. Photons could now travel freely without being absorbed. This is the radiation measured by COBE and WMAP. The universe, once opaque and filled with plasma, now became transparent. The sky, instead of being white, now became black.

6. 1 billion years—stars condense

The temperature dropped to 18 degrees. Quasars, galaxies, and galactic clusters began to condense, largely as a by-product of tiny quantum ripples in the original fireball. Stars began to "cook" the light elements, like carbon, oxygen, and nitrogen. Exploding stars spewed elements beyond iron into the heavens. This is the farthest era that can be probed by the Hubble space telescope.

7. 6.5 billion years—de Sitter expansion

The Friedmann expansion gradually ended, and the universe began to accelerate and enter an accelerating phase, called the de Sitter expansion, driven by a mysterious antigravity force that is still not understood.

8. 13.7 billion years—today

The present. The temperature has dropped to 2.7 degrees. We see the present universe of galaxies, stars, and planets. The universe is continuing to accelerate in a runaway mode.

THE FUTURE

Although inflation is the theory today that has the power to explain such a wide range of mysteries about the universe, this does not prove that it is correct. (In addition, rival theories have recently been proposed, as we see in chapter 7.) The supernova result has to be checked and rechecked, taking into account factors such as dust and anomalies in the production of supernovae. The "smoking gun" that would finally verify or disprove the inflationary scenario are "gravity waves" that were produced at the instant of the big bang. These gravity waves, like the microwave background, should still be reverberating throughout the universe and may actually be found by gravity wave detectors, as we see in chapter 9. Inflation makes specific predictions about the nature of these gravity waves, and these gravity wave detectors should find them.

But one of the most intriguing predictions of inflation cannot be directly tested, and that is the existence of "baby universes" existing in a multiverse of universes, each one obeying a slightly different set of physical laws. To understand the full implications of the multiverse, it is important to first understand that inflation takes full advantage of the bizarre consequences of both Einstein's equations and the quantum theory. In Einstein's theory, we have the possible existence of multiple universes, and in the quantum theory, we have the possible means of tunneling between them. And within a new framework called M-theory, we may have the final theory that can settle these questions about parallel universes and time travel, once and for all.

PART TWO

THE MULTIVERSE

CHAPTER FIVE

Dimensional Portals and Time Travel

> Inside every black hole that collapses may lie the seeds
> of a new expanding universe.
>
> —Sir Martin Rees

> Black holes may be apertures to elsewhen. Were we to
> plunge down a black hole, we would re-emerge, it is con-
> jectured, in a different part of the universe and in an-
> other epoch in time . . . Black holes may be entrances to
> Wonderlands. But are there Alices or white rabbits?
>
> —Carl Sagan

GENERAL RELATIVITY is like a Trojan horse. On the surface, the theory is magnificent. With a few simple assumptions, one can obtain the general features of the cosmos, including the bending of starlight and the big bang itself, all of which have been measured to astonishing accuracy. Even inflation can be accommodated if we insert a cosmological constant by hand into the early universe. These solutions give us the most compelling theory of the birth and death of the universe.

But lurking inside the horse, we find all sorts of demons and goblins, including black holes, white holes, wormholes, and even time machines, which defy common sense. These anomalies were considered so bizarre that even Einstein himself thought that they would never be found in nature. For years, he fought strenuously against these strange solutions. Today, we know that these anomalies cannot be easily dismissed. They are an integral part of general relativity. And in fact, they may even provide a salvation to any intelligent being confronting the big freeze.

But perhaps the strangest of these anomalies is the possibility of parallel universes and gateways connecting them. If we recall the metaphor introduced by Shakespeare that all the world is a stage, then general relativity admits the possibility of trapdoors. But instead of leading to the basement, we find that the trapdoors lead to parallel stages like the original. Imagine the stage of life consisting of multistory stages, one on top of the next. On each stage, the actors read their lines and wander around the set, thinking that their stage is the only one, oblivious of the possibilities of alternate realities. However, if one day they accidentally fall into a trapdoor, they find themselves thrust into an entirely new stage, with new laws, new rules, and a new script.

But if an infinite number of universes can exist, then is life possible in any of these universes with different physical laws? It is a question that Isaac Asimov posed in his classic science fiction tale *The Gods Themselves*, where he created a parallel universe with a nuclear force different from our own. New intriguing possibilities arise when the usual laws of physics are repealed and new ones are introduced.

The story begins in the year 2070, when a scientist, Frederick Hallam, notices that ordinary tungsten-186 is strangely being converted into a mysterious plutonium-186, which has too many protons and should be unstable. Hallam theorizes that this strange plutonium-186 comes from a parallel universe where the nuclear force is much stronger, so it overcomes the repulsion of the protons. Since this strange plutonium-186 gives off large amounts of energy in the form of electrons, it can be harnessed to give fabulous amounts of free energy. This makes possible the celebrated Hallam electron pump, which solves Earth's energy crisis, making him a wealthy man. But there is a price to pay. If enough alien plutonium-186 enters our universe, then the nuclear force in general will increase in intensity. This means more energy will be released from the fusion process, and the Sun will brighten and eventually explode, destroying the entire solar system!

Meanwhile, the aliens in the parallel universe have a different perspective. Their universe is dying. The nuclear force is quite strong in their universe, meaning that the stars have been consuming hydrogen at an enormous rate and will soon die. They set up the exchange whereby useless plutonium-186 is sent to our universe in exchange for valuable tungsten-186, which allows them to create the positron pump, which saves their dying world. Although they realize that the nuclear force will increase in strength in our universe, causing our stars to explode, they don't care.

Earth, it seems, is headed for disaster. Humanity has become addicted to Hallam's free energy, refusing to believe that the Sun will soon explode. Another scientist comes up with an ingenious solution to this conundrum. He is convinced that other parallel universes must exist. He successfully modifies a powerful atom smasher to create a hole in space that connects our universe to many others. Searching among them, he finally finds one parallel universe that is empty except for a "cosmic egg" containing unlimited amounts of energy, but with a weaker nuclear force.

By siphoning energy from this cosmic egg, he can create a new energy pump and, at the same time, weaken the nuclear force in our universe, thus preventing the Sun from exploding. There is, however, a price to be paid: this new parallel universe will have its nuclear force increased, causing it to explode. But he reasons that this explosion will merely cause the cosmic egg to "hatch," creating a new big bang. In effect, he realizes, he will become a midwife to a new expanding universe.

Asimov's science fiction tale is one of the few to actually use the laws of nuclear physics to spin a tale of greed, intrigue, and salvation. Asimov was correct in assuming that changing the strength of the forces in our universe would have disastrous consequences, that the stars in our universe would brighten and then explode if the nuclear force was increased in strength. This raises the inevitable question: are parallel universes consistent with the laws of physics? And if so, what would be required to enter one?

To understand these questions, we must first understand the nature of wormholes, negative energy, and, of course, those mysterious objects called black holes.

BLACK HOLES

In 1783, British astronomer John Michell was the first to wonder what would happen if a star became so large that light itself could not escape. Any object, he knew, had an "escape velocity," the velocity required to leave its gravitational pull. (For Earth, for example, the escape velocity is 25,000 miles per hour, the speed that any rocket must attain in order to break free of Earth's gravity.)

Michell wondered what might happen if a star became so massive that its escape velocity was equal to the speed of light. Its gravity would be so immense that nothing could escape it, not even light itself, and hence the object would appear black to the outside world. Finding such an object in space would in some sense be impossible, since it would be invisible.

The question of Michell's "dark stars" was largely forgotten for a century and a half. But the matter resurfaced in 1916, when Karl Schwarzschild, a German physicist serving the German army on the Russian front, found an exact solution of Einstein's equations for a massive star. Even today, the Schwarzschild solution is known to be the simplest and most elegant exact solution of Einstein's equations. Einstein was astonished that Schwarzschild could find a solution to his complex tensor equations while dodging artillery shells. He was equally astonished that Schwarzschild's solution had peculiar properties.

The Schwarzschild solution, from a distance, could represent the gravity of an ordinary star, and Einstein quickly used the solution to calculate the gravity surrounding the Sun and check his earlier calculations, in which he had made approximations. For this he was eternally thankful to Schwarzschild. But in Schwarzschild's second paper, he showed that surrounding a very massive star there was an imaginary "magic sphere" with bizarre properties. This "magic sphere" was the point of no return. Anyone passing through the "magic sphere" would be immediately sucked by gravity into the star, never to be seen again. Not even light could escape if it fell into this sphere. Schwarzschild did not realize that he was rediscovering Michell's dark star, through Einstein's equations.

He next calculated the radius for this magic sphere (called the Schwarzschild radius). For an object the size of our Sun, the magic sphere was about 3 kilometers (roughly 2 miles). (For Earth, its Schwarzschild radius was about a centimeter.) This meant that if one could compress the Sun down to 2 miles, then it would become a dark star and devour any object that passed this point of no return.

Experimentally, the existence of the magic sphere caused no problems, since it was impossible to squeeze the sun down to 2 miles. No mechanism was known to create such a fantastic star. But theoretically, it was a disaster. Although Einstein's general theory of relativity could yield brilliant results, like the bending of starlight around the Sun, the theory made no sense as you approached the magic sphere itself, where gravity became infinite.

A Dutch physicist, Johannes Droste, then showed that the solution was even crazier. According to relativity, light beams, he showed, would bend severely as they whipped around the object. In fact, at 1.5 times the Schwarzschild radius, light beams actually orbited in circles around the star. Droste showed that the distortions of time found in general relativity around these massive stars were much worse than those found in special relativity. He showed that, as you approached this magic sphere, someone from a distance would say that your clocks were getting slower and slower, until your clocks stopped totally when you hit the object. In fact, someone from the outside would say that you were frozen in time as you reached the magic sphere. Because time itself would stop at this point, some physicists believed that such a bizarre object could never exist in nature. To make matters even more interesting, mathematician Herman Weyl showed that if one investigated the world inside the magic sphere, there seemed to be another universe on the other side.

This was all so fantastic that even Einstein could not believe it. In 1922, during a conference in Paris, Einstein was asked by mathematician Jacques Hadamard what would happen if this "singularity" were real, that is, if gravity became infinite at the Schwarzschild radius. Einstein replied, "It would be a true disaster for the theory; and it would be very difficult to say *a priori* what could happen physically because the formula does not apply anymore." Einstein would later call this the "Hadamard disaster." But he thought that all this controversy around dark stars was pure speculation. First, no one had ever seen such a bizarre object, and perhaps they didn't exist, that is, they were unphysical. Moreover, you would be crushed to death if you ever fell into one. And since one could never pass through the magic sphere (since time has stopped), no one could never enter this parallel universe.

In the 1920s, physicists were thoroughly confused about this issue. But in 1932, an important breakthrough was made by Georges Lemaître, father of the big bang theory. He showed that the magic sphere was not a singularity at all where gravity became infinite; it was just a mathematical illusion caused by choosing an unfortunate set of mathematics. (If one chose a different set of coordinates or variables to examine the magic sphere, the singularity disappeared.)

Taking this result, the cosmologist H. P. Robertson then reexamined Droste's original result that time stops at the magic sphere. He found that time stopped only from the vantage point of an observer watching a rocket ship enter the magic sphere. From the vantage point of the rocket ship itself, it would only take a fraction of a second for gravity to suck you right past the magic sphere. In other words, a space traveler unfortunate enough to pass through the magic sphere would find himself crushed to death almost instantly, but to an observer watching from the outside, it would appear to take thousands of years.

This was an important result. It meant that the magic sphere was reachable and could no longer be dismissed as a mathematical monstrosity. One had to seriously consider what might happen if one passed through the magic sphere. Physicists then calculated what a journey through the magic sphere might look like. (Today, the magic sphere is called the event horizon. The horizon refers to the farthest point one can see. Here, it refers to the farthest point light can travel. The radius of the event horizon is called the Schwarzschild radius.)

As you approached the black hole in a rocket ship, you would see light that had been captured billions of years ago by the black hole, dating back to when the black hole itself was first created. In other words, the life history of the black hole would be revealed to you. As you got closer, tidal forces would gradually rip the atoms of your body apart, until even the nuclei of your atoms would look like spaghetti. The journey through the event horizon would be a one-way trip, because gravity would be so intense that you would inevitably be sucked right into the center, where you will be crushed to death. Once inside the event horizon, there could be no turning back. (To leave the event horizon, one would have to travel faster than light, which is impossible.)

In 1939, Einstein wrote a paper in which he tried to dismiss such dark stars, claiming that they cannot be formed by natural processes. He started by assuming that a star forms from a swirling collection of dust, gas, and debris rotating in a sphere, gradually coming together because of gravity. He then showed that this collection of swirling particles will never collapse to within its Schwarzschild radius, and hence will never become a black hole. At best, this swirling mass of particles will approach 1.5 times the Schwarzschild radius, and hence black holes will never form. (To go below 1.5 times the Schwarzschild radius, one would have to travel faster than the speed of light, which is impossible.) "The essential result of this investigation is a clear understanding of why the 'Schwarzschild singularities' do not exist in physical reality," Einstein wrote.

Arthur Eddington, too, had deep reservations about black holes and bore a lifelong suspicion that they could never exist. He once said that there should "be a law of Nature to prevent a star from behaving in this absurd way."

Ironically, that same year, J. Robert Oppenheimer (who would later build the atomic bomb) and his student Hartland Snyder showed that a black hole *could* indeed form, via another mechanism. Instead of assuming that a black hole came about from a swirling collection of particles collapsing under gravity, they used as their starting point an old, massive star that has used up its nuclear fuel and hence implodes under the force of gravity. For example, a dying, giant star forty times the mass of the Sun might exhaust its nuclear fuel and be compressed by gravity to within its Schwarzschild radius of 80 miles, in which case it would inevitably collapse into a black hole. Black holes, they suggested, were not only possible, they might be the natural end point for billions of dying giant stars in the galaxy. (Perhaps the idea of implosion, pioneered in 1939 by Oppenheimer, gave him the inspiration for the implosion mechanism used in the atomic bomb just a few years later.)

EINSTEIN-ROSEN BRIDGE

Although Einstein thought that black holes were too incredible to exist in nature, he then ironically showed that they were even stranger than anyone thought, allowing for the possibility of wormholes lying at the heart of a black hole. Mathematicians call them multiply connected spaces. Physicists call them wormholes because, like a worm drilling into the earth, they create an alternative shortcut between two points. They are sometimes called dimensional portals, or gateways. Whatever you call them, they may one day provide the ultimate means for interdimensional travel.

The first person to popularize wormholes was Charles Dodgson, who wrote under the pen name of Lewis Carroll. In *Through the Looking Glass*, he introduced the wormhole as the looking glass, which connected the countryside of Oxford to Wonderland. As a professional mathematician and Oxford don, Dodgson was familiar with these multiply connected spaces. By definition, a multiply connected space is one in which a lasso cannot be shrunk down to a point. Usually, any loop can effortlessly be collapsed to a point. But if we analyze a doughnut, then it's possible to place the lasso on its surface so that it encircles the doughnut hole. As we slowly collapse the loop, we find that it cannot be compressed to a point; at best, it can be shrunk to the circumference of the hole.

Mathematicians delighted in the fact that they had found an object that was totally useless in describing space. But in 1935, Einstein and his student Nathan Rosen introduced wormholes into the world of physics. They were trying to use the black hole solution as a model for elementary particles. Einstein never liked the idea, dating back to Newton, that a particle's gravity became infinite as you approached it. This "singularity," thought Einstein, should be removed because it made no sense.

Einstein and Rosen had the novel idea of representing an electron (which was usually thought of as a tiny point without any structure) as a black hole. In this way, general relativity could be used to explain the mysteries of the quantum world in a unified field theory. They started with the standard black hole solution, which resembles a large vase with a long throat. They then cut the throat, and merged it with another black hole solution that was flipped over. To Einstein, this strange but smooth configuration would be free of the singularity at the origin of the black hole and might act like an electron.

Unfortunately, Einstein's idea of representing an electron as a black hole failed. But today, cosmologists speculate that the Einstein-Rosen bridge can act as a gateway between two universes. We could move about freely in one universe until accidentally falling into a black hole, where we would be suddenly sucked through the hole to emerge on the other side (through a white hole).

To Einstein, any solution of his equations, if it began with a physically plausible starting point, should correspond to a physically possible object. But he wasn't worried about someone falling into a black hole and entering a parallel universe. The tidal forces would become infinite at the center, and anyone unfortunate enough to fall into a black hole would have their atoms ripped apart by the gravitational field. (The Einstein-Rosen bridge does open up momentarily, but it closes so fast that no object can pass through it in time to reach

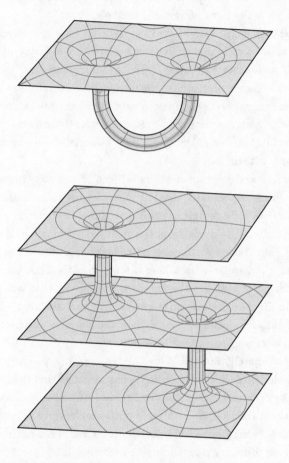

The Einstein-Rosen bridge. At the center of a black hole, there is a "throat" that connects space-time to another universe or another point in our universe. Although travel through a stationary black hole would be fatal, rotating black holes have a ringlike singularity, such that it may be possible to pass through the ring and through the Einstein-Rosen bridge, although this is still speculative.

the other side.) Einstein's attitude was that, while wormholes may exist, living creatures could never pass through one and live to tell about it.

ROTATING BLACK HOLES

In 1963, however, this view began to change, when New Zealand mathematician Roy Kerr found an exact solution of Einstein's equation describing perhaps the most realistic dying star, a spinning black hole. Because of the conservation of angular momentum, as a star collapses under gravity, it spins faster. (This is the same reason why spinning galaxies look like pinwheels, and why skaters spin faster when they bring their arms in.) A spinning star could collapse into a ring of neutrons, which would remain stable because of the intense centrifugal force pushing outward, canceling the inward force of gravity. The astonishing feature of such a black hole was that if you fell into the Kerr black hole, you would *not* be crushed to death. Instead, you would be sucked completely through the Einstein-Rosen bridge to a parallel universe. "Pass through this magic ring and—presto!—you're in a completely different universe where radius and mass are negative!" Kerr exclaimed to a colleague, when he discovered this solution.

The frame of Alice's looking glass, in other words, was like the spinning ring of Kerr. But any trip through the Kerr ring would be a one-way trip. If you were to pass through the event horizon surrounding the Kerr ring, the gravity would not be enough to crush you to death, but it would be sufficient to prevent a return trip back through the event horizon. (The Kerr black hole, in fact, has two event horizons. Some have speculated that you might need a second Kerr ring, connecting the parallel universe back to ours, in order to make a return trip.) In some sense, a Kerr black hole can be compared to an elevator inside a skyscraper. The elevator represents the Einstein-Rosen bridge, which connects different floors, where each floor is a different universe. In fact, there are an infinite number of floors in this skyscraper, each one different from the others. But the elevator can never go down. There is only an "up" button. Once you leave a floor, or universe, there would be no turning back because you would have passed an event horizon.

Physicists are divided about how stable a Kerr ring would be. Some calculations suggest that if one tried to pass through the ring, the person's very presence would destabilize the black hole, and the gateway would close. If a light beam, for example, were to pass into the Kerr black hole, it would gain enormously in energy as it fell toward the center and become blue-shifted—that is, it would increase in frequency and energy. As it approached the horizon, it would have so much energy that it would kill anyone trying to pass through the Einstein-Rosen bridge. It would also generate its own gravitational field, which would interfere with the original black hole, perhaps destroying the gateway.

In other words, while some physicists believe that the Kerr black hole is the most realistic of all black holes, and could indeed connect parallel universes, it is not clear how safe it would be to enter the bridge or how stable the doorway would be.

OBSERVING BLACK HOLES

Because of the bizarre properties of black holes, as late as the early 1990s their existence was still considered science fiction. "Ten years ago, if you found an object that you thought was a black hole in the center of a galaxy, half the field thought you were a little nuts," remarked astronomer Douglas Richstone of the University of Michigan in 1998. Since then, astronomers have identified several hundred black holes in outer space via the Hubble space telescope, the Chandra X-ray space telescope (which measures X-ray emissions from powerful stellar and galactic sources), and the Very Large Array Radio Telescope (which consists of a series of powerful radio telescopes in New Mexico). Many astronomers believe, in fact, that most of the galaxies in the heavens (which have central bulges at the center of their disks) have black holes at their centers.

As predicted, all of the black holes found in space are rotating very rapidly; some have been clocked by the Hubble space telescope rotating at about a million miles per hour. At the very center, one can see a flat, circular core often about a light-year across. Inside that core lies the event horizon and the black hole itself.

Because black holes are invisible, astronomers have to use indirect means to verify their existence. In photographs, they try to identify the "accretion disk" of swirling gas that surrounds the black hole. Astronomers have now collected beautiful photographs of these accretion disks. (These disks are almost universally found for most rapidly spinning objects in the universe. Even our own Sun probably had a similar disk surrounding it when it formed 4.5 billion years ago, which later condensed into the planets. The reason these disks form is that they represent the lowest state of energy for such a rapidly spinning object.) By using Newton's laws of motion, astronomers can calculate the mass of the central object by knowing the velocity of the stars orbiting around it. If the mass of the central object has an escape velocity equal to the speed of light, then even light itself cannot escape, providing indirect proof of the existence of a black hole.

The event horizon lies at the center of the accretion disk. (It is unfortunately too small to be identified with current technology. Astronomer Fulvio Melia claims that capturing the event horizon of a black hole on film is the "holy grail" of black hole science.) Not all the gas that falls toward a black hole passes through the event horizon. Some of it bypasses the event horizon and is hurled past it at huge velocities and ejected into space, forming two long jets of gas emanating from the black hole's north and south poles. This gives the black hole the appearance of a spinning top. (The reason jets are ejected like this is probably that the magnetic field lines of the collapsing star, as they become more intense, become concentrated above the north and south poles. As the star continues to collapse, these magnetic field lines condense into two tubes emanating from the north and south poles. As ionized particles fall into the collapsed star, they follow these narrow magnetic lines of force and are ejected as jets via the north and south polar magnetic fields.)

Two types of black holes have been identified. The first is the stellar black hole, in which gravity crushes a dying star until it implodes. The second, however, is more easily detected. These are galactic black holes, which lurk at the very centers of huge galaxies and quasars and weigh millions to billions of solar masses.

Recently, a black hole was conclusively identified in the center of our own Milky Way galaxy. Unfortunately, dust clouds obscure the galactic center; if not for that, a huge fireball would be visible to us on Earth every night coming from the direction of the constellation Sagittarius. Without the dust, the center of our Milky Way galaxy would probably outshine the Moon, making it the brightest object in the night sky. At the very center of this galactic nucleus lies a black hole that weighs about 2.5 million solar masses. In terms of its size, it is about a tenth of the radius of the orbit of Mercury. By galactic standards, this is not an especially massive black hole; quasars can have black holes that weigh several billion solar masses. The black hole in our backyard is rather quiescent at present.

The next closest galactic black hole lies at the center of the Andromeda galaxy, the closest galaxy to Earth. It weighs 30 million solar masses, and its Schwarzschild radius is about 60 million miles. (At the center of the Andromeda galaxy lie at least two massive objects, probably the leftovers of a previous galaxy that was devoured by Andromeda billions of years ago. If the Milky Way galaxy eventually collides with Andromeda billions of years from now, as appears likely, perhaps our galaxy will wind up in the "stomach" of the Andromeda galaxy.)

One of the most beautiful photographs of a galactic black hole is the one taken by the Hubble space telescope of the galaxy NGC 4261. In the past, radio telescope pictures of this galaxy showed two very graceful jets being shot out of the galaxy's north and south poles, but no one knew what the engine behind it was. The Hubble telescope photographed the very center of the galaxy, revealing a beautiful disk about 400 light-years across. At its very center was a tiny dot containing the accretion disk, about a light-year across. The black hole at the center, which could not be seen by the Hubble telescope, weighs approximately 1.2 billion solar masses.

Galactic black holes like this are so powerful they can consume entire stars. In 2004, NASA and the European Space Agency announced that they had detected a huge black hole in a distant galaxy devouring a star in a single gulp. The Chandra X-ray telescope and the European XMM-Newton satellite both observed the same event: a burst of X rays being emitted by the galaxy RX J1242–11, signaling that a star had been gobbled up by the huge black hole at the center. This black hole has been estimated to weigh 100 million times the mass of our Sun. Calculations have shown that, as a star comes perilously close to the event horizon of a black hole, the enormous gravity distorts and stretches the star until it breaks apart, emitting a telltale burst of X rays. "This star was stretched beyond its breaking point. This unlucky star just wandered into the wrong neighborhood," observed astronomer Stefanie Komossa of the Max Planck Institute in Garching, Germany.

The existence of black holes has helped to solve many old mysteries. The galaxy M-87, for example, was always a curiosity to astronomers because it looked like a massive ball of stars with a strange "tail" emerging from it. Because it emitted copious quantities of radiation, at one point astronomers thought that this tail represented a stream of antimatter. But today, astronomers have found that it is energized by a huge black hole weighing perhaps 3 billion solar masses. And that strange tail is now believed to be a gigantic jet of plasma which is streaming out of, not into, the galaxy.

One of the more spectacular discoveries concerning black holes occurred when the Chandra X-ray telescope was able to peer through a small gap in the dust in outer space to observe a collection of black holes near the edge of the visible universe. In all, six hundred black holes could be seen. Extrapolating from that, astronomers estimate there are at least 300 million black holes over the entire night sky.

GAMMA RAY BURSTERS

The black holes mentioned above are perhaps billions of years old. But astronomers now have the rare opportunity to see black holes being formed right before our eyes. Some of these are probably the mysterious gamma ray bursters which release the largest amount of energy in the universe. Huge gamma ray bursters are second only to the big bang itself in terms of the energy they release.

Gamma ray bursters have a curious history, dating back to the Cold War. In the late 1960s, the United States was worried that the Soviet Union or another country might secretly detonate a nuclear bomb, perhaps on a deserted part of the Earth or even on the Moon, violating existing treaties. So the United States launched the Vela satellite to specifically spot "nuke flashes," or unauthorized detonations of nuclear bombs. Because a nuclear detonation unfolds in distinct stages, microsecond by microsecond, each nuke flash gives off a characteristic double flash of light that can be seen by satellite. (The Vela satellite did pick up two such nuke flashes in the 1970s off the coast of Prince Edward Island near South Africa, in the presence of Israeli war ships, sightings that are still being debated by the intelligence community.)

But what startled the Pentagon was that the Vela satellite was picking up signs of huge nuclear explosions in space. Was the Soviet Union secretly detonating hydrogen bombs in deep space, using an unknown, advanced technology? Concerned that the Soviets might have leapfrogged over the U.S. in weapons technology, top scientists were brought in to analyze these deeply disturbing signals.

After the breakup of the Soviet Union, there was no need to classify this information, so the Pentagon dumped a mountain of astronomical data onto the world of astronomy, which was overwhelming. For the first time in decades, an entirely new astronomical phenomenon of immense power and scope had been revealed. Astronomers quickly realized that these gamma ray bursters, as they were called, were titanic in their power, releasing within seconds the entire energy output of our Sun over its entire life history (about 10 billion years). But these events were also fleeting; once detected by the Vela satellite, they had dimmed so much that by the time ground telescopes were pointed in their direction, nothing could be seen in their wake. (Most bursters last between 1 and 10 seconds, but the shortest one lasted 0.01 second, and some lasted as long as several minutes.)

Today, space telescopes, computers, and rapid response teams have changed our ability to spot gamma ray bursters. About three times a day, gamma ray bursters are detected, setting off a complex chain of events. As soon as the energy from one is detected by satellite, astronomers using computers rapidly locate its precise coordinates and aim more telescopes and sensors in its precise direction.

The data from these instruments has revealed truly astounding results. At the heart of these gamma ray bursters lies an object often only a few tens of miles across. In other words, the unimaginable cosmic power of gamma ray bursters is concentrated within an area the size of, say, New York City. For years, the leading candidates for such events were colliding neutron stars in a binary star system. According to this theory, as the orbit of these neutron stars decayed over time, and as they followed a death spiral, they would ultimately collide and create a mammoth release of energy. Such events are extremely rare, but because the universe is so large, and since these bursters light up the entire universe, they should be seen several times a day.

But in 2003, new evidence scientists collected suggested that gamma ray bursters are the result of a "hypernova" that creates a massive black hole. By rapidly focusing telescopes and satellites in the direction of gamma ray bursters, scientists found that they resembled a massive supernova. Since the exploding star has an enormous magnetic field and ejects radiation via its north and south polar directions, it might appear as if the supernova is more energetic than it actually is—that is, we observe these bursters only if they are pointed directly at Earth, giving the false impression that they are more powerful than they really are.

If indeed gamma ray bursters are black holes in formation, then the next generation of space telescopes should be able to analyze them in great detail and perhaps answer some of our deepest questions about space and time. Specifically, if black holes can bend space into a pretzel, can they also bend time?

VAN STOCKUM'S TIME MACHINE

Einstein's theory links space and time into an inseparable unity. As a result, any wormhole that connects two distant points in space might also connect two distant points in time. In other words, Einstein's theory allows for the possibility of time travel.

The concept of time itself has evolved over the centuries. To Newton, time was like an arrow; once fired, it never changed course and traveled unerringly and uniformly to its target. Einstein then introduced the concept of warped space, so time was more like a river that gently speeded up or slowed down as it meandered through the universe. But Einstein worried about the possibility that perhaps the river of time can bend back on itself. Perhaps there could be whirlpools or forks in the river of time.

In 1937, this possibility was realized when W. J. Van Stockum found a solution to Einstein's equations which permitted time travel. He began with an infinite, spinning cylinder. Although it's not physically possible to build an infinite object, he calculated that if such a cylinder spun around at or near the speed of light, it would drag the fabric of space-time along with it, much like molasses is dragged along with the blades of a blender. (This is called frame-dragging, and it has now been experimentally seen in detailed photographs of rotating black holes.)

Anyone brave enough to travel around the cylinder would be swept along, attaining fantastic speeds. In fact, to a distant observer, it would appear that the individual was exceeding the speed of light. Although Van Stockum himself did not realize it at the time, by making a complete trip around the cylinder, you could actually go back in time, returning before you left. If you left at noon, then by the time you returned to your starting point, say, it might be 6 p.m. the previous night. The faster the cylinder spun, the further back in time you would go (the only limitation being that you could not go further back in time than the creation of the cylinder itself).

Since the cylinder is like a maypole, every time you danced around the pole, you would wind up further and further back in time. Of course, one could dismiss such a solution because cylinders cannot be infinitely long. Also, if such a cylinder could be built, the centrifugal forces on the cylinder, because it spins near the speed of light, would be enormous, causing the material that made up the cylinder to fly apart.

GÖDEL UNIVERSE

In 1949, Kurt Gödel, the great mathematical logician, found an even stranger solution to Einstein's equations. He assumed that the entire universe was rotating. Like the Van Stockum cylinder, one is swept up by the molasses-like nature of space-time. By taking a rocket ship around the Gödel universe, you return to your starting point but shift back in time.

In Gödel's universe, a person can, in principle, travel between any two points in space and time in the universe. Every event, in any time period, can be visited, no matter how distant in the past. Because of gravity, there is a tendency for Gödel's universe to collapse on itself. Hence, the centrifugal force of rotation must balance this gravitational force. In other words, the universe must spin above a certain speed. The larger the universe, the greater the tendency to collapse, and the faster the universe would have to spin to prevent collapse.

For a universe our size, for example, Gödel calculated that it would have to rotate once every 70 billion years, and the minimum radius for time travel would be 16 billion light-years. To travel back in time, however, you would have to travel just below the speed of light.

Gödel was well aware of the paradoxes that could arise from his solution—the possibility of meeting yourself in the past and altering the course of history. "By making a round trip on a rocket ship in a sufficiently wide course, it is possible in these worlds to travel into any region of the past, present, and future, and back again, exactly as it is possible in other worlds to travel to distant parts of space," he wrote. "This state of affairs seems to imply an absurdity.

For it enables one to travel into the near past of those places where he has himself lived. There he would find a person who would be himself at some earlier period of life. Now he could do something to this person which, by his memory, he knows has not happened to him."

Einstein was deeply disturbed by the solution found by his friend and neighbor at the Institute for Advanced Study at Princeton. His response is quite revealing:

> Kurt Gödel's essay constitutes, in my opinion, an important contribution to the general theory of relativity, especially to the analysis of the concept of time. The problem here involved disturbed me already at the time of the building up of the general theory of relativity, without my having succeeded in clarifying it . . . The distinction "earlier-later" is abandoned for world-points which lie far apart in a cosmological sense, and those paradoxes, regarding the direction of the causal connection, arise, of which Mr. Gödel has spoken . . . It will be interesting to weigh whether these are not to be excluded on physical grounds.

Einstein's response is interesting for two reasons. First, he admitted that the possibility of time travel bothered him when he first formulated general relativity. Since time and space are treated like a piece of rubber that can bend and warp, Einstein worried that the fabric of space-time would warp so much that time travel might be possible. Second, he ruled out Gödel's solution on the basis of "physical grounds"—that is, the universe does not spin, it expands.

When Einstein died, it was widely known that his equations allowed for strange phenomena (time travel, wormholes). But no one gave them much thought because scientists felt they could not be realized in nature. The consensus was that these solutions had no basis in the real world; you would die if you tried to reach a parallel universe via a black hole; the universe did not spin; and you cannot make infinite cylinders, making time travel an academic question.

THORNE TIME MACHINE

The issue of time travel lay dormant for thirty-five years until 1985, when the astronomer Carl Sagan was writing his novel *Contact* and wanted to incorporate a way in which the heroine could travel to the star Vega. This would require a two-way journey, one in which the heroine would travel to Vega and then return to Earth, something that would not be allowed by black hole–type wormholes. He turned to the physicist Kip Thorne for advice. Thorne shocked the physics world by finding new solutions to Einstein's equations that allowed for time travel without many of the previous problems. In 1988, with colleagues Michael Morris and Ulvi Yurtsever, Thorne showed that it was possible to build a time machine if one could somehow obtain strange forms of matter and energy, such as "exotic negative matter" and "negative energy." Physicists were at first skeptical of this new solution, since no one had ever seen this exotic matter before, and negative energy only exists in minute quantities. But it represented a breakthrough in our understanding of time travel.

The great advantage of negative matter and negative energy is that they make a wormhole transversable, so you can make a two-way trip through it without having to worry about event horizons. In fact, Thorne's group found that a trip through such a time machine might be quite mild, compared to the stress found on a commercial airline.

One problem, however, is that exotic matter (or negative matter) is quite extraordinary in its properties. Unlike antimatter (which is known to exist and most likely falls to the ground under Earth's gravitational field), negative matter falls up, so it will float upward in Earth's gravity because it possesses antigravity. It is repelled, not attracted, by ordinary matter, and by other negative matter. This means that it is also quite difficult to find in nature, if it exists at all. When Earth was first formed 4.5 billion years ago, any negative matter on Earth would have floated away into deep space. So negative matter might possibly be floating in space, far away from any planets. (Negative matter will probably never strike a passing star or planet, since it is repelled by ordinary matter.)

While negative matter has never been seen (and quite possibly does not exist), negative energy is physically possible but extremely rare. In 1933, Henrik Casimir showed that two uncharged parallel metal plates can create negative energy. Normally, one would expect that two plates would remain stationary because they are uncharged. However, Casimir showed that there is a very small attractive force between these two uncharged parallel plates. In 1948, this tiny force was actually measured, showing that negative energy was a real possibility. The Casimir effect exploits a rather bizarre feature of the vacuum. According to the quantum theory, empty space is teeming with "virtual particles" which dance in and out of nothingness. This violation of the conservation of energy is possible because of the Heisenberg uncertainty principle, which allows for violations of cherished classical laws as long as they occur very briefly. For example, an electron and antielectron, due to uncertainty, have a certain small probability of being created out of nothing and then annihilating each other. Because the parallel plates are very close to each other, these virtual particles cannot easily come between the two plates. Thus, because there are more virtual particles surrounding the plates than there are between them, this creates an inward force from the outside that pushes the parallel plates together slightly. This effect was precisely measured in 1996 by Steven Lamoreaux at the Los Alamos National Laboratory. The attractive force he measured was tiny (equal to the weight of 1/30,000 of an insect like an ant). The smaller the separation of the plates, the greater the force of attraction.

So here is how the time machine Thorne dreamed up might operate. An advanced civilization would start with two parallel plates, separated by an extremely small gap. These parallel plates would then be reshaped into a sphere, so the sphere consists of an inner and outer shell. Then they would make two such spheres and somehow string a wormhole between them, so a tunnel in space connects both spheres. Each sphere now encloses a mouth of the wormhole.

Normally, time beats in synchronization for both spheres. But if we now put one sphere into a rocket ship that is sent speeding near the speed of light, time slows down for that rocket ship, so that the two spheres are no longer synchronized in time. The clock on the rocket ship beats much slower than the clock on Earth. Then if one jumps into the sphere on Earth, one may be sucked through the wormhole connecting them and wind up in the other rocket ship, sometime in the past. (This time machine, however, cannot take you back before the creation of the machine itself.)

PROBLEMS WITH NEGATIVE ENERGY

Although Thorne's solution was quite sensational when announced, there were severe obstacles to its actual creation, even for an advanced civilization. First, one must obtain large quantities of negative energy, which is quite rare. This type of wormhole depends on a huge amount of negative energy to keep the wormhole's mouth open. If one creates negative energy via the Casimir effect, which is quite small, then the size of the wormhole would have to be much smaller than an atom, making travel through the wormhole impractical. There are other sources of negative energy besides the Casimir effect, but all of them are quite difficult to manipulate. For example, physicists Paul Davies and Stephen Fulling have shown that a rapidly moving mirror can be shown to create negative energy, which accumulates in front of the mirror as it moves. Unfortunately, one has to move the mirror at near light speed in order to obtain negative energy. And like the Casimir effect, the negative energy created is small.

Another way to extract negative energy is to use high-powered laser beams. Within the energy states of the laser, there are "squeezed states" in which positive and negative energy coexist. However, this effect is also quite difficult to manipulate. A typical pulse of negative energy might last for 10^{-15} seconds, followed by a pulse of positive energy. Separating positive energy states from negative energy states is possible, although extremely difficult. I discuss this more in chapter 11.

Last, it turns out that a black hole also has negative energy, near its event horizon. As shown by Jacob Bekenstein and Stephen Hawking, a black hole is not perfectly black because it slowly evaporates energy. This is because the uncertainty principle makes possible the tunneling of radiation past the enormous gravity of a black hole. But because an evaporating black hole loses energy, the event horizon gradually gets smaller with time. Usually, if positive matter (like a star) is thrown into a black hole, the event horizon expands. But if we throw negative matter into the black hole, its event horizon will contract. Thus, black hole evaporation creates negative energy near the event horizon. (Some have advocated putting the mouth of the wormhole next to the event horizon in order to harvest negative energy. However, harvesting such negative energy would be extraordinarily difficult and dangerous, since you would have to be extremely close to the event horizon.)

Hawking has shown that in general negative energy is required to stabilize all wormhole solutions. The reasoning is quite simple. Usually, positive energy can create an opening of a wormhole that concentrates matter and energy. Thus, light rays converge as they enter the mouth of the wormhole. However, if these light rays emerge from the other side, then somewhere in the center of the wormhole light rays should defocus. The only way this can happen is if negative energy is present. Furthermore, negative energy is repulsive, which is required to keep the wormhole from collapsing under gravity. So the key to building a time machine or wormhole may be to find sufficient amounts of negative energy to keep the mouth open and stable. (A number of physicists have shown that, in the presence of large gravitational fields, negative energy fields are rather common. So perhaps one day gravitational negative energy may be used to drive a time machine.)

Another obstacle facing such a time machine is: where do we find a wormhole? Thorne relied upon the fact that wormholes occur naturally, in what is called the space-time foam. This goes back to a question asked by the Greek philosopher Zeno over two thousand years ago: what is the smallest distance one can travel?

Zeno once proved mathematically that it was impossible to cross a river. He first observed that the distance across a river can be sub-divided into an infinite number of points. But since it took an infi-nite amount of time to move across an infinite number of points, it was therefore impossible to cross the river. Or, for that matter, it was impossible for anything to move at all. (It would take another two thousand years, and the coming of calculus, to finally resolve this puzzle. It can be shown that an infinite number of points can be crossed in a finite amount of time, making motion mathematically possible after all.)

John Wheeler of Princeton analyzed Einstein's equations to find the smallest distance. Wheeler found that at incredibly small dis-tances, on the order of the Planck length (10^{-33} cm), Einstein's theory predicted that the curvature of space could be quite large. In other words, at the Planck length, space was not smooth at all but had large curvature—that is, it was kinky and "foamy." Space becomes lumpy and actually froths with tiny bubbles that dart in and out of the vacuum. Even empty space, at the tiniest distances, is constantly boiling with tiny bubbles of space-time, which are actually tiny wormholes and baby universes. Normally, "virtual particles" consist of electron and antielectron pairs that pop into existence momen-tarily before annihilating each other. But at the Planck distance, tiny bubbles representing entire universes and wormholes may spring into existence, only to vanish back into the vacuum. Our own universe may have started as one of these tiny bubbles floating in the space-time foam that suddenly inflated, for reasons we don't un-derstand.

Since wormholes are found naturally in the foam, Thorne as-sumed that an advanced civilization could somehow pick wormholes out of the foam and then expand and stabilize them with negative energy. Although this would be a very difficult process, it is within the realm of the laws of physics.

While Thorne's time machine seems theoretically possible, al-though exceedingly difficult to build from an engineering view-point, there is a third nagging question: does time travel violate a fundamental law of physics?

A UNIVERSE IN YOUR BEDROOM

In 1992, Stephen Hawking tried to resolve this question about time travel once and for all. Instinctively, he was against time travel; if journeys through time were as common as Sunday picnics, then we should see tourists from the future gawking at us and taking pictures.

But physicists often quote from T. H. White's epic novel *The Once and Future King*, where a society of ants declares, "Everything not forbidden is compulsory." In other words, if there isn't a basic principle of physics forbidding time travel, then time travel is necessarily a physical possibility. (The reason for this is the uncertainty principle. Unless something is forbidden, quantum effects and fluctuations will eventually make it possible if we wait long enough. Thus, unless there is a law forbidding it, it will eventually occur.) In response, Stephen Hawking proposed a "chronology protection hypothesis" that would prevent time travel and hence "make history safe for historians." According to this hypothesis, time travel is not possible because it violates specific physical principles.

Since wormhole solutions are extremely difficult to work with, Hawking began his argument by analyzing a simplified universe discovered by Charles Misner of the University of Maryland which had all the ingredients of time travel. Misner space is an idealized space in which your bedroom, for example, becomes the entire universe. Let's say that every point on the left wall of your bedroom is identical to the corresponding point on the right wall. This means that if you walk toward the left wall, you will not get a bloody nose, but will instead walk through the wall and reappear from the right wall. This means that the left and right wall are joined, in some sense, as in a cylinder.

In addition, the points on the front wall are identical to the points on the back wall, and the points on the ceiling are identical to the points on the floor. Thus, if you walk in any direction, you pass right through your bedroom walls and return back again to your bedroom. You cannot escape. In other words, your bedroom truly is the entire universe!

In a Misner space, the entire universe is contained in your bedroom. The opposite walls are all identified with each other, so entering one wall you immediately emerge from the opposite wall. The ceiling is likewise identified with the floor. Misner space is often studied because it has the same topology as a wormhole but is much simpler to handle mathematically. If the walls move, then time travel might be possible within the Misner universe.

What is really bizarre is that, if you look carefully at the left wall, you see that it is actually transparent and there is a carbon copy of your bedroom on the other side of this wall. In fact, there is an exact clone of yourself standing in the other bedroom, although you can only see your back side, never your front side. If you look below or above, you also see carbon copies of yourself. In fact, there is an infinite sequence of yourselves standing in front, behind, below, and above you.

Making contact with yourself is quite difficult. Every time you turn your head to catch a glimpse of the clones' faces, you find that they have also turned away, so you never see their faces. But if the bedroom is small enough, you might pass your hand through the wall and grab the shoulder of the clone in front of you. Then you might be shocked to find that the clone behind you has reached out and grabbed your shoulder as well. Also, you can reach out with your left and right hands, grabbing hold of the clones to your side, until there is an infinite sequence of yourselves holding hands. In effect, you have reached completely around the universe to grab ahold of yourself. (It is not advisable to harm your clones. If you take a gun and point it at the clone in front of you, you might reconsider pulling the trigger, because the clone behind you is pointing a gun at you as well!)

In Misner space, assume that the walls are collapsing around you. Now things become very interesting. Let's say the bedroom is being squeezed, with the right wall slowly coming toward you at 2 miles per hour. If you now walk through the left wall, you will return back from the moving right wall, but boosted by an additional 2 miles per hour, so you are now traveling at 4 miles per hour. In fact, each time you make a complete circuit into the left wall, you get an additional boost of 2 miles per hour emerging from the right wall, so you are now traveling at 6 miles per hour. After repeated trips around the universe, you travel 6, 8, 10 miles per hour, until you gradually approach incredible velocities close to the speed of light.

At a certain critical point, you are traveling so fast in this Misner universe that you travel back in time. In fact, you can visit any previous point in space-time. Hawking analyzed this Misner space carefully. He found that the left wall and right wall, mathematically speaking, are almost identical to the two mouths of a wormhole. In other words, your bedroom resembles a wormhole, where the left wall and the right wall are the same, similar to the two mouths of a wormhole, which are also identical.

Then he pointed out that this Misner space was unstable both classically and quantum mechanically. If you shine a flashlight at the left wall, for example, the light beam gains energy every time it emerges from the right wall. The light beam becomes blue-shifted—that is, it becomes more energetic, until it reaches infinite energy, which is impossible. Or, the light beam becomes so energetic that it creates a monstrous gravitational field of its own which collapses the bedroom/wormhole. Thus, the wormhole collapses if you try to walk through it. Also, one can show that something called the energy-momentum tensor, which measures the energy and matter content of space, becomes infinite because radiation can pass an infinite number of times through the two walls.

To Hawking, this was the coup de grâce for time travel—quantum radiation effects built up until they became infinite, creating a divergence, killing the time traveler and closing the wormhole.

Since Hawking's paper, the divergence question he raised has generated a lively discussion in the physics literature, with scientists taking both pro and con positions with regard to chronology protection. In fact, several physicists began to find loopholes in Hawking's proof by making suitable choices for wormholes, by changing their size, length, and so on. They found that in some wormhole solutions, the energy-momentum tensor did, in fact, diverge, but in others it was well defined. Russian physicist Sergei Krasnikov examined this divergence question for different types of wormholes and concluded that "there is not a grain of evidence to suggest that the time machine must be unstable."

The tide has swung so far in the other direction against Hawking that Princeton physicist Li-Xin Li even proposed an antichronology protection conjecture: "There is no law of physics preventing the appearance of closed timelike curves."

In 1998, Hawking was forced to make a retreat of sorts. He wrote, "The fact that the energy-momentum tensor fails to diverge [in certain cases] shows that the back reaction does not enforce chronology protection." This does not mean that time travel is possible, only that our understanding is still incomplete. Physicist Matthew Visser sees the failure of Hawking's conjecture is "not as a vindication for time travel enthusiasts, but rather as an indication that resolving issues of chronology protection requires a fully developed theory of quantum gravity."

Today, Hawking no longer says that time travel is absolutely impossible, only that it is highly unlikely and impractical. The odds are overwhelmingly against time travel. But one cannot rule it out entirely. If one can somehow harness large quantities of positive and negative energy and solve the stability problem, time travel may indeed be possible. (And perhaps the reason we are not flooded by tourists from the future is that the earliest time they can go back to is when the time machine was created, and perhaps time machines haven't been created yet.)

GOTT TIME MACHINE

In 1991, J. Richard Gott III of Princeton proposed yet another solution to Einstein's equations which allowed for time travel. His approach was interesting because he started from an entirely fresh approach, abandoning spinning objects, wormholes, and negative energy entirely.

Gott was born in Louisville, Kentucky, in 1947, and he still speaks in a gentle southern accent that seems a bit exotic in the rarefied, rough-and-tumble world of theoretical physics. He got his start in science as a child when he joined an amateur astronomy club and enjoyed stargazing.

While in high school, he won the prestigious Westinghouse Science Talent Search contest and has been associated with that contest ever since, acting as chairman of the judges for many years. After graduating from Harvard in mathematics, he went to Princeton, where he still works.

While doing research in cosmology, he became interested in "cosmic strings," a relic of the big bang that is predicted by many theories. Cosmic strings may have a width thinner than an atomic nucleus, but their mass may be stellar and they may extend for millions of light-years in space. Gott first found a solution to Einstein's equations which allowed for cosmic strings. But then he noticed something unusual about these cosmic strings. If you take two cosmic strings and send them toward each other, then, just before they collide, it is possible to use this as a time machine. First, he found that if you made the round-trip around the colliding cosmic strings, space was contracted, giving it strange properties. We know that if we move around a table, for example, and return to where we started, we have traveled 360 degrees. But when a rocket travels around the two cosmic strings as they pass each other, it actually travels through less than 360 degrees, because space has shrunk. (This has the topology of a cone. If we move completely around a cone, we also find that we travel less than 360 degrees.) Thus, by going rapidly around both strings, you could actually exceed the speed of light (as seen by a distant observer) since the total distance was less than expected. This does not violate special relativity, however, because in your own frame of reference your rocket never exceeds light speed.

But this also means that if you travel around the colliding cosmic strings, you can take a trip to the past. Gott recalls, "When I found this solution, I was quite excited. The solution used only positive-density matter, moving at speeds slower than the speed of light. By contrast, wormhole solutions require more exotic negative-energy-density material (stuff that weighs less than nothing)."

But the energy necessary for a time machine is enormous. "To allow time travel to the past, cosmic strings with a mass-per-unit length of about 10 million billion tons per centimeter must each move in opposite directions at speeds of at least 99.999999996 percent of the speed of light. We have observed high-energy protons in the universe moving at least this fast, so such speeds are possible," he observes.

Some critics have pointed out that cosmic strings are rare, if they exist at all, and colliding cosmic strings are even rarer. So Gott proposed the following. An advanced civilization may find a single cosmic string in outer space. Using gigantic spaceships and huge tools, they might reshape the string into a rectangular loop that is slightly bent (resembling the shape of a reclining chair). The loop, he hypothesized, might collapse under its own gravity, so that two straight pieces of the cosmic string might fly past each other near the speed of light, briefly creating a time machine. Nevertheless, Gott admits, "A collapsing loop of string large enough to allow you to circle it once and go back in time a year would have to be more than half the mass-energy of an entire galaxy."

TIME PARADOXES

Traditionally, another reason physicists dismissed the idea of time travel was because of time paradoxes. For example, if you go back in time and kill your parents before you are born, then your birth is impossible. Hence you could never go back in time to kill your parents to begin with. This is important, because science is based on logically consistent ideas; a genuine time paradox would be enough to completely rule out time travel.

These time paradoxes can be grouped into several categories:

Grandfather paradox. In this paradox, you alter the past in a way that makes the present impossible. For example, by going back into the distant past to meet the dinosaurs, you accidentally step on a small, furry mammal that is the original ancestor of humanity. By destroying your ancestor, you cannot logically exist.

Information paradox. In this paradox, information comes from the future, which means that it may have no origin. For example, let's say a scientist creates a time machine and then goes back in time to give the secret of time travel to himself as a youth. The secret of time travel would have no origin, since the time machine the youthful scientist possesses was not created by him but was handed to him by his older self.

Bilker's paradox. In this kind of paradox, a person knows what the future will be and does something that makes the future impossible. For example, you make a time machine to take you to the future, and you see that you are destined to marry a woman named Jane. However, on a lark, you decide to marry Helen instead, thereby making your own future impossible.

The sexual paradox. In this kind of paradox, you father yourself, which is a biological impossibility. In a tale written by the British philosopher Jonathan Harrison, the hero of the story not only fathers himself, but he also cannibalizes himself. In Robert Heinlein's classic tale "All You Zombies," the hero is simultaneously his mother, father, daughter, and son—that is, a family tree unto himself. (See the notes for details. Unraveling the sexual paradox is actually rather delicate, requiring knowledge of both time travel and the mechanics of DNA.)

In *The End of Eternity*, Isaac Asimov envisions a "time police" that is responsible for preventing these paradoxes. The *Terminator* movies hinge on an information paradox—a microchip recovered from a robot from the future is studied by scientists, who then create a race of robots that become conscious and take over the world. In other words, the design for these super robots was never created by an inventor; it simply came from a piece of debris left over from one of the robots of the future. In the movie *Back to the Future*, Michael J. Fox struggles to avoid a grandfather paradox when he goes back in time and meets his mother as a teenager, who promptly falls in love with him. But if she spurns the advances of Fox's future father, then his very existence is threatened.

Scriptwriters willingly violate the laws of physics in making Hollywood blockbusters. But in the physics community, such paradoxes are taken very seriously. Any solution to these paradoxes must be compatible with relativity and the quantum theory. For example, to be compatible with relativity, the river of time simply cannot end. You cannot dam the river of time. Time, in general relativity, is represented by a smooth, continuous surface and cannot be torn or ripped. It may change topology, but it cannot simply stop. This means that if you kill your parents before you are born, you cannot simply disappear. This would violate the laws of physics.

Currently, physicists are congregating around two possible solutions to these time paradoxes. First, Russian cosmologist Igor Novikov believes that we are forced to act in a way so that no paradoxes occur. His approach is called the self-consistency school. If the river of time smoothly bends back on itself and creates a whirlpool, he suggests that an "invisible hand" of some sort would intervene if we were to jump back into the past and were about to create a time paradox. But Novikov's approach presents problems with free will. If we go back in time and meet our parents before we are born, we might think that we have free will in our actions; Novikov believes that an undiscovered law of physics prevents any action that will change the future (such as killing your parents or preventing your birth). He notes, "We cannot send a time traveler back to the Garden of Eden to ask Eve not to pick the apple from the tree."

What is this mysterious force that prevents us from altering the past and creating a paradox? "Such a constraint on our free will is unusual and mysterious but not completely without parallel," he writes. "For example, it can be my will to walk on the ceiling without the aid of any special equipment. The law of gravity prevents me from doing this; I will fall down if I try, so my free will is restricted."

But time paradoxes can occur when inanimate matter (with no free will at all) is cast into the past. Let's suppose that just before the historic battle between Alexander the Great and Darius III of Persia in 330 B.C., you send machine guns back into time, giving instructions on how to use them. We would potentially change all subsequent European history (and might find ourselves speaking a version of the Persian language rather than a European language).

In fact, even the tiniest disturbance into the past may cause unexpected paradoxes in the present. Chaos theory, for example, uses the metaphor of the "butterfly effect." At critical times in the formation of Earth's weather, even the fluttering of the wings of a butterfly sends ripples that can tip the balance of forces and set off a powerful storm. Even the smallest inanimate objects sent back into the past will inevitably change the past in unpredictable ways, resulting in a time paradox.

A second way to resolve the time paradox is if the river of time smoothly forks into two rivers, or branches, forming two distinct universes. In other words, if you were to go back in time and shoot your parents before you were born, you would have killed people who are genetically the same as your parents in an alternate universe, one in which you will never be born. But your parents in your original universe will be unaffected.

This second hypothesis is called the "many worlds theory"—the idea that all possible quantum worlds might exist. This eliminates the infinite divergences found by Hawking, since radiation does not repeatedly go through the wormhole as in Misner space. It only goes through once. Each time it passes through the wormhole, it enters a new universe. And this paradox goes to perhaps the deepest question in the quantum theory: how can a cat be dead and alive at the same time?

To answer this question, physicists have been forced to entertain two outrageous solutions: either there is a cosmic consciousness that watches over us all, or else there are an infinite number of quantum universes.

CHAPTER SIX

Parallel Quantum Universes

I think I can safely say that nobody understands quantum mechanics.

—Richard Feynman

Anyone who is not shocked by the quantum theory does not understand it.

—Niels Bohr

The Infinite Improbability Drive is a wonderful new method of crossing vast interstellar distances in a mere nothingth of a second, without all that tedious mucking about in hyperspace.

—Douglas Adams

IN THE *Hitchhiker's Guide to the Galaxy*, the bestselling, irreverent, wacky science fiction novel by Douglas Adams, the hero stumbles upon a most ingenious method of traveling to the stars. Instead of using wormholes, hyperdrives, or dimensional portals to travel between galaxies, he conceives of harnessing the uncertainty principle to dart across the vastness of intergalactic space. If we can somehow control the probability of certain improbable events, then anything, including faster-than-light travel, and even time travel, is possible. Reaching the distant stars in seconds is highly unlikely, but when one can control quantum probabilities at will, then even the impossible may become commonplace.

The quantum theory is based on the idea that there is a probability that all possible events, no matter how fantastic or silly, might occur. This, in turn, lies at the heart of the inflationary universe theory—when the original big bang took place, there was a quantum transition to a new state in which the universe suddenly inflated by an enormous amount. Our entire universe, it appears, may have sprung out of a highly unlikely quantum leap. Although Adams wrote in jest, we physicists realize that if we could somehow control these probabilities, one could perform feats that would be indistinguishable from magic. But for the present time, altering the probabilities of events is far beyond our technology.

I sometimes ask our Ph.D. students at the university simpler questions, such as, calculate the probability that they will suddenly dissolve and rematerialize on the other side of a brick wall. According to the quantum theory, there is a small but calculable probability that this could take place. Or, for that matter, that we will dissolve in our living room and wind up on Mars. According to the quantum theory, one could in principle suddenly rematerialize on the red planet. Of course, the probability is so small that we would have to wait longer than the lifetime of the universe. As a result, in our everyday life, we can dismiss such improbable events. But at the subatomic level, such probabilities are crucial for the functioning of electronics, computers, and lasers.

Electrons, in fact, regularly dematerialize and find themselves rematerialized on the other side of walls inside the components of your PC and CD. Modern civilization would collapse, in fact, if electrons were not allowed to be in two places at the same time. (The molecules of our body would also collapse without this bizarre principle. Imagine two solar systems colliding in space, obeying Newton's laws of gravity. The colliding solar systems would collapse into a chaotic jumble of planets and asteroids. Similarly, if the atoms obeyed Newton's laws, they would disintegrate whenever they bumped into another atom. What keeps two atoms locked in a stable molecule is the fact that electrons can simultaneously be in so many places at the same time that they form an electron "cloud" which binds the atoms together. Thus, the reason why molecules are stable and the universe does not disintegrate is that electrons can be many places at the same time.)

But if electrons can exist in parallel states hovering between existence and nonexistence, then why can't the universe? After all, at one point the universe was smaller than an electron. Once we introduce the possibility of applying the quantum principle to the universe, we are forced to consider parallel universes.

It is exactly this possibility that is explored in Philip K. Dick's disturbing science fantasy tale *The Man in the High Castle*. In the book, there is an alternate universe separated from ours because of a single pivotal event. In 1933, in that universe, world history is changed when an assassin's bullet kills President Roosevelt during his first year in office. Vice President Garner takes over and establishes an isolationist policy that weakens the United States militarily. Unprepared for the attack on Pearl Harbor, and unable to recover from the destruction of the entire U.S. fleet, by 1947 the United States is forced to surrender to the Germans and the Japanese. The United States is eventually cut up into three pieces, with the German Reich controlling the east coast, the Japanese controlling the west coast, and an uneasy buffer, the Rocky Mountain states, in between. In this parallel universe, a mysterious individual writes a book, called *The Grasshopper Lies Heavy*, based on a line in the Bible, which is banned by the Nazis. It talks about an alternate universe in which Roosevelt was not assassinated, and the United States and Britain defeated the Nazis. The mission of the heroine in the story is to see if there is any truth in an alternate universe in which democracy and freedom prevail, rather than tyranny and racism.

TWILIGHT ZONE

The world of *The Man in the High Castle* and our world are separated by only the tiniest of accidents, a single assassin's bullet. However, it is also possible that a parallel world may be separated from ours by the smallest possible event: a single quantum event, a cosmic ray impact.

In one episode of the *Twilight Zone* television series, a man wakes up only to find that his wife does not recognize him. She screams at him to leave before she calls the police. When he wanders around town, he finds that his lifelong friends also fail to recognize him, as if he never existed. Finally, he visits his parents' house and is shaken to the core. His parents claim that they have never seen him before and that they never had a son. Without friends, family, or a home, he drifts aimlessly around town, eventually falling asleep on a park bench, like a homeless man. When he wakes up the next day, he finds himself comfortably back in bed with his wife. However, when his wife turns around, he is shocked to find that she is not his wife at all, but a strange woman that he has never seen before.

Are such preposterous stories possible? Perhaps. If the protagonist in *The Twilight Zone* had asked some revealing questions of his mother, he might have found that she had a miscarriage and hence never had a son. Sometimes a single cosmic ray, a single particle from outer space, can strike deep in the DNA within an embryo and cause a mutation that will eventually lead to a miscarriage. In such a case, a single quantum event can separate two worlds, one in which you live as a normal, productive citizen, and another that is exactly identical, except that you were never born.

To slip between these worlds *is* within the laws of physics. But it is extremely unlikely; the probability of it happening is astronomically small. But as you can see, the quantum theory gives us a picture of the universe much stranger than the one given to us by Einstein. In relativity, the stage of life on which we perform may be made of rubber, with the actors traveling in curved paths as they move across the set. As in Newton's world, the actors in Einstein's world parrot their lines from a script that was written beforehand. But in a quantum play, the actors suddenly throw away the script and act on their own. The puppets cut their strings. Free will has been established. The actors may disappear and reappear from the stage. Even stranger, they may find themselves appearing in two places at the same time. The actors, when delivering their lines, never know for sure whether or not they are speaking to someone who might suddenly disappear and reappear in another place.

MONSTER MIND: JOHN WHEELER

Except perhaps for Einstein and Bohr, no man has wrestled more with the absurdities and successes of the quantum theory than John Wheeler. Is all physical reality just an illusion? Do parallel quantum universes exist? In the past, when he was not mulling over these intractable quantum paradoxes, Wheeler was applying these probabilities to build the atomic and hydrogen bombs and was pioneering the study of black holes. John Wheeler is the last of the giants, or "monster minds," as his student Richard Feynman once called them, who have grappled with the insane conclusions of the quantum theory.

It was Wheeler who coined the term "black hole" in 1967 at a conference at NASA's Goddard Institute for Space Studies in New York City after the discovery of the first pulsars.

Wheeler was born in 1911 in Jacksonville, Florida. His father was a librarian, but engineering was in his family's blood. Three of his uncles were mining engineers and often used explosives in their work. The idea of using dynamite fascinated him, and he loved to watch explosions. (One day, he was carelessly experimenting with a piece of dynamite and it accidentally exploded in his hand, blowing off part of his thumb and the end of one finger. Coincidentally, when Einstein was a college student, a similar explosion took place in his hand due to carelessness, requiring several stitches.)

Wheeler was a precocious kid, mastering calculus and devouring every book he could find on the new theory that his friends were buzzing about: quantum mechanics. Right before his eyes, a new theory was being developed in Europe by Niels Bohr, Werner Heisenberg, and Erwin Schrödinger that suddenly unlocked the secrets of the atom. Only a few years before, followers of the philosopher Ernst Mach had scoffed at the existence of atoms, stating that atoms had never been observed in the laboratory and probably were a fiction. What couldn't be seen probably did not exist, they claimed.

The great German physicist Ludwig Boltzmann, who laid down the laws of thermodynamics, committed suicide in 1906, in part because of the intense ridicule he faced while promoting the concept of atoms.

Then, in few momentous years, from 1925 to 1927, the secrets of the atom came tumbling out. Never in modern history (except for the year 1905, with the work of Einstein) had breakthroughs of this magnitude been accomplished in so short a time. Wheeler wanted to be part of this revolution. But he realized that the United States was in the backwash of physics; there was not a single world-class physicist among its ranks. Like J. Robert Oppenheimer before him, Wheeler left the United States and journeyed to Copenhagen to learn from the master himself, Niels Bohr.

Previous experiments on electrons demonstrated that they acted both as a particle and as a wave. This strange duality between particles and waves was finally unraveled by the quantum physicists: the electron, in its dance around the atom, was shown to be a particle, but it was accompanied by a mysterious wave. In 1925, Austrian physicist Erwin Schrödinger proposed an equation (the celebrated Schrödinger wave equation) that accurately described the motion of the wave that accompanies the electron. This wave, represented by the Greek letter *psi*, gave breathtakingly precise predictions for the behavior of atoms which sparked a revolution in physics. Suddenly, almost from first principles, one could peer inside the atom itself to calculate how electrons danced in their orbits, making transitions and bonding atoms together in molecules.

As quantum physicist Paul Dirac boasted, physics would soon reduce all of chemistry to mere engineering. He proclaimed, "The underlying physical laws necessary for the mathematical theory of a larger part of physics and the whole of chemistry are thus completely known, and the difficulty is only that the application of these laws leads to equations much too complicated to be soluble." As spectacular as this *psi* function was, it was still a mystery as to what it really represented.

Finally, in 1928, physicist Max Born proposed the idea that this wave function represented the probability of finding the electron at any given point. In other words, you could never know for sure precisely where an electron was; all you could do was calculate its wave function, which told you the probability of it being there. So, if atomic physics could be reduced to waves of probability of an electron being here or there, and if an electron could seemingly be in two places at the same time, how do we finally determine where the electron really is?

Bohr and Heisenberg eventually formulated the complete set of recipes in a quantum cookbook that has worked beautifully in atomic experiments with magnificent precision. The wave function only tells you the probability that the electron is located here or there. If the wave function is large at a certain point, it means that there is a high likelihood that the electron is located there. (If it is small there, then it is unlikely that the electron can be found there.) For example, if we could "see" the wave function of a person, it would look remarkably like the person himself. However, the wave function also gently seeps out into space, meaning that there is a small probability that the person can be found on the moon. (In fact, the person's wave function actually spreads out throughout the universe.)

This also means that the wave function of a tree can tell you the probability that it is either standing or falling, but it cannot definitively tell you in which state it actually is. But common sense tells us that objects are in definite states. When you look at a tree, the tree is definitely in front of you—it is either standing or fallen, but not both.

To resolve the discrepancy between waves of probability and our commonsense notion of existence, Bohr and Heisenberg assumed that after a measurement is made by an outside observer, the wave function magically "collapses," and the electron falls into a definite state—that is, after looking at the tree, we see that it is truly standing. *In other words, the process of observation determines the final state of the electron.* Observation is vital to existence. After we look at the electron, its wave function collapses, so the electron is now in a definite state and there is no more need for wave functions.

So the postulates of Bohr's Copenhagen school, loosely speaking, can be summarized as follows:

a. All energy occurs in discrete packets, called quanta. (The quantum of light, for example, is the photon. The quanta of the weak force are called the W- and Z-boson, the quantum for the strong force is called the gluon, and the quantum for gravity is called the graviton, which has yet to be seen in the laboratory.)

b. Matter is represented by point particles, but the probability of finding the particle is given by a wave. The wave, in turn, obeys a specific wave equation (such as Schrödinger's wave equation).

c. Before an observation is made, an object exists in all possible states simultaneously. To determine which state the object is in, we have to make an observation, which "collapses" the wave function, and the object goes into a definite state. The act of observation destroys the wave function, and the object now assumes a definite reality. The wave function as served its purpose: it has given us the precise probability of finding the object in that particular state.

DETERMINISM OR UNCERTAINTY?

The quantum theory is the most successful physical theory of all time. The highest formulation of the quantum theory is the Standard Model, which represents the fruit of decades of experiments with particle accelerators. Parts of this theory have been tested to 1 part in 10 billion. If one includes the mass of the neutrino, then the Standard Model is consistent with all experiments on subatomic particles, without exception.

But no matter how successful the quantum theory is, experimentally it is based on postulates that have unleashed storms of philosophical and theological controversy for the past eighty years. The second postulate, in particular, has raised the ire of religions because it asks who decides our fate. Throughout the ages, philosophers, theologians, and scientists have been fascinated by the future and whether somehow our destinies are knowable. In Shakespeare's *Macbeth*, Banquo, desperate to lift the veil that clouds our destiny, delivers the memorable lines:

If you can look into the seeds of time
And say which grain will grow and which will not,
Speak then to me . . .
(act I, scene 3)

Shakespeare wrote these words in 1606. Eighty years later, another Englishman, Isaac Newton, had the audacity to claim that he knew the answer to this ancient question. Both Newton and Einstein believed in the concept called determinism, which states that all future events can be determined in principle. To Newton, the universe was a gigantic clock wound up by God at the beginning of time. Ever since then, it's been ticking, obeying his three laws of motion, in a precisely predictable way. The French mathematician Pierre Simon de Laplace, who was a scientific advisor to Napoleon, wrote that, using Newton's laws, one could predict the future with the same precision that one views the past. He wrote that if a being could know the position and velocity of all the particles in the universe, "for such an intellect, nothing could be uncertain; and the future just like the past would be present before his eyes." When Laplace presented Napoleon with a copy of his masterwork, *Celestial Mechanics*, the emperor said, "You have written this huge work on the heavens without once mentioning God." Laplace replied, "Sire, I had no need of that hypothesis."

To Newton and Einstein, the notion of free will, that we are masters of our destiny, was an illusion. This commonsense notion of reality, that concrete objects that we touch are real and exist in definite states, Einstein called "objective reality." He most clearly presented his position as follows:

I am a determinist, compelled to act as if free will existed, because if I wish to live in a civilized society, I must act responsibly. I know

philosophically a murderer is not responsible for his crimes, but I pre-
fer not to take tea with him. My career has been determined by various forces over which I have no control, primarily those mysterious
glands in which nature prepares the very essence of life. Henry Ford
may call it is his Inner Voice, Socrates referred to it as his daemon:
each man explains in his own way the fact that the human will is not
free . . . Everything is determined . . . by forces over which we have
no control . . . for the insect as well as for the star. Human beings,
vegetables, or cosmic dust, we all dance to a mysterious time, intoned
in the distance by an invisible player.

Theologians have also wrestled with this question. Most religions
of the world believe in some form of predestination, the idea that
God is not only omnipotent (all-powerful) and omnipresent (exists
everywhere), but also omniscient (knows everything, even the future). In some religions, this means that God knows whether we will
go to heaven or hell, even before we are born. In essence, there is a
"book of destiny" somewhere in heaven with all of our names listed,
including our birth date, our failures and triumphs, our joys and our
defeats, even our death date, and whether we will live in paradise or
eternal damnation.

(This delicate theological question of predestination, in part,
helped to split the Catholic Church in half in 1517, when Martin
Luther tacked the ninety-five theses on the church at Wittenberg. In
it, he attacked the church's practice of selling indulgences—essentially bribes that paved the journey to heaven for the rich. Perhaps,
Luther seemed to say, God does know our future ahead of time and
our fates are predestined, but God cannot be persuaded to change his
mind by our making a handsome donation to the church.)

But to physicists who accept the concept of probability, the most
controversial postulate by far is the third postulate, which has given
headaches to generations of physicists and philosophers. "Observation" is a loose, ill-defined concept. Moreover, it relies on the
fact that there are actually two types of physics: one for the
bizarre subatomic world, where electrons can seemingly be in two
places at the same time, and the other for the macroscopic world

that we live in, which appears to obey the commonsense laws of Newton.

According to Bohr, there is an invisible "wall" separating the atomic world from the everyday, familiar macroscopic world. While the atomic world obeys the bizarre rules of the quantum theory, we live out our lives outside that wall, in the world of well-defined planets and stars where the waves have already collapsed.

Wheeler, who learned quantum mechanics from its creators, liked to summarize the two schools of thought on this question. He gives the example of three umpires at a baseball game discussing the finer points of baseball. In making a decision, the three umpires say:

> Number 1: I calls 'em like I see 'em.
> Number 2: I calls 'em the way they *are*.
> Number 3: They ain't *nothing* till I calls 'em.

To Wheeler, the second umpire is Einstein, who believed there was an absolute reality outside human experience. Einstein called this "objective reality," the idea that objects can exist in definite states without human intervention. The third umpire is Bohr, who argued that reality existed only after an observation was made.

TREES IN THE FOREST

Physicists sometimes view philosophers with a certain disdain, quoting from the Roman Cicero, who once said, "There is nothing so absurd that it has not been said by philosophers." The mathematician Stanislaw Ulam, who took a dim view of giving lofty names to silly concepts, once said, "Madness is the ability to make fine distinctions on different kinds of nonsense." Einstein himself once wrote of philosophy, "Is not all of philosophy as if written in honey? It looks wonderful when one contemplates it, but when one looks again it is all gone. Only mush remains."

Physicists also like to tell the apocryphal story supposedly told by a university president who became exasperated looking at the budget for the physics, math, and philosophy departments. He supposedly said, "Why is it that you physicists always require so much expensive equipment? Now the Department of Mathematics requires nothing but money for paper, pencils, and waste paper baskets and the Department of Philosophy is better still. It doesn't even ask for waste paper baskets."

However, philosophers may yet get the last laugh. The quantum theory is incomplete and rests on shaky philosophical grounds. This quantum controversy forces one to reexamine the work of philosophers like Bishop Berkeley, who in the eighteenth century claimed that objects exist only because humans are there to observe them, a philosophy called solipsism or idealism. If a tree falls in the forest but no one is there to see it, then it does not really fall, they claim.

Now we have a quantum reinterpretation of trees falling in the forest. Before an observation is made, you don't know whether it has fallen or not. In fact, the tree exists in all possible states simultaneously: it might be burnt, fallen, firewood, sawdust, and so on. Once an observation is made, then the tree suddenly springs into a definite state, and we see that it has fallen, for instance.

Comparing the philosophical difficulty of relativity and the quantum theory, Feynman once remarked, "There was a time when the newspapers said that only twelve men understood the theory of relativity. I do not believe there was ever such a time . . . On the other hand, I think I can safely say that nobody understands quantum mechanics." He writes that quantum mechanics "describes nature as absurd from the point of view of common sense. And it fully agrees with experiment. So I hope you can accept nature as she is—absurd." This has created an uneasy feeling among many practicing physicists, who feel as if they are creating entire worlds based on shifting sands. Steven Weinberg writes, "I admit to some discomfort in working all my life in a theoretical framework that no one fully understands."

In traditional science, the observer tries to keep as dispassionately detached from the world as possible. (As one wag said, "You can always spot the scientist at a strip club, because he is the only one examining the audience.") But now, for the first time, we see that it is impossible to separate the observer from the observed. As Max Planck once remarked, "Science cannot solve the ultimate mystery of Nature. And it is because in the last analysis we ourselves are part of the mystery we are trying to solve."

THE CAT PROBLEM

Erwin Schrödinger, who introduced the wave equation in the first place, thought that this was going too far. He confessed to Bohr that he regretted ever proposing the wave concept if it introduced the concept of probability into physics.

To demolish the idea of probabilities, he proposed an experiment. Imagine a cat sealed in a box. Inside the box, there is a bottle of poison gas, connected to a hammer, which in turn is connected to a Geiger counter placed near a piece of uranium. No one disputes that the radioactive decay of the uranium atom is purely a quantum event that cannot be predicted ahead of time. Let's say there is a 50 percent chance that a uranium atom will decay in the next second. But if a uranium atom decays, it sets off the Geiger counter, which sets off the hammer that breaks the glass, killing the cat. Before you open the box, it is impossible to tell whether the cat is dead or alive. In fact, in order to describe the cat, physicists add the wave function of the live cat and the dead cat—that is, we put the cat in a nether world of being 50 percent dead and 50 percent alive simultaneously.

Now open the box. Once we peer into the box, an observation is made, the wave function collapses, and we see that the cat is, say, alive. To Schrödinger, this was silly. How can a cat be both dead and alive at the same time, just because we haven't looked at it? Does it suddenly spring into existence as soon as we observe it? Einstein was also displeased with this interpretation. Whenever guests came over to his house, he would say: look at the moon. Does it suddenly spring into existence when a mouse looks at it? Einstein believed the answer was no. But in some sense, the answer might be yes.

Things finally came to a head in 1930 in a historic clash at the Solvay Conference between Einstein and Bohr. Wheeler would later remark that it was the greatest debate in intellectual history that he knew about. In thirty years, he had never heard of a debate between two greater men on a deeper issue with deeper consequences for an understanding of the universe.

Einstein, always bold, daring, and supremely eloquent, proposed a barrage of "thought experiments" to demolish the quantum theory. Bohr, who mumbled incessantly, was reeling after each attack. Physicist Paul Ehrenfest observed, "It was wonderful for me to be present at the dialogues between Bohr and E. E, like a chess player, with ever new examples. A kind of perpetuum mobile of the second kind, intent on breaking through uncertainty. Bohr always, out of a cloud of philosophical smoke, seeking the tools for destroying one example after another. Einstein like a jack-in-a-box, popping up fresh every morning. Oh, it was delightful. But I am almost unreservedly pro Bohr and contra E. He now behaves toward Bohr exactly as the champions of absolute simultaneity had behaved toward him."

Finally, Einstein proposed an experiment that he thought would give the coup de grâce to the quantum theory. Imagine a box containing a gas of photons. If the box has a shutter, it can briefly release a single photon. Since one can measure the shutter speed precisely, and also measure the photon's energy, one can therefore determine the state of the photon with infinite precision, thereby violating the uncertainty principle.

Ehrenfest wrote, "To Bohr, this was a heavy blow. At the moment he saw no solution. He was extremely unhappy all through the evening, walked from one person to another, trying to persuade them all that this could not be true, because if E was right this would mean the end of physics. But he could think of no refutation. I will never forget the sight of the two opponents leaving the university club. Einstein, a majestic figure, walking calmly with a faint ironical smile, and Bohr trotting along by his side, extremely upset."

When Ehrenfest later encountered Bohr, he was speechless; all he could do was mumble the same word over and over again, "Einstein . . . Einstein . . . Einstein."

The next day, after an intense, sleepless night, Bohr was able to find a tiny flaw in Einstein's argument. After emitting the photon, the box was slightly lighter, since matter and energy were equivalent. This meant that the box rose slightly under gravity, since energy has weight, according to Einstein's own theory of gravity. But this introduced uncertainty in the photon's energy. If one then calculated the uncertainty in the weight and uncertainty in the shutter speed, one found that the box obeyed the uncertainty principle exactly. In effect, Bohr had used Einstein's own theory of gravity to refute Einstein! Bohr had emerged victorious. Einstein was defeated.

When Einstein later complained that "God does not play dice with the world," Bohr reportedly fired back, "Stop telling God what to do." Ultimately, Einstein admitted that Bohr had successfully refuted his arguments. Einstein would write, "I am convinced that this theory undoubtedly contains a piece of definitive truth." (Einstein, however, had disdain for physicists who failed to appreciate the subtle paradoxes inherent in the quantum theory. He once wrote, "Of course, today every rascal thinks he knows the answer, but he is deluding himself.")

After these and other fierce debates with quantum physicists, Einstein finally gave in, but took a different approach. He conceded that the quantum theory was correct, but only within a certain domain, only as an approximation to the real truth. In the same way that relativity generalized (but did not destroy) Newton's theory, he wanted to absorb the quantum theory into a more general, more powerful theory, the unified field theory.

(This debate, between Einstein and Schrödinger on one side, and Bohr and Heisenberg on the other, cannot be easily dismissed, since these "thought experiments" can now be performed in the laboratory. Although scientists cannot make a cat appear both dead and alive, they can now manipulate individual atoms with nanotechnology. Recently, these mind-bending experiments were done with a Buckyball containing sixty carbon atoms, so the "wall" envisioned by Bohr separating large objects from quantum objects is rapidly crumbling. Experimental physicists are now even contemplating what would be required to show that a virus, consisting of thousands of atoms, can be in two places at the same time.)

THE BOMB

Unfortunately, discussions over these delicious paradoxes were interrupted with the rise of Hitler in 1933 and the rush to build an atomic bomb. It was known for years, via Einstein's famous equation $E = mc^2$, that there was a vast storehouse of energy locked in the atom. But most physicists pooh-poohed the idea of ever being able to harness this energy. Even Ernest Rutherford, the man who discovered the nucleus of the atom, said, "The energy produced by the breaking down of the atom is a very poor kind of thing. Anyone who expects a source of power from the transformation of these atoms is talking moonshine."

In 1939, Bohr made a fateful trip to the United States, landing in New York to meet his student John Wheeler. He was bearing ominous news: Otto Hahn and Lise Meitner had shown that the uranium nucleus could be split in half, releasing energy, in a process called fission. Bohr and Wheeler began to work out the quantum dynamics of nuclear fission. Since everything in the quantum theory is a matter of probability and chance, they estimated the probability that a neutron will break apart the uranium nucleus, releasing two or more neutrons, which then fission even more uranium nuclei, which then release ever more neutrons, and so on, setting off a chain reaction capable of devastating a modern city. (In quantum mechanics, you can never know if any particular neutron will fission a uranium atom, but you can compute with incredible accuracy the probability that billions of uranium atoms will fission in a bomb. That is the power of quantum mechanics.)

Their quantum computations indicated that an atomic bomb might be possible. Two months later, Bohr, Eugene Wigner, Leo Szilard, and Wheeler met at Einstein's old office at Princeton to discuss the prospects for an atomic bomb. Bohr believed it would take the resources of an entire nation to build the bomb. (A few years later, Szilard would persuade Einstein to write the fateful letter to President Franklin Roosevelt, urging him to build the atomic bomb.)

That same year, the Nazis, aware that the catastrophic release of energy from the uranium atom could give them an unbeatable weapon, ordered Bohr's student, Heisenberg, to create the atomic bomb for Hitler. Overnight, the discussions over the quantum probability of fission became deadly serious, with the fate of human history at stake. Discussions of the probability of finding live cats would soon be replaced by discussions of the probability of fissioning uranium.

In 1941, with the Nazis overrunning most of Europe, Heisenberg made a secret journey to meet his old mentor, Bohr, in Copenhagen. The precise nature of the meeting is still shrouded in mystery, and award-winning plays have been written about it, with historians still debating its content. Was Heisenberg offering to sabotage the Nazi atomic bomb? Or was Heisenberg trying to recruit Bohr for the Nazi bomb? Six decades later, in 2002, much of the mystery over Heisenberg's intentions was finally lifted, when the Bohr family released a letter written by Bohr to Heisenberg in the 1950s but never mailed. In that letter, Bohr recalled that Heisenberg had said at that meeting that a Nazi victory was inevitable. Since there was no stopping the Nazi juggernaut, it was only logical that Bohr work for the Nazis.

Bohr was appalled, shaken to the core. Trembling, he refused to allow his work on the quantum theory to fall into Nazi hands. Because Denmark was under Nazi control, Bohr planned a secret escape by plane, and he was almost suffocated due to lack of oxygen on the plane trip to freedom.

Meanwhile, at Columbia University, Enrico Fermi had shown that a nuclear chain reaction was feasible. After he reached this conclusion, he peered out over New York City and realized that a single bomb could destroy everything he saw of the famed skyline. Wheeler, realizing how high the stakes had become, voluntarily left Princeton and joined Fermi in the basement of Stagg Field at the University of Chicago, where together they built the first nuclear reactor, officially inaugurating the nuclear age.

Over the next decade, Wheeler witnessed some of the most momentous developments in atomic warfare. During the war, he helped supervise the construction of the mammoth Hanford Reservation in Washington State, which created the raw plutonium necessary to build the bombs that would devastate Nagasaki. A few years later, he worked on the hydrogen bomb, witnessing the first hydrogen bomb blast in 1952 and the devastation caused when a piece of the Sun was unleashed on a small island in the Pacific. But after being at the forefront of world history for over a decade, he finally returned to his first love, the mysteries of the quantum theory.

SUM OVER PATHS

One of Wheeler's legion of students after the war was Richard Feynman, who stumbled on perhaps the simplest yet most profound way of summarizing the intricacies of the quantum theory. (One consequence of this idea would win Feynman the Nobel Prize in 1965.) Let's say that you want to walk across the room. According to Newton, you would simply take the shortest path, from point A to point B, called the classical path. But according to Feynman, first you would have to consider all possible paths connecting points A and B. This means considering paths that take you to Mars, Jupiter, the nearest star, even paths that go backward in time, back to the big bang. No matter how crazy and utterly bizarre the paths are, you must consider them. Then Feynman assigned a number for each path, giving a precise set of rules by which to calculate this number. Miraculously, by adding up these numbers from all possible paths, you found the probability of walking from point A to point B given by standard quantum mechanics. This was truly remarkable.

Feynman found that the sum of these numbers over paths that were bizarre and violated Newton's laws of motion usually canceled out to give a small total. This was the origin of quantum fluctuations—that is, they represented paths whose sum was very small. But he also found that the commonsense Newtonian path was the one that did not cancel out and hence had the largest total; it was the path with the greatest probability. Thus, our commonsense notion of the physical universe is simply the most probable state among an infinite number of states. But we coexist with all possible states, some of which take us back to the dinosaur era, to the nearest supernova, and to the edges of the universe. (These bizarre paths create tiny deviations from the commonsense Newtonian sense path but fortunately have a very low probability associated with them.)

In other words, as odd as it may seem, every time you walk across the room, somehow your body "sniffs out" all possible paths ahead of time, even those extending to the distant quasars and the big bang, and then adds them up. Using powerful mathematics called functional integrals, Feynman showed that the Newtonian path is simply the most probable path, not the only path. In a mathematical tour de force, Feynman was able to prove that this picture, as astounding as it may seem, is exactly equivalent to ordinary quantum mechanics. (In fact, Feynman was able to give a derivation of the Schrödinger wave equation using this approach.)

The power of Feynman's "sum over paths" is that today, when we formulate GUT theories, inflation, even string theory, we use Feynman's "path integral" point of view. This method is now taught in every graduate school in the world and is by far the most powerful and convenient way of formulating the quantum theory.

(I use the Feynman path integral approach every day in my own research. Every equation I write is written in terms of these sum over paths. When I first learned of Feynman's point of view as a graduate student, it changed my entire mental picture of the universe. Intellectually, I understood the abstract mathematics of the quantum theory and general relativity, but it was the idea that I am in some sense sniffing out paths that take me to Mars or the distant stars as I walk across the room that altered my worldview. Suddenly, I had a strange new mental picture of myself living in a quantum world. I began to realize that quantum theory is much more alien than the mind-bending consequences of relativity.)

When Feynman developed this bizarre formulation, Wheeler, who was at Princeton University, rushed over next door to the Institute for Advanced Study to visit Einstein to convince him of the elegance and power of this new picture. Wheeler excitedly explained to Einstein Feynman's new theory of path integrals. Wheeler did not fully realize how utterly crazy this must have sounded to Einstein.

Afterward, Einstein shook his head and repeated that he still did not believe that God played dice with the world. Einstein admitted to Wheeler that he could be wrong, but he also insisted that he had earned the right to be wrong.

WIGNER'S FRIEND

Most physicists shrug their shoulders and throw up their hands when confronted with the mind-bending paradoxes of quantum mechanics. To most practicing scientists, quantum mechanics is a set of cookbook rules that yields the right probabilities with uncanny accuracy. As the physicist-turned-priest John Polkinghorne has said, "The average quantum mechanic is no more philosophical than the average motor mechanic."

However, some of the deepest thinkers in physics have struggled with these questions. For example, there are several ways of resolving the Schrödinger cat problem. The first, advocated by Nobel laureate Eugene Wigner and others, is that *consciousness determines existence.* Wigner has written that it "was not possible to formulate the laws of quantum mechanics in a fully consistent way, without reference to the consciousness [of the observer] . . . the very study of the external world led to the conclusion that the content of the consciousness is the ultimate reality." Or, as the poet John Keats once wrote, "Nothing ever becomes real till it is experienced."

But if I make an observation, what is to determine which state I am in? This means that someone else has to observe me to collapse my wave function. This is sometimes called "Wigner's friend." But it also means that someone has to observe Wigner's friend, and Wigner's friend's friend, and so on. Is there a cosmic consciousness that determines the entire sequence of friends by observing the entire universe?

One physicist who tenaciously believes in the central role of consciousness is Andrei Linde, one of the founders of the inflationary universe.

For me as a human being, I do not know any sense in which I could claim that the universe is here in the absence of observers. We are together, the universe and us. The moment you say that the universe exists without any observers, I cannot make any sense out of that. I cannot imagine a consistent theory of everything that ignores consciousness. A recording device cannot play the role of an observer, because who will read what is written on this recording device. In order for us to see that something happens, and say to one another that something happens, you need to have a universe, you need to have a recording device, and you need to have us . . . In the absence of observers, our universe is dead.

According to Linde's philosophy, dinosaur fossils don't really exist until you look at them. But when you do look at them, they spring into existence as if they had existed millions of years ago. (Physicists who hold to this point of view are careful to point out that this picture is experimentally consistent with a world in which dinosaur fossils really are millions of years old.)

(Some people, who dislike introducing consciousness into physics, claim that a camera can make an observation of an electron, hence wave functions can collapse without resorting to conscious beings. But then who is to say if the camera exists? Another camera is necessary to "observe" the first camera and collapse its wave function. Then a second camera is necessary to observe the first camera, and a third camera to observe the second camera, ad infinitum. So introducing cameras does not answer the question of how wave functions collapse.)

DECOHERENCE

A way to partially resolve some of these thorny philosophical questions, one gaining popularity among physicists, is called decoherence. It was first formulated by German physicist Dieter Zeh in 1970. He noticed that in the real world you cannot separate the cat from the environment. The cat is in constant contact with the molecules of air, the box, and even cosmic rays that pass through the experiment. These interactions, no matter how small, radically affect the wave function: if the wave function is disturbed to the slightest degree, then the wave function suddenly splits into two distinct wave functions of the dead cat or the live cat, which no longer interact. Zeh showed that a collision with a single air molecule was enough to collapse it, forcing the permanent separation of the dead cat and live cat wave functions, which can no longer communicate with each other. In other words, even before you open the box, the cat has been in contact with air molecules and hence is already dead or alive.

Zeh made the key observation that had been overlooked: for the cat to be both dead and alive, the wave function of the dead cat and the wave function of the live cat must be vibrating in almost exact synchronization, a state called coherence. But experimentally, this is almost impossible. Creating coherent objects vibrating in unison in the laboratory is extraordinarily difficult. (In practice, it is difficult to get more than a handful of atoms to vibrate coherently because of interference from the outside world.) In the real world, objects interact with the environment, and the slightest interaction with the outside world can disturb the two wave functions, and then they start to "decohere"—that is, fall out of synchronization and separate. Once the two wave functions are no longer vibrating in phase with each other, Zeh showed, the two wave functions no longer interact with each other.

MANY WORLDS

At first, decoherence sounds very satisfying, since the wave function now collapses not via consciousness but by random interactions with the outside world. But this still doesn't solve the fundamental question that bothered Einstein: how does nature "choose" which state to collapse into? When an air molecule hits the cat, who or what determines the final state of the cat? On this question, decoherence theory simply states that the two wave functions separate and no longer interact, but it does not answer the original question: is the cat dead or alive? In other words, decoherence makes consciousness unnecessary in quantum mechanics, but it does not resolve the key question that disturbed Einstein: how does nature "choose" the final state of the cat? On this question, decoherence theory is silent.

There is, however, a natural extension of decoherence that resolves this question that is gaining wide acceptance today among physicists. This second approach was pioneered by another of Wheeler's students, Hugh Everett III, who discussed the possibility that perhaps the cat can be both dead and alive at the same time but in two different universes. When Everett's Ph.D. thesis was finished in 1957, it was barely noticed. Over the years, however, interest in the "many worlds" interpretation began to grow. Today, it has unleashed a tidal wave of renewed interest in the paradoxes of the quantum theory.

In this radically new interpretation, the cat is both dead and alive because the universe has split into two. In one universe, the cat is dead; in another universe, the cat is alive. In fact, at each quantum juncture, the universe splits in half, in a never-ending sequence of splitting universes. All universes are possible in this scenario, each as real as the other. People living in each universe might vigorously protest that *their* universe is the real one, and that all the others are imaginary or fake. These parallel universes are not ghost worlds with an ephemeral existence; within each universe, we have the appearance of solid objects and concrete events as real and as objective as any.

The advantage of this interpretation is that we can drop condition number three, the collapse of the wave function. Wave functions never collapse, they just continue to evolve, forever splitting into other wave functions, in a never-ending tree, with each branch representing an entire universe. The great advantage of the many worlds theory is that it is simpler than the Copenhagen interpretation: it requires no collapse of the wave function. The price we pay is that now we have universes that continually split into millions of branches. (Some find it difficult to understand how to keep track of all these proliferating universes. However, the Schrödinger wave equation does this automatically. By simply tracing the evolution of the wave equation, one immediately finds all the numerous branches of the wave.)

If this interpretation is correct, then at this very instant your body coexists with the wave functions of dinosaurs engaged in mortal combat. Coexisting in the room you are in is the wave function of a world where the Germans won World War II, where aliens from outer space roam, where you were never born. The worlds of *The Man in the High Castle* and *The Twilight Zone* are among the universes existing in your living room. The catch is that we can no longer interact with them, since they have decohered from us.

As Alan Guth has said, "There is a universe where Elvis is still alive." Physicist Frank Wilczek has written, "We are haunted by the awareness that infinitely many slightly variant copies of ourselves are living out their parallel lives and that every moment more duplicates spring into existence and take up our many alternative futures." He notes that the history of Greek civilization, and hence the Western world, might have been different had Helen of Troy not been such a captivating beauty, if instead she had an ugly wart on her nose. "Well, warts can arise from mutations in single cells, often triggered by exposure to the ultraviolet rays of the sun." He goes on, "Conclusion: there are many, many worlds in which Helen of Troy *did* have a wart at the tip of her nose."

I am reminded of the passage from Olaf Stapledon's classic work of science fiction, *Star Maker:* "Whenever a creature was faced with several possible courses of action, it took them all, thereby creating many . . . distinct histories of the cosmos. Since in every evolutionary sequence of the cosmos there were many creatures and each was constantly faced with many possible courses, and the combinations of all their courses were innumerable, an infinity of distinct universes exfoliated from every moment of every temporal sequence."

The mind reels when we realize that, according to this interpretation of quantum mechanics, all possible worlds coexist with us. Although wormholes might be necessary to reach such alternate worlds, these quantum realities exist in the very same room that we live in. They coexist with us wherever we go. The key question is: if this is true, why don't we see these alternate universes filling up our living room? This is where decoherence comes in: our wave function has decohered with these other worlds (that is, the waves are no longer in phase with each other). We are no longer in contact with them. This means that even the slightest contamination with the environment will prevent the various wave functions from interacting with each other. (In chapter 11, I mention a possible exception to this rule, in which intelligent beings may be able to travel between quantum realities.)

Does this seem too strange to be possible? Nobel laureate Steven Weinberg likens this multiple universe theory to radio. All around you, there are hundreds of different radio waves being broadcast from distant stations. At any given instant, your office or car or living room is full of these radio waves. However, if you turn on a radio, you can listen to only one frequency at a time; these other frequencies have decohered and are no longer in phase with each other. Each station has a different energy, a different frequency. As a result, your radio can only be turned to one broadcast at a time.

Likewise, in our universe we are "tuned" into the frequency that corresponds to physical reality. But there are an infinite number of parallel realities coexisting with us in the same room, although we cannot "tune into" them. Although these worlds are very much alike, each has a different energy. And because each world consists of trillions upon trillions of atoms, this means that the energy difference can be quite large. Since the frequency of these waves is proportional to their energy (by Planck's law), this means that the waves of each world vibrate at different frequencies and cannot interact anymore. For all intents and purposes, the waves of these various worlds do not interact or influence each other.

Surprisingly, scientists, by adopting this strange point of view, can rederive all the results of the Copenhagen approach without ever having to collapse the wave function. In other words, experiments done with the Copenhagen interpretation, or the many worlds interpretation, will yield precisely the same experimental results. Bohr's collapse of the wave function is mathematically equivalent to contamination with the environment. In other words, Schrödinger's cat can be dead and alive at the same time if we can somehow isolate the cat from possible contamination from every atom or cosmic ray. Of course, this is practically impossible. Once the cat is in contact with a cosmic ray, the dead cat and live cat wave functions decohere, and it appears as if the wave function has collapsed.

IT FROM BIT

With all this renewed interest in the measurement problem in the quantum theory, Wheeler has become science's grand old man of quantum physics, appearing at numerous conferences in his honor. He has even been hailed as a guru of sorts by New Age advocates who are fascinated by the question of consciousness in physics. (However, he is not always pleased with such associations. Once, he was distressed to find himself on the same program with three parapsychologists. He quickly put out a statement that included the sentence "Where there's smoke, there's smoke.")

After seventy years of contemplating the paradoxes of the quantum theory, Wheeler is the first one to admit that he does not have all the answers. He continues to always question his assumptions. When asked about the measurement problem in quantum mechanics, he says, "I am just driven crazy by that question. I confess that sometimes I do take 100 percent seriously the idea that the world is a figment of the imagination and, other times, that the world does exist out there independent of us. However, I subscribe wholeheartedly to those words of Leibniz, 'This world may be a phantasm and existence may be merely a dream, but this dream or phantasm to me is real enough if using reason well we are never deceived by it.'"

Today, the many worlds/decoherence theory is gaining popularity among physicists. But Wheeler is bothered that it requires "too much excess baggage." He is toying with yet another explanation of the Schrödinger cat problem. He calls his theory "It from bit." It's an unorthodox theory, which starts with the assumption that information is at the root of all existence. When we look at the moon, a galaxy, or an atom, their essence, he claims, is in the information stored within them. But this information sprang into existence when the universe observed itself. He draws a circular diagram, representing the history of the universe. At the beginning of the universe, it sprang into being because it was observed. This means that "it" (matter in the universe) sprang into existence when information ("bit") of the universe was observed. He calls this the "participatory universe"—the idea that the universe adapts to us in the same way that we adapt to the universe, that our very presence makes the universe possible. (Since there is no universal consensus on the measurement problem in quantum mechanics, most physicists take a wait-and-see attitude toward It from Bit.)

QUANTUM COMPUTING AND TELEPORTATION

Such philosophical discussions may seem hopelessly impractical, devoid of any practical application in our world. Instead of debating how many angels can dance on the head of a pin, quantum physicists seem to be debating how many places an electron can be at the same time.

However, these are not the idle musings of ivory-tower academics. One day they may have the most practical application of all: to drive the economies of the world. One day, the wealth of entire nations may depend on the subtleties of Schrödinger's cat. At that time, perhaps our computers will be computing in parallel universes. Almost all of our computer infrastructure today is based on silicon transistors. Moore's law, which states that computer power doubles every eighteen months, is possible because of our ability to etch smaller and smaller transistors onto silicon chips via beams of ultraviolet radiation. Although Moore's law has revolutionized the technological landscape, it cannot continue forever. The most advanced Pentium chip has a layer twenty atoms across. Within fifteen to twenty years, scientists may be calculating on layers perhaps five atoms across. At these incredibly small distances, we have to abandon Newtonian mechanics and adopt the quantum mechanics, where the Heisenberg uncertainty principle takes over. As a consequence, we no longer know precisely where the electron is. This means that short circuits will take place as electrons drift outside insulators and semiconductors instead of staying within them.

In the future, we will reach the limits of etching on silicon wafers. The Age of Silicon will soon be coming to a close. Perhaps it will usher in the quantum era. Silicon Valley could become a Rust Belt. One day we may be forced to compute on atoms themselves, introducing a new architecture for computation. Computers today are based on the binary system—every number is based on zeros and ones. Atoms, however, can have their spin pointed up, down, or sideways, simultaneously. Computer bits (os and 1s) could be replaced by "qubits" (anything between 0 and 1), making quantum computation much more powerful than ordinary computers.

A quantum computer, for example, could shake the foundations of international security. Today, large banks, multinational corporations, and industrial nations code their secrets with complex computer algorithms. Many secret codes are based on factorizing huge numbers. It would take centuries, for example, for an ordinary computer to factorize a number containing a hundred digits. But for a quantum computer, such calculations may be effortless; they could break the secret codes of the nations of the world.

To see how a quantum computer would work, let's say that we align a series of atoms, with their spins pointing in one direction in a magnetic field. Then we shine a laser beam on them, so many of the spins flip as the laser beam reflects off the atoms. By measuring the reflected laser light, we have recorded a complex mathematical operation, the scattering of light off atoms. If we calculate this process using the quantum theory, following Feynman, we must add together all possible positions of the atoms, spinning in all possible directions. Even a simple quantum calculation, which would take a fraction of a second, would be almost impossible to perform on a standard computer, no matter how much time is allotted.

In principle, as David Deutsch of Oxford has stressed, this means that when we use quantum computers, we would have to sum over all possible parallel universes. Although we cannot directly make contact with these alternate universes, an atomic computer could calculate them using the spin states existing in parallel universes. (While we are no longer coherent with the other universes in our living room, the atoms in a quantum computer are, by construction, vibrating coherently in unison.)

Although the potential of quantum computers is truly staggering, in practice, the problems are equally enormous. At present, the world record for the number of atoms used in a quantum computer is seven. At best, we can multiply three by five, to get fifteen on a quantum computer, hardly impressive. For a quantum computer to be competitive with even an ordinary laptop, we would need hundreds, perhaps millions of atoms vibrating coherently. Because even the collision with a single air molecule could make the atoms decohere, one would have to have extraordinarily clean conditions to isolate the test atoms from the environment. (To construct a quantum computer that would exceed the speed of modern computers would require thousands to millions of atoms, so quantum computing is still decades away.)

QUANTUM TELEPORTATION

There may ultimately be another practical application to physicists' seemingly pointless discussion of parallel quantum universes: quantum teleportation. The "transporter" used in Star Trek and other science fiction programs to transport people and equipment through space seems like a marvelous way to zip across vast distances. But as tantalizing as it seems, the idea of teleportation has stumped physicists because it seems to violate the uncertainty principle. By making a measurement on an atom, you disturb the state of the atom, and hence an exact copy cannot be made.

But scientists found a loophole in this argument in 1993, through something called quantum entanglement. This is based on an old experiment proposed in 1935 by Einstein and his colleagues Boris Podolsky and Nathan Rosen (the so-called EPR paradox) to show how crazy the quantum theory really is. Let's say that there is an explosion, and two electrons fly apart in opposite directions, traveling at near light speed. Since electrons can spin like a top, assume that the spins are correlated—that is, if one electron has its spin axis pointing up, the other electron is spinning down (such that the total spin is zero). Before we make a measurement, however, we do not know which direction each electron is spinning.

Now wait several years. By then, the two electrons are many light-years apart. If we now make a measurement of the spin of one electron and find that its axis of spin points up, then we instantly know that the other electron is spinning down (and vice versa). In fact, the fact that the electron is found to be spinning up *forces* the other electron to spin down. This means that we now know something about an electron many light-years away, instantly. (Information, it seems, has traveled faster than the speed of light, in apparent violation of Einstein's special relativity.) By subtle reasoning, Einstein could show that, by making successive measurements on one pair, one could violate the uncertainty principle. More important, he showed that quantum mechanics is more bizarre than anyone had previously thought.

Up to then, physicists believed the universe was local, that disturbances in one part of the universe only spread out locally from the source. Einstein showed that quantum mechanics is essentially *nonlocal*—disturbances from one source can instantly affect distant parts of the universe. Einstein called it a "spooky action-at-a-distance," which he thought was absurd. Thus, thought Einstein, the quantum theory must be wrong.

(The critics of quantum mechanics could resolve the Einstein-Podolsky-Rosen paradox by assuming that, if our instruments were only sensitive enough, they could really determine which way the electrons were spinning. The apparent uncertainty in the spin and position of an electron was a fiction, due to the fact that our instruments were too crude. They introduced the concept called hidden variables—that is, there must be a hidden *sub*quantum theory, in which there is no uncertainty at all, based on new variables called hidden variables.)

The stakes were raised enormously in 1964, when physicist John Bell put the EPR paradox and hidden variables to the acid test. He showed that if one performed the EPR experiment, there should be a numerical correlation between the spins of the two electrons, depending on which theory one used. If the hidden variable theory was correct, as the skeptics believed, then the spins should be correlated in one way. If quantum mechanics was correct, the spins should be correlated in another way. In other words, quantum mechanics (the foundation of all modern atomic physics) would rise and fall on the basis of a single experiment.

But experiments have conclusively proved Einstein wrong. In the early 1980s, Alan Aspect and colleagues in France performed the EPR experiment with two detectors 13 meters apart, which measured the spins of photons emitted from calcium atoms. In 1997, the EPR experiment was performed with detectors separated by 11 kilometers. Each time the quantum theory won. A certain form of knowledge *does* travel faster than light. (Although Einstein was wrong on the EPR experiment, he was right on the larger question of faster-than-light communication. The EPR experiment, although it does allow you to know something instantly about the other side of the galaxy, does not allow you to send a message in this way. You cannot, for example, send Morse code. In fact, an "EPR transmitter" would send only random signals, since the spins you measure are random each time you measure them. The EPR experiment allows you to acquire information about the other side of the galaxy, but it does not allow you to transmit information that is useful—that is, not random.)

Bell liked to describe the effect by using the example of a mathematician called Bertelsman. He had the strange habit of every day wearing a green sock on one foot and a blue sock on the other, in random order. If one day you notice that he is wearing a blue sock on his left foot, you now know, faster than light, that his other sock is green. But knowing that does not allow you to communicate information in this fashion. Revealing information is different from sending it. The EPR experiment does not mean that we can communicate information through telepathy, faster-than-light travel, or time travel. But it does mean that it is impossible to completely separate ourselves from the oneness of the universe.

It forces us to hold a different picture of our universe. There is a cosmic "entanglement" between every atom of our body and atoms that are light-years distant. Since all matter came from a single explosion, the big bang, in some sense the atoms of our body are linked with some atoms on the other side of the universe in some kind of cosmic quantum web. Entangled particles are somewhat like twins still joined by an umbilical cord (their wave function) which can be light-years across. What happens to one member automatically affects the other, and hence knowledge concerning one particle can instantly reveal knowledge about its pair. Entangled pairs act as if they were a single object, although they may be separated by a large distance. (More precisely, since the wave functions of the particles in the big bang were once connected and coherent, their wave functions might still be partially connected billions of years after the big bang, so that disturbances in one part of the wave function can influence another distant part of the wave function.)

In 1993, scientists proposed using the concept of EPR entanglement to provide a mechanism for quantum teleportation. In 1997 and 1998, the scientists at Cal Tech, Aarhus University in Denmark, and the University of Wales made the first experimental demonstration of quantum teleportation when a single photon was teleported across a tabletop. Samuel Braunstein of the University of Wales, who was part of this team, has compared entangled pairs to lovers "who know each other so well that they could answer for their lover even if separated by long distances."

(Quantum teleportation experiments require three objects, called A, B, and C. Let B and C be two twins that are entangled. Although B and C may be separated by a large distance, they are still entangled with each other. Now let B come in contact with A, which is the object to be teleported. B "scans" A, so the information contained in A is transferred to B. This information is then transferred automatically to the twin C. Thus, C becomes an exact replica of A.)

Progress in quantum teleportation is moving rapidly. In 2003, scientists at the University of Geneva in Switzerland were able to teleport photons a distance of 1.2 miles through fiber optic cable. Photons of light (at 1.3-mm wavelength) in one laboratory were teleported into photons of light of a different wavelength (1.55 mm) in another laboratory connected by this long cable. Nicolas Gisin, a physicist on this project, has said, "Possibly, larger objects like a molecule will be teleported in my lifetime, but really large objects are not teleportable using foreseeable technologies."

Another significant breakthrough was made in 2004, when scientists at the National Institute of Standards and Technology (NIST) teleported not just a quantum of light but an entire atom. They successfully entangled three beryllium atoms and were able to transfer the characteristics of one atom into another, a major accomplishment.

The practical applications of quantum teleportation are potentially enormous. However, one should point out that there are several practical problems to quantum teleportation. First, the original object is destroyed in the process, so that you cannot make carbon copies of the object being teleported. Only one copy is possible. Second, you cannot teleport an object faster than light. Relativity still holds, even for quantum teleportation. (To teleport object A into object C, you still need an intermediate object B connecting the two, so information transfer takes place slower than light.) Third, perhaps the most important limitation on quantum teleportation is the same one facing quantum computing: the objects in question must be coherent. The slightest contamination with the environment will destroy quantum teleportation. But it is conceivable that within the twenty-first century the first virus may be teleported.

Teleporting a human being may pose other problems. Braunstein observes, "The key thing for now is the sheer amount of information involved. Even with the best communication channels we could conceive of at the moment, transferring all that info would take the age of the universe."

WAVE FUNCTION OF THE UNIVERSE

But perhaps the ultimate realization of the quantum theory may come when we apply quantum mechanics not just to individual photons but to the entire universe. Stephen Hawking has quipped that whenever he hears the cat problem, he reaches for his gun. He has proposed his own solution to the problem—to have a wave function of the entire universe. If the entire universe is part of the wave function, then there is no necessity for an observer (who must exist outside the universe).

In the quantum theory, every particle is associated with a wave. The wave, in turn, tells you the probability of finding the particle at any point. However, the universe, when it was very young, was smaller than a subatomic particle. Therefore, perhaps the universe itself has a wave function. Since the electron can exist in many states at the same time, and since the universe was smaller than an electron, perhaps the universe also existed simultaneously in many states, described by a super wave function.

This is a variation of the many worlds theory: there is no need to invoke a cosmic observer that can observe the entire universe all at once. But Hawking's wave function is quite different from Schrödinger's wave function. In Schrödinger's wave function, at every point in space-time, there is a wave function. In Hawking's wave function, for every universe, there is a wave. Instead of Schrödinger's *psi* function describing all possible states of the electron, Hawking introduces a *psi* function that represents all possible states of the universe. In ordinary quantum mechanics, the electron exists in ordinary space. However, in the wave function of the universe, the wave function exists in "super space," the space of all possible universes, introduced by Wheeler.

This master wave function (the mother of all wave functions) obeys not the Schrödinger equation (which only works for single electrons) but the Wheeler-DeWitt equation, which works for all possible universes. In the early 1990s, Hawking wrote that he was able to partially solve his wave function of the universe and show that the most likely universe was one with a vanishing cosmological constant. This paper provoked quite a bit of controversy because it depended on summing over all possible universes. Hawking performed this sum by including wormholes connecting our universe with all possible universes. (Imagine an infinite sea of soap bubbles floating in air, all connected by thin filaments or wormholes, and then adding them all together.)

Ultimately, doubts were raised about Hawking's ambitious method. It was pointed out that the sum of all possible universes was a mathematically unreliable one, at least until we had a "theory of everything" to guide us. Until a theory of everything is constructed, critics have argued that one cannot really trust any of the calculations about time machines, wormholes, the instant of the big bang, and wave functions of the universe.

Today, however, scores of physicists believe that we have finally found the theory of everything, although it is not yet in its final form: string theory, or M-theory. Will it allow us to "read the Mind of God," as Einstein believed?

CHAPTER SEVEN

M-Theory: The Mother of All Strings

> To someone who could grasp the Universe from a unified standpoint the entire creation would appear as a unique truth and necessity.
>
> —J. D'Alembert

> I feel that we are so close with string theory that—in my moments of greatest optimism—I imagine that any day, the final form of the theory might drop out of the sky and land in someone's lap. But more realistically, I feel that we are now in the process of constructing a much deeper theory than anything we have had before and that well into the twenty-first century, when I am too old to have any useful thoughts on the subject, younger physicists will have to decide whether we have in fact found the final theory.
>
> —Edward Witten

H. G. Wells's classic novel of 1897, *The Invisible Man*, begins with a strange tale. One cold wintry day, a stranger comes in from the darkness dressed in a bizarre fashion. His face is completely covered; he is wearing dark blue glasses, and a white bandage blankets his entire face.

At first, the villagers take pity on him, thinking that he was in a horrible accident. But strange things happen around the village. One day, his landlady entered his empty room and screamed when she saw clothing moving about by itself. Hats were whirling across the room, the bedclothes leaped into the air, chairs moved, and "the furniture went mad," she recalled in horror.

Soon, the entire village is buzzing with rumors of these unusual occurrences. Finally, a group of villagers gathers and confronts the mysterious stranger. To their amazement, he slowly begins to unwrap his bandages. The crowd is aghast. Without the bandages, the stranger's face is completely missing. In fact, he is invisible. Chaos erupts, as people yell and scream. The villagers try to chase the invisible man, who easily fights them off.

After committing a string of petty crimes, the invisible man seeks out an old acquaintance and recounts his remarkable story. His true name is Mr. Griffen of University College. Although he started out learning medicine, he stumbled upon a revolutionary way in which to change the refractive and reflective properties of flesh. His secret is the fourth dimension. He exclaims to Dr. Kemp, "I found a general principle . . . a formula, a geometrical expression involving four dimensions."

Sadly, instead of using this great discovery to help humanity, his thoughts are of robbery and private gain. He proposes to recruit his friend as an accomplice. Together, he claims, they can plunder the world. But the friend is horrified and reveals Mr. Griffen's presence to the police. This leads to a final manhunt, in which the invisible man is mortally wounded.

As with the best science fiction novels, there is a germ of science in many of H. G. Wells's tales. Anyone who can tap into the fourth spatial dimension (or what is today called the fifth dimension, with time being the fourth dimension) can indeed become invisible, and can even assume the powers normally ascribed to ghosts and gods. Imagine, for the moment, that a race of mythical beings can inhabit the two-dimensional world of a tabletop, as in Edwin Abbot's 1884 novel *Flatland*. They conduct their affairs unaware that an entire universe, the third dimension, surrounds them.

But if a Flatland scientist could perform an experiment that allows him to hover inches off the table, he would become invisible, because light would pass below him as if he didn't exist. Floating just above Flatland, he could see events unfolding below on the tabletop. Hovering in hyperspace has decided advantages, for anyone looking down from hyperspace would have the powers of a god.

Not only would light pass beneath him, making him invisible, he could also pass over objects. In other words, he could disappear at will and walk through walls. By simply leaping into the third dimension, he would vanish from the universe of Flatland. And if he jumped back onto the tabletop, he would suddenly rematerialize out of nowhere. He could therefore escape from any jail. A prison in Flatland would consist of a circle drawn around a prisoner, so it would be easy to simply jump into the third dimension and be outside.

It would be impossible to keep secrets away from a hyperbeing. Gold that is locked in a vault could be easily seen from the vantage point of the third dimension, since the vault is just an open rectangle. It would be child's play to reach into the rectangle and lift the gold out without ever breaking into the vault. Surgery would be possible without cutting the skin.

Similarly, H. G. Wells wanted to convey the idea that in a four-dimensional world, we are the Flatlanders, oblivious of the fact that higher planes of existence might hover right above ours. We believe that our world consists of all that we can see, unaware that there may be entire universes right above our noses. Although another universe might be hovering just inches above us, floating in the fourth dimension, it would appear to be invisible.

Because a hyperbeing would possess superhuman powers usually ascribed to a ghost or a spirit, in another science fiction story, H. G. Wells pondered the question of whether supernatural beings might inhabit higher dimensions. He raised a key question that is today the subject of great speculation and research: could there be new laws of physics in these higher dimensions? In his 1895 novel *The Wonderful Visit*, a vicar's gun accidentally hits an angel, who happens to be passing through our dimension. For some cosmic reason, our dimension and a parallel universe temporarily collided, allowing this angel to fall into our world. In the story, Wells writes, "There may be any number of three-dimensional Universes packed side by side." The vicar questions the wounded angel. He is shocked to find that our laws of nature no longer apply in the angel's world. In his universe, for example, there are no planes, but rather cylinders, so space itself is curved. (Fully twenty years before Einstein's general theory of relativity, Wells was entertaining thoughts about universes existing on curved surfaces.) As the vicar puts it, "Their geometry is different because their space has a curve in it so that all their planes are cylinders; and their law of Gravitation is not according to the law of inverse squares, and there are four-and-twenty primary colours instead of only three." More than a century after Wells wrote his tale, physicists today realize that new laws of physics, with different sets of subatomic particles, atoms, and chemical interactions, might indeed exist in parallel universes. (As we see in chapter 9, several experiments are now being conducted to detect the presence of parallel universes that might be hovering just above ours.)

The concept of hyperspace has intrigued artists, musicians, mystics, theologians, and philosophers, especially near the beginning of the twentieth century. According to art historian Linda Dalrymple Henderson, Pablo Picasso's interest in the fourth dimension influenced the creation of cubism. (The eyes of the women he painted look directly at us, even though their noses face to the side, allowing us to view the women in their entirety. Similarly, a hyperbeing looking down on us will see us in our entirety: front, back, and sides simultaneously.) In his famous painting *Christus Hypercubus*, Salvador Dalí painted Jesus Christ crucified in front of an unraveled four-dimensional hypercube, or a tesseract. In his painting *The Persistence of Memory*, Dalí tried to convey the idea of time as the fourth dimension with melted clocks. In Marcel Duchamp's painting *Nude Descending a Staircase (No. 2)*, we see a nude in time-lapse motion walking down the stairs, in another attempt to capture the fourth dimension of time on a two-dimensional surface.

M-THEORY

Today, the mystery and lore surrounding the fourth dimension are being resurrected for an entirely different reason: the development of string theory and its latest incarnation, M-theory. Historically, the concept of hyperspace has been resisted strenuously by physicists; they scoffed that higher dimensions were the province of mystics and charlatans. Scientists who seriously proposed the existence of unseen worlds were subject to ridicule.

With the coming of M-theory, all that has changed. Higher dimensions are now in the center of a profound revolution in physics because physicists are forced to confront the greatest problem facing physics today: the chasm between general relativity and the quantum theory. Remarkably, these two theories comprise the sum total of all physical knowledge about the universe at the fundamental level. At present, only M-theory has the ability to unify these two great, seemingly contradictory theories of the universe into a coherent whole, to create a "theory of everything." Of all the theories proposed in the past century, the only candidate that can potentially "read the Mind of God," as Einstein put it, is M-theory.

Only in ten- or eleven-dimensional hyperspace do we have "enough room" to unify all the forces of nature in a single elegant theory. Such a fabulous theory would be able to answer the eternal questions: What happened before the beginning? Can time be reversed? Can dimensional gateways take us across the universe? (Although its critics correctly point out that testing this theory is beyond our present experimental ability, there are a number of experiments currently being planned that may change this situation, as we shall see in chapter 9.)

All attempts for the past fifty years to create a truly unified description of the universe have ended in ignominious failure. Conceptually, this is easy to understand. General relativity and the quantum theory are diametrical opposites in almost every way. General relativity is a theory of the very large: black holes, big bangs, quasars, and the expanding universe. It is based on the mathematics of smooth surfaces, like bed sheets and trampoline nets. The quantum theory is precisely the opposite—it describes the world of the very tiny: atoms, protons and neutrons, and quarks. It is based on a theory of discrete packets of energy called quanta. Unlike relativity, the quantum theory states that only the probability of events can be calculated, so we can never know for sure precisely where an electron is located. These two theories are based on different mathematics, different assumptions, different physical principles, and different domains. It is not surprising that all attempts to unify them have floundered.

The giants of physics—Erwin Schrödinger, Werner Heisenberg, Wolfgang Pauli, and Arthur Eddington—who have followed Einstein have tried their hand at a unified field theory, only to fail miserably. In 1928, Einstein accidentally created a media stampede with an early version of his unified field theory. The *New York Times* even published parts of the paper, including his equations. Over a hundred reporters swarmed outside his house. Writing from England, Eddington commented to Einstein, "You may be amused to hear that one of our great department stores in London (Selfridges) has posted on its window your paper (the six pages pasted up side by side) so that passers-by can read it all through. Large crowds gather around to read it."

In 1946, Erwin Schrödinger also caught the bug and discovered what he thought was the fabled unified field theory. Hurriedly, he did something rather unusual for his time (but which is not so unusual today): he called a press conference. Even Ireland's prime minister, Eamon De Valera, showed up to listen to Schrödinger. When asked how certain he was that he had finally bagged the unified field theory, he replied, "I believe I am right. I shall look like an awful fool if I am wrong." (The *New York Times* eventually found out about this press conference and mailed the manuscript to Einstein and others for comment. Sadly, Einstein realized that Schrödinger had rediscovered an old theory that he had proposed years earlier and had rejected. Einstein was polite in his response, but Schrödinger was humiliated.)

In 1958, physicist Jeremy Bernstein attended a talk at Columbia

University where Wolfgang Pauli presented his version of the unified field theory, which he developed with Werner Heisenberg. Niels Bohr, who was in the audience, was not impressed. Finally, Bohr rose up and said, "We in the back are convinced that your theory is crazy. But what divides us is whether your theory is crazy enough."

Pauli immediately knew what Bohr meant—that the Heisenberg-Pauli theory was too conventional, too ordinary to be the unified field theory. To "read the Mind of God" would mean introducing radically different mathematics and ideas.

Many physicists are convinced that there is a simple, elegant, and compelling theory behind everything that nonetheless is crazy and absurd enough to be true. John Wheeler of Princeton points out that, in the nineteenth century, explaining the immense diversity of life found on Earth seemed hopeless. But then Charles Darwin introduced the theory of natural selection, and a single theory provided the architecture to explain the origin and diversity of all life on Earth.

Nobel laureate Steven Weinberg uses a different analogy. After Columbus, the maps detailing the daring exploits of the early European explorers strongly indicated that there must exist a "north pole," but there was no direct proof of its existence. Because every map of Earth showed a huge gap where the north pole should be located, the early explorers simply assumed that a north pole should exist, although none of them had ever visited it. Similarly, the physicists of today, like the early explorers, find ample indirect evidence pointing to the existence of a theory of everything, although at present there is no universal consensus on what that theory is.

HISTORY OF STRING THEORY

One theory that clearly is "crazy enough" to be the unified field theory is string theory, or M-theory. String theory has perhaps the most bizarre history in the annals of physics. It was discovered quite by accident, applied to the wrong problem, relegated to obscurity, and suddenly resurrected as a theory of everything. And in the final analysis, because it is impossible to make small adjustments without destroying the theory, it will either be a "theory of everything" or a "theory of nothing."

The reason for this strange history is that string theory has been evolving backward. Normally, in a theory like relativity, one starts with fundamental physical principles. Later, these principles are honed down to a set of basic classical equations. Last, one calculates quantum fluctuations to these equations. String theory evolved backward, starting with the accidental discovery of its quantum theory; physicists are still puzzling over what physical principles may guide the theory.

The origin of string theory dates back to 1968, when two young physicists at the nuclear laboratory at CERN, Geneva, Gabriele Veneziano and Mahiko Suzuki, were independently flipping through a math book and stumbled across the Euler Beta function, an obscure eighteenth-century mathematical expression discovered by Leonard Euler, which strangely seemed to describe the subatomic world. They were astonished that this abstract mathematical formula seemed to describe the collision of two π meson particles at enormous energies. The Veneziano model soon created quite a sensation in physics, with literally hundreds of papers attempting to generalize it to describe the nuclear forces.

In other words, the theory was discovered by pure accident. Edward Witten of the Institute for Advanced Study (whom many believe to be the creative engine behind many of the stunning breakthroughs in the theory) has said, "By rights, twentieth-century physicists shouldn't have had the privilege of studying this theory. By rights, string theory shouldn't have been invented."

I vividly remember the stir string theory created. I was still a graduate student in physics at the University of California at Berkeley at that time, and I recall seeing physicists shaking their heads and stating that physics was not supposed to be this way. In the past, physics was usually based on making painfully detailed observations of nature, formulating some partial hypothesis, carefully testing the idea against the data, and then tediously repeating the process, over and over again. String theory was a seat-of-your-pants method based on simply guessing the answer. Such breathtaking shortcuts were not supposed to be possible.

Because subatomic particles cannot be seen even with our most powerful instruments, physicists have resorted to a brutal but effective way to analyze them, by smashing them together at enormous energies. Billions of dollars have been spent building huge "atom smashers," or particle accelerators, which are many miles across, creating beams of subatomic particles that collide into each other. Physicists then meticulously analyze the debris from the collision. The goal of this painful and arduous process is to construct a series of numbers, called the scattering matrix, or S-matrix. This collection of numbers is crucial because it encodes within it all the information of subatomic physics—that is, if one knows the S-matrix, one can deduce all the properties of the elementary particles.

One of the goals of elementary particle physics is to predict the mathematical structure of the S-matrix for the strong interactions, a goal so difficult that some physicists believed it was beyond any known physics. One can then imagine the sensation caused by Veneziano and Suzuki when they simply guessed the S-matrix by flipping through a math book.

The model was a completely different kind of animal from anything we had ever seen before. Usually, when someone proposes a new theory (such as quarks), physicists try to tinker with the theory, changing simple parameters (like the particles' masses or coupling strengths). But the Veneziano model was so finely crafted that even the slightest disturbance in its basic symmetries ruined the entire formula. As with a delicately crafted piece of crystal, any attempt to alter its shape would shatter it.

Of the hundreds of papers that trivially modified its parameters, thereby destroying its beauty, none have survived today. The only ones that are still remembered are those that sought to understand why the theory worked at all—that is, those that tried to reveal its symmetries. Eventually, physicists learned that the theory had no adjustable parameters whatsoever.

The Veneziano model, as remarkable as it was, still had several problems. First, physicists realized that it was just a first approximation to the final S-matrix and not the whole picture. Bunji Sakita, Miguel Virasoro, and Keiji Kikkawa, then at the University of Wisconsin, realized that the S-matrix could be viewed as an infinite series of terms, and that the Veneziano model was just the first and most important term in the series. (Crudely speaking, each term in the series represents the number of ways in which particles can bump into each other. They postulated some of the rules by which one could construct the higher terms in their approximation. For my Ph.D. thesis, I decided to rigorously complete this program and construct all possible corrections to the Veneziano model. Along with my colleague L. P. Yu, I calculated the infinite set of correction terms to the model.)

Finally, Yoichiro Nambu of the University of Chicago and Tetsuo Goto of Nihon University identified the key feature that made the model work—a vibrating string. (Work along these lines was also done by Leonard Susskind and Holger Nielsen.) When a string collided with another string, it created an S-matrix described by the Veneziano model. In this picture, each particle is nothing but a vibration or note on the string. (I discuss this concept in detail later.)

Progress was very rapid. In 1971, John Schwarz, André Neveu, and Pierre Ramond generalized the string model so that it included a new quantity called spin, making it a realistic candidate for particle interactions. (All subatomic particles, as we shall see, appear to be spinning like a miniature top. The amount of spin of each subatomic particle, in quantum units, is either an integer like 0, 1, 2 or a half integer like 1/2, 3/2. Remarkably, the Neveu-Schwarz-Ramond string gave precisely this pattern of spins.)

I was, however, still unsatisfied. The dual resonance model, as it was called back then, was a loose collection of odd formulas and rules of thumb. All physics for the previous 150 years had been based on "fields," since they were first introduced by British physicist Michael Faraday. Think of the magnetic field lines created by a bar magnet. Like a spiderweb, the lines of force permeate all space. At every point in space, you can measure the strength and direction of the magnetic lines of force. Similarly, a field is a mathematical object that assumes different values at every point in space. Thus, the field measures the strength of the magnetic, electrical, or the nuclear force at any point in the universe. Because of this, the fundamental description of electricity, magnetism, the nuclear force, and gravity is based on fields. Why should strings be different? What was required was a "field theory of strings" that would allow one to summarize the entire content of the theory into a single equation.

In 1974, I decided to tackle this problem. With my colleague Keiji Kikkawa of Osaka University, I successfully extracted the field theory of strings. In an equation barely an inch and a half long, we could summarize all the information contained within string theory. Once the field theory of strings was formulated, I had to convince the larger physics community of its power and beauty. I attended a conference in theoretical physics at the Aspen Center in Colorado that summer and gave a seminar to a small but select group of physicists. I was quite nervous: in the audience were two Nobel laureates, Murray Gell-Mann and Richard Feynman, who were notorious for asking sharp, penetrating questions that often left the speaker flustered. (Once, when Steven Weinberg was giving a talk, he wrote down on the blackboard an angle, labeled by the letter W, which is called the Weinberg angle in his honor. Feynman then asked what the W on the blackboard represented. As Weinberg began to answer, Feynman shouted "Wrong!" which broke up the audience. Feynman may have amused the audience, but Weinberg got the last laugh. This angle represented a crucial part of Weinberg's theory which united the electromagnetic and weak interactions, and which would eventually win him the Nobel Prize.)

In my talk, I emphasized that string field theory would produce the simplest, most comprehensive approach to string theory, which was largely a motley collection of disjointed formulas. With string field theory, the entire theory could be summarized in a single equation about an inch and a half long—all the properties of the Veneziano model, all the terms of the infinite perturbation approximation, and all the properties of spinning strings could be derived from an equation that would fit onto a fortune cookie. I emphasized the symmetries of string theory that gave it its beauty and power. When strings move in space-time, they sweep out two-dimensional surfaces, resembling a strip. The theory remains the same no matter what coordinates we use to describe this two-dimensional surface. I will never forget that, afterward, Feynman came up to me and said, "I may not agree totally with string theory, but the talk you gave is one of the most beautiful I have ever heard."

TEN DIMENSIONS

But just as string theory was taking off, it quickly unraveled. Claude Lovelace of Rutgers discovered that the original Veneziano model had a tiny mathematical flaw that could only be eliminated if space-time had twenty-six dimensions. Similarly, the superstring model of Neveu, Schwarz, and Ramond could only exist in ten dimensions. This shocked physicists. This had never been seen before in the entire history of science. Nowhere else do we find a theory that selects out its own dimensionality. Newton's and Einstein's theories, for example, can be formulated in any dimension. The famed inverse-square law of gravity, for example, can be generalized to an inverse-cube law in four dimensions. String theory, however, could only exist in specific dimensions.

From a practical point of view, this was a disaster. Our world, it was universally believed, existed in three dimensions of space (length, width, and breadth) and one of time. To admit a ten-dimensional universe meant that the theory bordered on science fiction. String theorists became the butt of jokes. (John Schwarz remembers riding in the elevator with Richard Feynman, who jokingly said to him, "Well, John, and how many dimensions do you live in today?") But no matter how string physicists tried to salvage the model, it quickly died. Only the die-hards continued to work on the theory. It was a lonely effort during this period.

Two die-hards who continued to work on the theory during those bleak years were John Schwarz of Cal Tech and Joël Scherk of the École Normale Supérieure in Paris. Until then, the string model was supposed to describe just the strong nuclear interactions. But there was a problem: the model predicted a particle that did not occur in the strong interactions, a curious particle with zero mass that possessed 2 quantum units of spin. All attempts to get rid of this pesky particle had failed. Every time one tried to eliminate this spin-2 particle, the model collapsed and lost its magical properties. Somehow, this unwanted spin-2 particle seemed to hold the secret of the entire model.

Then Scherk and Schwarz made a bold conjecture. Perhaps the flaw was actually a blessing. If they reinterpreted this worrisome spin-2 particle as the graviton (a particle of gravity arising from Einstein's theory), then the theory actually incorporated Einstein's theory of gravity! (In other words, Einstein's theory of general relativity simply emerges as the lowest vibration or note of the superstring.) Ironically, while in other quantum theories physicists strenuously try to avoid including any mention of gravity, string theory demands it. (That, in fact, is one of the attractive features of string theory—that it must include gravity or else the theory is inconsistent.) With this daring leap, scientists realized that the string model was incorrectly being applied to the wrong problem. It was not meant to be a theory of just the strong nuclear interactions; it was instead a theory of everything. As Witten has emphasized, one attractive feature of string theory is that it demands the presence of gravity. While standard field theories have failed for decades to incorporate gravity, gravity is actually obligatory in string theory.

Scherk and Schwarz's seminal idea, however, was universally ignored. For string theory to describe both gravity and the subatomic world, it meant that the strings would have to be only 10^{-33} cm long (the Planck length); in other words, they were a billion billion times smaller than a proton. This was too much for most physicists to accept.

But by the mid-1980s, other attempts at a unified field theory had floundered. Theories that tried to naively attach gravity to the Standard Model were drowning in a morass of infinities (which I shall explain shortly). Every time someone tried to artificially marry gravity with the other quantum forces, it led to mathematical inconsistencies that killed the theory. (Einstein believed that perhaps God had no choice in creating the universe. One reason for this might be that only a single theory is free of all these mathematical inconsistencies.)

There were two such kinds of mathematical inconsistencies. The first was the problem of infinities. Usually, quantum fluctuations are tiny. Quantum effects are usually only a small correction to Newton's laws of motion. This is why we can, for the most part, ignore them in our macroscopic world—they are too small to be noticed. However, when gravity is turned into a quantum theory, these quantum fluctuations actually become infinite, which is nonsense. The second mathematical inconsistency has to do with "anomalies," small aberrations in the theory that arise when we add quantum fluctuations to a theory. These anomalies spoil the original symmetry of the theory, thereby robbing it of its original power.

For example, think of a rocket designer who must create a sleek, streamlined vehicle to slice through the atmosphere. The rocket must possess great symmetry in order to reduce air friction and drag (in this case, cylindrical symmetry, so the rocket remains the same when we rotate it around its axis). This symmetry is called O(2). But there are two potential problems. First, because the rocket travels at such great velocity, vibrations can occur in the wings. Usually, these vibrations are quite small in subsonic airplanes. However, traveling at hypersonic velocities, these fluctuations can grow in intensity and eventually tear the wing off. Similar divergences plague any quantum theory of gravity. Normally, they are so small they can be ignored, but in a quantum theory of gravity they blow up in your face.

The second problem with the rocket ship is that tiny cracks may occur in the hull. These flaws ruin the original O(2) symmetry of the rocket ship. Tiny as they are, these flaws can eventually spread and rip the hull apart. Similarly, such "cracks" can kill the symmetries of a theory of gravity.

There are two ways to solve these problems. One is to find Band-Aid solutions, like patching up the cracks with glue and bracing the wings with sticks, hoping that the rocket won't explode in the atmosphere. This is the approach historically taken by most physicists in trying to marry quantum theory with gravity. They tried to brush these two problems under the rug. The second way to proceed is to start all over again, with a new shape and new, exotic materials that can withstand the stresses of space travel.

Physicists had spent decades trying to patch up a quantum theory of gravity, only to find it hopelessly riddled with new divergences and anomalies. Gradually, they realized the solution might be to abandon the Band-Aid approach and adopt an entirely new theory.

STRING BANDWAGON

In 1984, the tide against string theory suddenly turned. John Schwarz of Cal Tech and Mike Green, then at Queen Mary's College in London, showed that string theory was devoid of all the inconsistencies that had killed off so many other theories. Physicists already knew that string theory was free of mathematical divergences. But Schwarz and Green showed that it was also free of anomalies. As a result, string theory became the leading (and today, the only) candidate for a theory of everything.

Suddenly, a theory that had been considered essentially dead was resurrected. From a theory of nothing, string theory suddenly became a theory of everything. Scores of physicists desperately tried to read the papers on string theory. An avalanche of papers began to pour out of research laboratories around the world. Old papers that were gathering dust in the library suddenly became the hottest topic in physics. The idea of parallel universes, once considered too outlandish to be true, now came center stage in the physics community, with hundreds of conferences and literally tens of thousands of papers devoted to the subject.

(At times, things got out of hand, as some physicists got "Nobel fever." In August, 1991, *Discover* magazine even splashed on its cover the sensational title: "The New Theory of Everything: A Physicist Tackles the Ultimate Cosmic Riddle." The article quoted one physicist who was in hot pursuit of fame and glory: "I'm not one to be modest. If this works out, there will be a Nobel Prize in it," he boasted. When faced with the criticism that string theory was still in its infancy, he shot back, "The biggest string guys are saying it would take four hundred years to prove strings, but I say they should shut up.")

The gold rush was on.

Eventually, there was a backlash against the "superstring bandwagon." One Harvard physicist has sneered that string theory is not really a branch of physics at all, but actually a branch of pure mathematics, or philosophy, if not religion. Nobel laureate Sheldon Glashow of Harvard led the charge, comparing the superstring bandwagon to the Star Wars program (which consumes vast resources yet can never be tested). Glashow has said that he is actually quite happy that so many young physicists work on string theory, because, he says, it keeps them out of his hair. When asked about Witten's comment that string theory may dominate physics for the next fifty years, in the same way that quantum mechanics dominated the last fifty years, he replies that string theory will dominate physics the same way that Kaluza-Klein theory (which he considers "kooky") dominated physics for the last fifty years, which is not at all. He tried to keep string theorists out of Harvard. But as the next generation of physicists shifted to string theory, even the lone voice of a Nobel laureate was soon drowned out. (Harvard has since hired several young string theorists.)

COSMIC MUSIC

Einstein once said that if a theory did not offer a physical picture that even a child could understand, then it was probably useless. Fortunately, behind string theory there is a simple physical picture, a picture based on music.

According to string theory, if you had a supermicroscope and could peer into the heart of an electron, you would see not a point particle but a vibrating string. (The string is extremely tiny, at the Planck length of 10^{-33} cm, a billion billion times smaller than a proton, so all subatomic particles appear pointlike.) If we were to pluck this string, the vibration would change; the electron might turn into a neutrino. Pluck it again and it might turn into a quark. In fact, if you plucked it hard enough, it could turn into any of the known subatomic particles. In this way, string theory can effortlessly explain why there are so many subatomic particles. They are nothing but different "notes" that one can play on a superstring. To give an analogy, on a violin string the notes A or B or C sharp are not fundamental. By simply plucking the string in different ways, we can generate all the notes of the musical scale. B flat, for example, is not more fundamental than G. All of them are nothing but notes on a violin string. In the same way, electrons and quarks are not fundamental, but the string is. In fact, all the subparticles of the universe can be viewed as nothing but different vibrations of the string. The "harmonies" of the string are the laws of physics.

Strings can interact by splitting and rejoining, thus creating the interactions we see among electrons and protons in atoms. In this way, through string theory, we can reproduce all the laws of atomic and nuclear physics. The "melodies" that can be written on strings correspond to the laws of chemistry. The universe can now be viewed as a vast symphony of strings.

Not only does string theory explain the particles of the quantum theory as the musical notes of the universe, it explains Einstein's relativity theory as well—the lowest vibration of the string, a spin-two particle with zero mass, can be interpreted as the graviton, a particle or quantum of gravity. If we calculate the interactions of these gravitons, we find precisely Einstein's old theory of gravity in quantum form. As the string moves and breaks and reforms, it places enormous restrictions on space-time. When we analyze these constraints, we again find Einstein's old theory of general relativity. Thus, string theory neatly explains Einstein's theory with no additional work. Edward Witten has said that if Einstein had never discovered relativity, his theory might have been discovered as a by-product of string theory. General relativity, in some sense, is for free.

The beauty of string theory is that it can be likened to music. Music provides the metaphor by which we can understand the nature of the universe, both at the subatomic level and at the cosmic level. As the celebrated violinist Yehudi Menuhin once wrote, "Music creates order out of chaos; for rhythm imposes unanimity upon the divergent; melody imposes continuity upon the disjointed; and harmony imposes compatibility upon the incongruous."

Einstein would write that his search for a unified field theory would ultimately allow him to "read the Mind of God." If string theory is correct, we now see that the Mind of God represents cosmic music resonating through ten-dimensional hyperspace. As Gottfried Leibniz once said, "Music is the hidden arithmetic exercise of a soul unconscious that it is calculating."

Historically, the link between music and science was forged as early as the fifth century B.C., when the Greek Pythagoreans discovered the laws of harmony and reduced them to mathematics. They found that the tone of a plucked lyre string corresponded to its length. If one doubled the length of a lyre string, then the note went down by a full octave. If the length of a string was reduced by two-thirds, then the tone changed by a fifth. Hence, the laws of music and harmony could be reduced to precise relations between numbers. Not surprisingly, the Pythagoreans' motto was "All things are numbers." Originally, they were so pleased with this result that they dared to apply these laws of harmony to the entire universe. Their effort failed because of the enormous complexity of matter. However, in some sense, with string theory, physicists are going back to the Pythagorean dream.

Commenting on this historic link, Jamie James once said, "Music and science were [once] identified so profoundly that anyone who suggested that there was any essential difference between them would have been considered an ignoramus, [but now] someone proposing that they have anything in common runs the risk of being labeled a philistine by one group and a dilettante by the other—and, most damning of all, a popularizer by both."

PROBLEMS IN HYPERSPACE

But if higher dimensions actually exist in nature and not only in pure mathematics, then string theorists have to face the same problem that dogged Theodr Kaluza and Felix Klein back in 1921 when they formulated the first higher-dimensional theory: where are these higher dimensions?

Kaluza, a previously obscure mathematician, wrote a letter to Einstein proposing to formulate Einstein's equations in five dimensions (one dimension of time and four dimensions of space). Mathematically, this was no problem, since Einstein's equations can be trivially written in any dimension. But the letter contained a startling observation: if one manually separated out the fourth-dimensional pieces contained within the five-dimensional equations, you would automatically find, almost by magic, Maxwell's theory of light! In other words, Maxwell's theory of the electromagnetic force tumbles right out of Einstein's equations for gravity if we simply add a fifth dimension. Although we cannot see the fifth dimension, ripples can form on the fifth dimension, which correspond to light waves! This is a gratifying result, since generations of physicists and engineers have had to memorize Maxwell's difficult equations for the past 150 years. Now, these complex equations emerge effortlessly as the simplest vibrations one can find in the fifth dimension.

Imagine fish swimming in a shallow pond, just below the lily pads, thinking that their "universe" is only two-dimensional. Our three-dimensional world may be beyond their ken. But there is a way in which they can detect the presence of the third dimension. If it rains, they can clearly see the shadows of ripples traveling along the surface of the pond. Similarly, we cannot see the fifth dimension, but ripples in the fifth dimension appear to us as light.

(Kaluza's theory was a beautiful and profound revelation concerning the power of symmetry. It was later shown that if we add even more dimensions to Einstein's old theory and make them vibrate, then these higher-dimensional vibrations reproduce the W- and Z-bosons and gluons found in the weak and strong nuclear forces! If the program advocated by Kaluza was correct, then the universe was apparently much simpler than previously thought. Simply vibrating higher and higher dimensions reproduced many of the forces that ruled the world.)

Although Einstein was shocked by this result, it was too good to be true. Over the years, problems were discovered that rendered Kaluza's idea useless. First, the theory was riddled with divergences and anomalies, which is typical of quantum gravity theories. Second, there was the much more disturbing physical question: why don't we see the fifth dimension? When we shoot arrows into the sky, we don't see them disappear into another dimension. Think of smoke, which slowly permeates every region of space. Since smoke is never observed to disappear into a higher dimension, physicists realized that higher dimensions, if they exist at all, must be smaller than an atom. For the past century, mystics and mathematicians have entertained the idea of higher dimensions, but physicists scoffed at the idea, since no one had ever seen objects enter a higher dimension.

To salvage the theory, physicists had to propose that these higher dimensions were so small that they could not be observed in nature. Since our world is a four-dimensional world, it meant that the fifth dimension has to be rolled up into a tiny circle smaller than an atom, too small to be observed by experiment.

String theory has to confront this same problem. We have to curl up these unwanted higher dimensions into a tiny ball (a process called compactification). According to string theory, the universe was originally ten-dimensional, with all the forces unified by the string. However, ten-dimensional hyperspace was unstable, and six of the ten dimensions began to curl up into a tiny ball, leaving the other four dimensions to expand outward in a big bang. The reason we can't see these other dimensions is that they are much smaller than an atom, and hence nothing can get inside them. (For example, a garden hose and a straw, from a distance, appear to be one-dimensional objects defined by their length. But if one examines them closely, one finds that they are actually two-dimensional surfaces or cylinders, but the second dimension has been curled up so that one does not see it.)

WHY STRINGS?

Although previous attempts at a unified field theory have failed, string theory has survived all challenges. In fact, it has no rival. There are two reasons why string theory has succeeded where scores of other theories have failed.

First, being a theory based on an extended object (the string), it avoids many of divergences associated with point particles. As Newton observed, the gravitational force surrounding a point particle becomes infinite as we approach it. (In Newton's famous inverse square law, the force of gravity grows as $1/r^2$, so that it soars to infinity as we approach the point particle—that is, as r goes to zero, the gravitational force grows as $1/0$, which is infinite.)

Even in a quantum theory, the force remains infinite as we approach a quantum point particle. Over the decades, a series of arcane rules have been invented by Feynman and many others to brush these and many other types of divergences under the rug. But for a quantum theory of gravity, even the bag of tricks devised by Feynman is not sufficient to remove all the infinites in the theory. The problem is that point particles are infinitely small, meaning that their forces and energies are potentially infinite.

But when we analyze string theory carefully, we find two mechanisms that can eliminate these divergences. The first mechanism is due to the topology of strings; the second, due to its symmetry, is called supersymmetry.

The topology of string theory is entirely different from the topology of point particles, and hence the divergences are much different. (Roughly speaking, because the string has a finite length, it means that the forces do not soar to infinity as we approach the string. Near the string, forces only grow as $1/L^2$, where L is the length of the string, which is on the order of the Planck length of 10^{-33} cm. This length L acts to cut off the divergences.) Because a string is not a point particle but has a definite size, one can show that the divergences are "smeared out" along the string, and hence all physical quantities become finite.

Although it seems intuitively obvious that the divergences of string theory are smeared out and hence finite, the precise mathematical expression of this fact is quite difficult and is given by the "elliptic modular function," one of the strangest functions in mathematics, with a history so fascinating it played a key role in a Hollywood movie. *Good Will Hunting* is the story of a rough working-class kid from the backstreets of Cambridge, played by Matt Damon, who exhibits astounding mathematical abilities. When he is not getting into fistfights with neighborhood toughs, he works as a janitor at MIT. The professors at MIT are shocked to find that this street tough is actually a mathematical genius who can simply write down the answers to seemingly intractable mathematical problems. Realizing that this street tough has learned advanced mathematics on his own, one of them blurts out that he is the "next Ramanujan."

In fact, *Good Will Hunting* is loosely based on the life of Srinivasa Ramanujan, the greatest mathematical genius of the twentieth century, a man who grew up in poverty and isolation near Madras, India, at the turn of the last century. Living in isolation, he had to derive much of nineteenth-century European mathematics on his own. His career was like a supernova, briefly lighting up the heavens with his mathematical brilliance. Tragically, he died of tuberculosis in 1920 at the age of thirty-seven. Like Matt Damon in *Good Will Hunting*, he dreamed of mathematical equations, in this case the elliptic modular function, which possesses strange but beautiful mathematical properties, but only in twenty-four dimensions. Mathematicians are still trying to decipher the "lost notebooks of Ramanujan" found after his death. Looking back at Ramanujan's work, we see that it can be generalized to eight dimensions, which is directly applicable to string theory. Physicists add two extra dimensions in order to construct a physical theory. (For example, polarized sunglasses use the fact that light has two physical polarizations; it can vibrate left-right or up-down. But the mathematical formulation of light in Maxwell's equation is given with four components. Two of these four vibrations are actually redundant.) When we add two more dimensions to Ramanujan's functions, the "magic numbers" of mathematics become 10 and 26, precisely the "magic numbers" of string theory. So in some sense, Ramanujan was doing string theory before World War I!

The fabulous properties of these elliptic modular functions explain why the theory must exist in ten dimensions. Only in that precise number of dimensions do most of the divergences that plague other theories disappear, as if by magic. But the topology of strings, by itself, is not powerful enough to eliminate all the divergences. The remaining divergences of the theory are removed by a second feature of string theory, its symmetry.

SUPERSYMMETRY

The string possesses some of the largest symmetries known to science. In chapter 4, in discussing inflation and the Standard Model, we see that symmetry gives us a beautiful way in which to arrange the subatomic particles into pleasing and elegant patterns. The three types of quarks can be arranged according to the symmetry $SU(3)$, which interchanges three quarks among themselves. It is believed that in GUT theory, the five types of quarks and leptons might be arranged according to the symmetry $SU(5)$.

In string theory, these symmetries cancel the remaining divergences and anomalies of the theory. Since symmetries are among the most beautiful and powerful tools at our disposal, one might expect that the theory of the universe must possess the most elegant and powerful symmetry known to science. The logical choice is a symmetry that interchanges not just the quarks but all the particles found in nature—that is, the equations remain the same if we reshuffle all the subatomic particles among themselves. This precisely describes the symmetry of the superstring, called supersymmetry. *It is the only symmetry that interchanges all the subatomic particles known to physics.* This makes it the ideal candidate for the symmetry that arranges all the particles of the universe into a single, elegant, unified whole.

If we look at the forces and particles of the universe, all of them fall into two categories: "fermions" and "bosons," depending on their spin. They act like tiny spinning tops that can spin at various rates. For example, the photon, a particle of light that mediates the electromagnetic force, has spin 1. The weak and strong nuclear forces are mediated by W-bosons and gluons, which also have spin 1. The graviton, a particle of gravity, has spin 2. All these with integral spin are called bosons. Similarly, the particles of matter are described by subatomic particles with half-integral spin—1/2, 3/2, 5/2, and so on. (Particles of half-integral spins are called fermions and include the electron, the neutrino, and the quarks.) Thus, supersymmetry elegantly represents the duality between bosons and fermions, between forces and matter.

In a supersymmetric theory, all the subatomic particles have a partner: each fermion is paired with a boson. Although we have never seen these supersymmetric partners in nature, physicists have dubbed the partner of the electron the "selectron," with spin 0. (Physicists add an "s" to describe the superpartner of a particle.) The weak interactions include particles called leptons; their superpartners are called sleptons. Likewise, the quark may have a spin-0 partner called the squark. In general, the partners of the known particles (the quarks, leptons, gravitons, photons, and so on) are called sparticles, or superparticles. These sparticles have yet to be found in our atom smashers (probably because our machines are not powerful enough to create them).

But since all subatomic particles are either fermions or bosons, a supersymmetric theory has the potential of unifying all known subatomic particles into one simple symmetry. *We now have a symmetry large enough to include the entire universe.*

Think of a snowflake. Let each of the six prongs of the snowflake represent a subatomic particle, with every other prong being a boson, and the one that follows being a fermion. The beauty of this "super snowflake" is that when we rotate it, it remains the same. In this way, the super snowflake unifies all the particles and their sparticles. So if we were to try to construct a hypothetical unified field theory with just six particles, a natural candidate would be the super snowflake.

Supersymmetry helps to eliminate the remaining infinities that are fatal to other theories. We mentioned earlier that most divergences are eliminated because of the topology of the string—that is, because the string has a finite length, the forces do not soar to infinity as we approach it. When we examine the remaining divergences, we find that they are of two types, from the interactions of bosons and fermions. However, these two contributions always occur with the opposite signs, hence the boson contribution precisely cancels the fermion contribution! In other words, since fermionic and bosonic contributions always have opposite signs, the remaining infinities of the theory cancel against each other. So supersymmetry is more than window dressing; not only is it an aesthetically pleasing symmetry because it unifies all the particles of nature, it is also essential in canceling the divergences of string theory.

Recall the analogy of designing a sleek rocket, in which vibrations in the wings may eventually grow and tear the wings off. One solution is to exploit the power of symmetry, to redesign the wings so that vibrations in one wing cancel against vibrations in another. When one wing vibrates clockwise, the other wing vibrates counterclockwise, canceling the first vibration. Thus the symmetry of the rocket, instead of being just an artificial, artistic device, is crucial to canceling and balancing the stresses on the wings. Similarly, supersymmetry cancels divergences by having the bosonic and fermionic parts cancel out against each other.

(Supersymmetry also solves a series of highly technical problems that are actually fatal to GUT theory. Intricate mathematical inconsistencies in GUT theory require supersymmetry to eliminate them.)

Although supersymmetry represents a powerful idea, at present there is absolutely no experimental evidence to support it. This may be because the superpartners of the familiar electrons and protons are simply too massive to be produced in today's particle accelerators. However, there is one tantalizing piece of evidence that points the way to supersymmetry. We know now that the strengths of the three quantum forces are quite different. In fact, at low energies, the strong force is thirty times stronger than the weak force, and a hundred times more powerful than the electromagnetic force. However, this was not always so. At the instant of the big bang, we suspect that

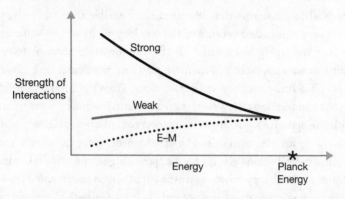

The strengths of the weak, strong, and electromagnetic forces are quite different in our everyday world. However, at energies found near the big bang, the strengths of these forces should converge perfectly. This convergence takes place if we have a supersymmetric theory. Thus, supersymmetry may be a key element in any unified field theory.

all three forces were equal in strength. Working backward, physicists can calculate what the strengths of the three forces would have been at the beginning of time. By analyzing the Standard Model, physicists find that the three forces seem to converge in strength near the big bang. But they are not precisely equal. When one adds supersymmetry, however, all three forces fit perfectly and are of equal strength, precisely what a unified field theory would suggest. Although this is not direct proof of supersymmetry, it shows at least that supersymmetry is consistent with known physics.

DERIVING THE STANDARD MODEL

Although superstrings have no adjustable parameters at all, string theory can offer solutions that are astonishingly close to the Standard Model, with its motley collections of bizarre subatomic particles and nineteen free parameters (such as the masses of the particles and their coupling strengths). In addition, the Standard

Model has three identical and redundant copies of all the quarks and leptons, which seems totally unnecessary. Fortunately, string theory can derive many of the qualitative features of the Standard Model effortlessly. It's almost like getting something for nothing. In 1984, Philip Candelas of the University of Texas, Gary Horowitz and Andrew Strominger of the University of California at Santa Barbara, and Edward Witten showed that if you wrapped up six of the ten dimensions of string theory and still preserved supersymmetry in the remaining four dimensions, the tiny, six-dimensional world could be described by what mathematicians called a Calabi-Yau manifold. By making a few simple choices of the Calabi-Yau spaces, they showed that the symmetry of the string could be broken down to a theory remarkably close to the Standard Model.

In this way, string theory gives us a simple answer as to why the Standard Model has three redundant generations. In string theory, the number of generations or redundancies in the quark model is related to the number of "holes" we have in the Calabi-Yau manifold. (For example, a doughnut, an inner tube, and a coffee cup are all surfaces with one hole. Eyeglass frames have two holes. Calabi-Yau surfaces can have an arbitrary number of holes.) Thus, by simply choosing the Calabi-Yau manifold that has a certain number of holes, we can construct a Standard Model with different generations of redundant quarks. (Since we never see the Calabi-Yau space because it is so small, we also never see the fact that this space has doughnut holes in it.) Over the years, teams of physicists have arduously tried to catalog all the possible Calabi-Yau spaces, realizing that the topology of this six-dimensional space determines the quarks and leptons of our four-dimensional universe.

M-THEORY

The excitement surrounding string theory unleashed back in 1984 could not last forever. By the mid-1990s, the superstring bandwagon was gradually losing steam among physicists. The easy problems the theory posed were picked off, leaving the hard ones behind. One such problem was that billions of solutions of the string equations were being discovered. By compactifying or curling up space-time in different ways, string solutions could be written down in any dimension, not just four. Each of the billions of string solutions corresponded to a mathematically self-consistent universe.

Physicists were suddenly drowning in string solutions. Remarkably, many of them looked very similar to our universe. With a suitable choice of a Calabi-Yau space, it was relatively easy to reproduce many of the gross features of the Standard Model, with its strange collection of quarks and leptons, even with its curious set of redundant copies. However, it was exceedingly difficult (and remains a challenge even today) to find precisely the Standard Model, with the specific values of its nineteen parameters and three redundant generations. (The bewildering number of string solutions was actually welcomed by physicists who believe in the multiverse idea, since each solution represents a totally self-consistent parallel universe. But it was distressing that physicists had trouble finding precisely our own universe among this jungle of universes.)

One reason that this is so difficult is that one must eventually break supersymmetry, since we do not see supersymmetry in our low-energy world. In nature, for example, we do not see the selectron, the superpartner of the electron. If supersymmetry is unbroken, then the mass of each particle should equal the mass of its superparticle. Physicists believe that supersymmetry is broken, with the result that the masses of the superparticles are huge, beyond the range of current particle accelerators. But at present no one has come up with a credible mechanism to break supersymmetry.

David Gross of the Kavli Institute for Theoretical Physics in Santa Barbara has remarked that there are millions upon millions of solutions to string theory in three spatial dimensions, which is slightly embarrassing since there is no good way of choosing among them.

There were other nagging questions. One of the most embarrassing was the fact that there were five self-consistent string theories. It was hard to imagine that the universe could tolerate five distinct unified field theories. Einstein believed that God had no choice in creating the universe, so why should God create five of them?

The original theory based on the Veneziano formula describes what is called type I superstring theory. Type I theory is based on both open strings (strings with two ends) as well as closed strings (circular strings). This is the theory that was most intensely studied in the early 1970s. (Using string field theory, Kikkawa and I were able

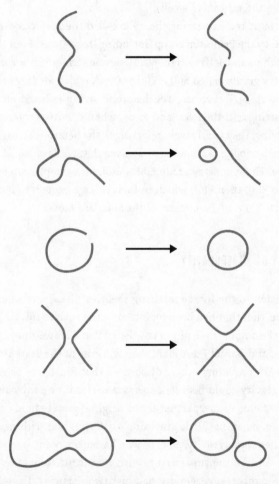

Type I strings undergo five possible interactions, in which strings can break, join, and fission. For closed strings, only the last interaction is necessary (resembling the mitosis of cells).

to catalog the complete set of type I string interactions. We showed that type I strings require five interactions; for closed strings, we showed that only one interaction term is necessary.)

Kikkawa and I also showed that it is possible to construct fully self-consistent theories with only closed strings (those resembling a loop). Today, these are called type II string theories, where strings interact via pinching a circular string into two smaller strings (resembling the mitosis of a cell).

The most realistic string theory is called the heterotic string, formulated by the Princeton group (including David Gross, Emil Martinec, Ryan Rohm, and Jeffrey Harvey). Heterotic strings can accommodate symmetry groups called E(8) × E(8) or O(32), which are large enough to swallow up GUT theories. The heterotic string is based entirely on closed strings. In the 1980s and 1990s, when scientists referred to the superstring, they tacitly were referring to the heterotic string, because it was rich enough to allow one to analyze the Standard Model and GUT theories. The symmetry group E(8) × E(8), for example, can be broken down to E(8), then E(6), which in turn is large enough to include the SU(3) × SU(2) × U(1) symmetry of the Standard Model.

MYSTERY OF SUPERGRAVITY

In addition to the five superstring theories, there was another nagging question that had been forgotten in the rush to solve string theory. Back in 1976, three physicists, Peter Van Nieuwenhuizen, Sergio Ferrara, and Daniel Freedman, then working at the State University of New York at Stony Brook, discovered that Einstein's original theory of gravity could become supersymmetric if one introduced just one new field, a superpartner to the original gravity field (called the gravitino, meaning "little graviton," with spin 3/2). This new theory was called supergravity, and it was based on point particles, not strings. Unlike the superstring, with its infinite sequence of notes and resonances, supergravity had just two particles. In 1978, it was shown by Eugene Cremmer, Joël Scherk, and Bernard Julia of the École Normale Supérieure that the most general supergravity could be written down in eleven dimensions. (If we tried to write down supergravity theory in twelve or thirteen dimensions, mathematical inconsistencies would arise.) In the late 1970s and early 1980s, it was thought that supergravity might be the fabled unified field theory. The theory even inspired Stephen Hawking to speak of "the end of theoretical physics" being in sight when he gave his inaugural lecture upon taking the Lucasian Chair of Mathematics at Cambridge University, the same chair once held by Isaac Newton. But supergravity soon ran into the same difficult problems that had killed previous theories. Although it had fewer infinities than ordinary field theory, in the final analysis supergravity was not finite and was potentially riddled with anomalies. Like all other field theories (except for string theory), it blew up in scientists' faces.

Another supersymmetric theory that can exist in eleven dimensions is supermembrane theory. Although the string has just one dimension that defines its length, the supermembrane can have two or more dimensions because it represents a surface. Remarkably, it was shown that two types of membranes (a two-brane and five-brane) are self-consistent in eleven dimensions, as well.

However, supermembranes also had problems; they are notoriously difficult to work with, and their quantum theories actually diverge. While violin strings are so simple that the Greek Pythagoreans worked out their laws of harmony two thousand years ago, membranes are so difficult that even today no one has a satisfactory theory of the music based on them. Plus, it was shown that these membranes are unstable and eventually decay into point particles.

So, by the mid 1990s, physicists had several mysteries. Why were there five string theories in ten dimensions? And why were there two theories in eleven dimensions, supergravity and supermembranes? Moreover, all of them possessed supersymmetry.

ELEVENTH DIMENSION

In 1994, a bombshell was dropped. Another breakthrough took place that once again changed the entire landscape. Edward Witten and

Paul Townsend of Cambridge University mathematically found that ten-dimensional string theory was actually an approximation to a higher, mysterious, eleven-dimensional theory of unknown origin. Witten, for example, showed that if we take a membranelike theory in eleven dimensions and curl up one dimension, then it becomes ten-dimensional type IIa string theory!

Soon afterward, it was found that all five string theories could be shown to be the same—just different approximations of the same mysterious eleven-dimensional theory. Since membranes of different sorts can exist in eleven dimensions, Witten called this new theory M-theory. But not only did it unify the five different string theories, as a bonus it also explained the mystery of supergravity.

Supergravity, if you'll recall, was an eleven-dimensional theory that contained just two particles with zero mass, the original Einstein graviton, plus its supersymmetric partner (called the gravitino). M-theory, however, has an infinite number of particles with different masses (corresponding to the infinite vibrations that can ripple on some sort of eleven-dimensional membrane). But M-theory can explain the existence of supergravity if we assume that a tiny portion of M-theory (just the massless particles) is the old supergravity theory. In other words, supergravity theory is a tiny subset of M-theory. Similarly, if we take this mysterious eleven-dimensional membranelike theory and curl up one dimension, the membrane turns into a string. In fact, it turns into precisely type II string theory! For example, if we look at a sphere in eleven dimensions and then curl up one dimension, the sphere collapses, and its equator becomes a closed string. We see that string theory can be viewed as a slice of a membrane in eleven dimensions if we curl up the eleventh dimension into a small circle.

Thus, we find a beautiful and simple way of unifying all ten-dimensional and eleven-dimensional physics into a single theory! It was a conceptual tour de force.

I still remember the shock generated by this explosive discovery. I was giving a talk at Cambridge University at that time. Paul Townsend was gracious enough to introduce me to the audience. But before my talk, he explained with great excitement this new result,

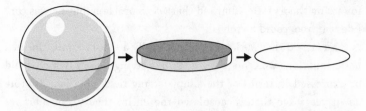

A ten-dimensional string can emerge from an eleven-dimensional membrane by slicing or curling up one dimension. The equator of a membrane becomes the string after one dimension is collapsed. There are five ways in which this reduction can take place, giving rise to five different superstring theories in ten dimensions.

that in the eleventh dimension, the various string theories can be unified into a single theory. The title of my talk mentioned the tenth dimension. He told me before I spoke that, if this proved to be successful, then the title of my talk would be obsolete.

I thought silently to myself, "Uh oh." Either he was raving mad, or the physics community was going to be turned completely upside down.

I could not believe what I was hearing, so I fired a barrage of questions at him. I pointed out that eleven-dimensional supermembranes, a theory he helped to formulate, were useless because they were mathematically intractable, and worse, they were unstable. He admitted this was a problem, but he was confident that these questions would be solved in the future.

I also said that eleven-dimensional supergravity was not finite; it blew up, like all the other theories except string theory. That was no longer a problem, he replied calmly, because supergravity was nothing but an approximation of a larger, still mysterious theory, M-theory, which *was* finite—it was actually string theory reformulated in the eleventh dimension in terms of membranes.

Then I said that supermembranes were unacceptable because no one had ever been able to explain how membranes interact as they collide and re-form (as I had done in my own Ph.D. thesis years ago for string theory). He admitted that was a problem, but he was confident it, too, could be solved.

Last, I said that M-theory was not really a theory at all, since its basic equations were not known. Unlike string theory (which could be expressed in terms of the simple string field equations I wrote down years ago that encapsulated the entire theory), membranes had no field theory at all. He conceded this point as well. But he remained confident that the equations for M-theory would eventually be found.

My mind was sent swimming. If he was right, string theory was once again about to undergo a radical transformation. Membranes, which were once relegated to the dustbin of physics history, suddenly were being resurrected.

The origin of this revolution is that string theory is still evolving backward. Even today, no one knows the simple physical principles that underlie the entire theory. I like to visualize this as walking in the desert and accidentally stumbling upon a small, beautiful pebble. When we brush away the sand, we find that the pebble is actually the top of a gigantic pyramid buried under tons of sand. After decades of painfully excavating the sand, we find mysterious hieroglyphics, hidden chambers, and tunnels. One day, we will find the ground floor and finally open up the doorway.

BRANE WORLD

One of the novel features of M-theory is that it introduces not only strings but a whole menagerie of membranes of different dimensions. In this picture, point particles are called "zero-branes," because they are infinitely small and have no dimension. A string is then a "one-brane," because it is a one-dimensional object defined by its length. A membrane is a "two-brane," like the surface of a basketball, defined by length and width. (A basketball can float in three dimensions, but its surface is only two-dimensional.) Our universe might be some kind of "three-brane," a three-dimensional object that has length, width, and breadth. (As one wit noted, if space has *p* dimensions, *p* being an integer, then our universe is a *p*-brane, pronounced "pea-brain." A chart showing all these pea-brains is called a "brane-scan.")

There are several ways in which we can take a membrane and collapse it down to a string. Instead of wrapping up the eleventh dimension, we can also slice off the equator of an eleven-dimensional membrane, creating a circular ribbon. If we let the thickness of the ribbon shrink, then the ribbon becomes a ten-dimensional string. Petr Horava and Edward Witten showed that we derive the heterotic string in this fashion.

In fact, it can be shown that there are five ways in which to reduce eleven-dimensional M-theory down to ten dimensions, thereby yielding the five superstring theories. M-theory gives us a quick, intuitive answer to the mystery of why there are five different string theories. Imagine standing on a large hilltop and looking down on the plains. From the vantage point of the third dimension, we can see the different parts of the plain unified into a single coherent picture. Likewise, from the vantage point of the eleventh dimension, looking down on the tenth dimension, we see the crazy quilt of five superstring theories as nothing more than different patches of the eleventh dimension.

DUALITY

Although Paul Townsend could not answer most of the questions I asked him at that time, what ultimately convinced me of the correctness of this idea was the power of yet another symmetry. Not only does M-theory have the largest set of symmetries known to physics, it has yet another trick up its sleeve: duality, which gives M-theory the uncanny ability to absorb all five superstring theories into one theory.

Consider electricity and magnetism, which are governed by Maxwell's equations. It was noticed long ago that if you simply interchange the electric field with the magnetic field, the equations look almost the same. This symmetry can be made exact if you can add monopoles (single poles of magnetism) into Maxwell's equations. The revised Maxwell's equations remain precisely the same if we exchange the electric field with the magnetic field and interchange the electric charge e with the inverse of the magnetic charge g. This means that electricity (if the electric charge is low) is precisely equivalent to magnetism (if the magnetic charge is high). This equivalence is called duality.

In the past, this duality was considered nothing more than a scientific curiosity, a parlor trick, since no one has ever seen a monopole, even today. However, physicists found it remarkable that Maxwell's equations had a hidden symmetry that nature apparently does not use (at least in our sector of the universe).

Similarly, the five string theories are all dual to each other. Consider type I and the heterotic $SO(32)$ string theory. Normally, these two theories don't even look alike. The type I theory is based on closed and open strings that can interact in five different ways, with strings splitting and joining. The $SO(32)$ string, on the other hand, is based entirely on closed strings that have one possible way of interacting, undergoing mitosis like a cell. The type I string is defined entirely in ten-dimensional space, while the $SO(32)$ string is defined with one set of vibrations defined in twenty-six-dimensional space.

Normally, you cannot find two theories that seem so dissimilar. However, just as in electromagnetism, the theories possess a powerful duality: if you let the strength of the interactions increase, type I strings change into $SO(32)$ heterotic strings, as if by magic. (This result is so unexpected that when I first saw this result, I had to shake my head in amazement. In physics, we rarely see two theories that appear totally dissimilar in all respects being shown to be mathematically equivalent.)

LISA RANDALL

Perhaps the greatest advantage that M-theory has over string theory is that these higher dimensions, instead of being quite small, may actually be quite large and even observable in the laboratory. In string theory, six of the higher dimensions must be wrapped up into a tiny ball, a Calabi-Yau manifold, too small to be observed with today's instruments. These six dimensions have all been compactified, so that entering a higher dimension is impossible—more than a little disappointing to those who would one day hope to soar into an infinite hyperspace rather than merely take brief short-cuts through compactified hyperspace via wormholes.

However, M-theory also features membranes; it is possible to view our entire universe as a membrane floating in a much larger universe. As a result, not all of these higher dimensions have to be wrapped up in a ball. Some of them, in fact, can be huge, infinite in extent.

One physicist who has tried to exploit this new picture of the universe is Lisa Randall of Harvard. Resembling the actress Jodie Foster a bit, Randall seems out of place in the fiercely competitive, testosterone-driven, intensely male profession of theoretical physics. She is pursuing the idea that if the universe is really a three-brane floating in higher-dimensional space, perhaps that explains why gravity is so much weaker than the other three forces.

Randall grew up in Queens, New York (the same borough immortalized by Archie Bunker). While she showed no particular interest in physics as a child, she adored mathematics. Although I believe we are all born scientists as children, not all of us manage to continue our love of science as adults. One reason is that we hit the brick wall of mathematics.

Whether we like it or not, if we are to pursue a career in science, eventually we have to learn the "language of nature": mathematics. Without mathematics, we can only be passive observers to the dance of nature rather than active participants. As Einstein once said, "Pure mathematics is, in its way, the poetry of logical ideas." Let me offer an analogy. One may love French civilization and literature, but to truly understand the French mind, one must learn the French language and how to conjugate French verbs. The same is true of science and mathematics. Galileo once wrote, "[The universe] cannot be read until we have learnt the language and become familiar with the characters in which it is written. It is written in mathematical language, and the letters are triangles, circles, and other geometrical figures, without which means it is humanly impossible to understand a single word."

But mathematicians often pride themselves at being the most impractical of all scientists. The more abstract and useless the mathematics, the better. What set Randall off into a different direction while an undergraduate at Harvard in the early 1980s was the fact that she loved the idea that physics can create "models" of the universe. When we physicists first propose a new theory, it is not simply based on a bunch of equations. New physical theories are usually based on simplified, idealized models which approximate a phenomenon. These models are usually graphic, pictorial, and simple to grasp. The quark model, for example, is based on the idea that within a proton there are three small constituents, the quarks. Randall was impressed that simple models, based on physical pictures, could adequately explain much of the universe.

In the 1990s, she became interested in M-theory, in the possibility that the entire universe was a membrane. She zeroed in on perhaps the most puzzling feature of gravity, that its strength is astronomically small. Neither Newton nor Einstein had addressed this fundamental but mysterious question. While the other three forces of the universe (electromagnetism, the weak nuclear force, and the strong nuclear force) are roughly all of the same strength, gravity is wildly different.

In particular, the masses of the quarks are so much smaller than the mass associated with quantum gravity. "The discrepancy is not small; the two mass scales are separated by sixteen orders of magnitude! Only theories that explain this huge ratio are likely candidates for theories underlying the Standard Model," says Randall.

The fact that gravity is so weak explains why the stars are so big. Earth, with its oceans, mountains, and continents, is nothing but a tiny speck when compared to the massive size of the Sun. But because gravity is so weak, it takes the mass of an entire star to squeeze hydrogen so that it can overcome the proton's electrical force of repulsion. So stars are so massive because gravity is so weak compared to the other forces.

With M-theory generating so much excitement in physics, several groups have tried to apply this theory to our universe. Assume the universe is a three-brane floating in a five-dimensional world. This time, the vibrations on the surface of the three-brane correspond to the atoms we see around us. Thus, these vibrations never leave the three-brane and hence cannot drift off into the fifth dimension. Even though our universe floats in the fifth dimension, our atoms cannot leave our universe because they represent vibrations on the surface of the three-brane. This then can answer the question Kaluza and Einstein asked in 1921: where is the fifth dimension? The answer is: we are floating in the fifth dimension, but we cannot enter it because our bodies are stuck on the surface of a three-brane.

But there is a potential flaw in this picture. Gravity represents the curvature of space. Thus, naively we might expect that gravity can fill up all five-dimensional space, rather than just the three-brane; in doing so, gravity would be diluted as it leaves the three-brane. This weakens the force of gravity. This is a good thing in supporting the theory, because gravity, we know, is so much weaker than the other forces. But it weakens gravity too much: Newton's inverse square law would be violated, yet the inverse square law works perfectly well for planets, stars, and galaxies. Nowhere in space do we find an inverse cube law for gravity. (Imagine a lightbulb illuminating a room. The light spreads out in a sphere. The strength of the light is diluted across this sphere. Thus, if you double the radius of the sphere, then the light is spread out over the sphere with four times the area. In general, if a lightbulb exists in n dimensional space, then its light is diluted across a sphere whose area increases as the radius is raised to the $n - 1$ power.)

To answer this question, a group of physicists, including N. Arkani-Hamed, S. Dimopoulos, and G. Dvali, have suggested that perhaps the fifth dimension is not infinite but is a millimeter away from ours, floating just above our universe, as in H. G. Wells's science fiction story. (If the fifth dimension were farther than a millimeter away, then it might create measurable violations of Newton's inverse square law.) If the fifth dimension is only a millimeter away, this prediction could be tested by looking for tiny deviations to Newton's law of gravity over very small distances. Newton's law of gravity works fine over astronomical distances, but it has never been tested down to the size of a millimeter. Experimentalists are now rushing to test for tiny deviations from Newton's inverse square law. This result is currently the subject of several ongoing experiments, as we see in chapter 9.

Randall and her colleague Raman Sundrum decided to take a new approach, to reexamine the possibility that the fifth dimension was not a millimeter away but perhaps even infinite. To do this, they had to explain how the fifth dimension could be infinite without destroying Newton's law of gravity. This is where Randall found a potential answer to the puzzle. She found that the three-brane has a gravitational pull of its own that prevents gravitons from drifting freely into the fifth dimension. The gravitons have to cling to the three-brane (like flies trapped on flypaper) because of the gravity exerted by the three-brane. Thus, when we try to measure Newton's law, we find that it is approximately correct in our universe. Gravity is diluted and weakened as it leaves the three-brane and drifts into the fifth dimension, but it doesn't get very far: the inverse square law is still roughly maintained because gravitons are still attracted to the three-brane. (Randall also introduced the possibility of a second membrane existing parallel to ours. If we calculate the subtle interaction of gravity across the two membranes, it can be adjusted so that we can numerically explain the weakness of gravity.)

"There was a lot of excitement when it was first suggested that extra dimensions provide alternative ways to address the origin of the [hierarchy problem]," Randall says. "Additional spatial dimensions may seem like a wild and crazy idea at first, but there are powerful reasons to believe that there really are extra dimensions of space."

If these physicists are correct, then gravity is just as strong as the other forces, except that gravity is attenuated because some of it leaks into higher-dimensional space. One profound consequence of this theory is that the energy at which these quantum effects become measurable may not be the Planck energy (10^{19} billion electron volts), as previously thought. Perhaps only trillions of electron volts are necessary, in which case the Large Hadron Collider (scheduled for completion by 2007) may be able to pick up quantum gravitational effects within this decade. This has stimulated considerable interest among experimental physicists to hunt for exotic particles beyond the Standard Model of subatomic particles. Perhaps quantum gravitational effects are just within our reach.

Membranes also give a plausible, though speculative, answer to the riddle of dark matter. In H. G. Wells's novel *The Invisible Man*, the protagonist hovered in the fourth dimension and hence was invisible. Similarly, imagine that there is a parallel world hovering just above our own universe. Any galaxy in that parallel universe would be invisible to us. But because gravity is caused by the bending of hyperspace, gravity could hop across universes. Any large galaxy in that universe would be attracted across hyperspace to a galaxy in our universe. Thus, when we measure the properties of our galaxies, we would find that their gravitational pull was much stronger than expected from Newton's laws because there is another galaxy hiding right behind it, floating on a nearby brane. This hidden galaxy perched behind our galaxy would be totally invisible, floating in another dimension, but it would give the appearance of a halo surrounding our galaxy containing 90 percent of the mass. Thus, dark matter may be caused by the presence of a parallel universe.

COLLIDING UNIVERSES

It may be a bit premature to apply M-theory to serious cosmology. Nonetheless, physicists have tried to apply "brane physics" to make a new twist on the usual inflationary approach to the universe. Three possible cosmologies have attracted some attention.

The first cosmology tries to answer the question: why do we live in four space-time dimensions? In principle, M-theory can be formulated in all dimensions up to eleven, so it seems like a mystery that four dimensions are singled out. Robert Brandenberger and Cumrun Vafa have speculated that this may be due to the particular geometry of strings.

In their scenario, the universe started perfectly symmetrically, with all higher dimensions tightly curled up at the Planck scale. What kept the universe from expanding were loops of strings that tightly coiled around the various dimensions. Think of a compressed coil that cannot expand because it is tightly wrapped by strings. If the strings somehow break, the coil suddenly springs free and expands.

In these tiny dimensions, the universe is prevented from expanding because we have windings of both strings and antistrings (roughly speaking, antistrings wind in the opposite direction from strings). If a string and antistring collide, then they can annihilate and disappear, like the unraveling of a knot. In very large dimensions, there is so much "room" that strings and antistrings rarely collide and never unravel. However, Brandenberger and Vafa showed that in three or fewer spatial dimensions, it is more likely that strings will collide with antistrings. Once these collisions take place, the strings unravel, and the dimensions spring rapidly outward, giving us the big bang. The appealing feature of this picture is that the topology of strings explains roughly why we see the familiar four-dimensional space-time around us. Higher-dimensional universes are possible but less likely to be seen because they are still wrapped up tightly by strings and antistrings.

But there are other possibilities in M-theory as well. If universes can pinch or bud off each other, spawning new universes, then perhaps the reverse can happen: universes can collide, creating sparks in the process, spawning new universes. In such a scenario, perhaps the big bang occurred because of a collision of two parallel brane-universes rather than the budding of a universe.

This second theory was proposed by physicists Paul Steinhardt of Princeton, Burt Ovrut of the University of Pennsylvania, and Neil Turok of Cambridge University, who created the "ekpyrotic" universe (meaning "conflagration" in Greek) to incorporate the novel features of the M-brane picture, in which some of the extra dimensions could be large and even infinite in size. They begin with two flat, homogenous, and parallel three-branes that represent the lowest energy state. Originally, they start as empty, cold universes, but gravity gradually pulls them together. They eventually collide, and the vast kinetic energy of the collision is converted into the matter and radiation making up our universe. Some call this the "big splat" theory rather than the big bang theory, because the scenario involves the collision of two branes.

The force of the collision pushes the two universes apart. As these two membranes separate from each other, they cool rapidly, giving us the universe we see today. The cooling and expansion continue for trillions of years, until the universes approach absolute zero in temperature, and the density is only one electron per quadrillion cubic light-years of space. In effect, the universe becomes empty and inert. But gravity continues to attract the two membranes, until, trillions of years later, they collide once again, and the cycle repeats all over again.

This new scenario is able to obtain the good results of inflation (flatness, uniformity). It solves the question of why the universe is so flat—because the two branes were flat to begin with. The model can also explain the horizon problem—that is, why the universe seems so remarkably uniform in all directions. It is because the membrane has a long time to slowly reach equilibrium. Thus, while inflation explains the horizon problem by having the universe inflate abruptly, this scenario solves the horizon problem in the opposite way, by having the universe reach equilibrium in slow motion.

(This also means that there are possibly other membranes floating out there in hyperspace that may collide with ours in the future, creating another big splat. Given the fact that our universe is accelerating, another collision may in fact be likely. Steinhardt adds, "Maybe the acceleration of the expansion of the universe is a precursor of such a collision. It is not a pleasant thought.")

Any scenario that dramatically challenges the prevailing picture of inflation is bound to elicit heated replies. In fact, within a week of the paper being placed on the Web, Andrei Linde and his wife, Renata Kallosh (herself a string theorist), and Lev Kofman of the University of Toronto issued a critique of this scenario. Linde criticized this model because anything so catastrophic as the collision of two universes might create a singularity, where temperatures and densities approach infinity. "That would be like throwing a chair into a black hole, which would vaporize the particles of the chair, and saying it somehow preserves the shape of the chair," Linde protested.

Steinhardt fired back, saying, "What looks like a singularity in four dimensions may not be one in five dimensions . . . When the branes crunch together, the fifth dimension disappears temporarily, but the branes themselves don't disappear. So the density and temperature don't go to infinity, and time continues right through. Although general relativity goes berserk, string theory does not. And what once looked like a disaster in our model now seems manageable."

Steinhardt has on his side the power of M-theory, which is known to eliminate singularities. In fact, that is the reason theoretical physicists need a quantum theory of gravity to begin with, to eliminate all infinities. Linde, however, points out a conceptual vulnerability of this picture, that the branes existed in a flat, uniform state at the beginning. "If you start with perfection, you might be able to explain what you see . . . but you still haven't answered the question: Why must the universe start out perfect?" Linde says. Steinhardt answers back, "Flat plus flat equals flat." In other words, you have to assume that the membranes started out in the lowest energy state of being flat.

Alan Guth has kept an open mind. "I don't think Paul and Neil come close to proving their case. But their ideas are certainly worth looking at," he says. He turns the tables and challenges string theorists to explain inflation: "In the long run, I think it's inevitable that string theory and M-theory will need to incorporate inflation, since inflation seems to be an obvious solution to the problems it was designed to address—that is, why is the universe so uniform and flat." So he asks the question: can M-theory derive the standard picture of inflation?

Last, there is another competing theory of cosmology that employs string theory, the "pre–big bang" theory of Gabriele Veneziano, the physicist who helped start string theory back in 1968. In his theory, the universe actually started out as a black hole. If we want to know what the inside of a black hole looks like, all we have to do is look outside.

In this theory, the universe is actually infinitely old and started out in the distant past as being nearly empty and cold. Gravity began to create clumps of matter throughout the universe, which gradually condensed into regions so dense that they turned into black holes. Event horizons began to form around each black hole, permanently separating the exterior of the event horizon from the interior. Within each event horizon, matter continued to be compressed by gravity, until the black hole eventually reached the Planck length.

At this point, string theory takes over. The Planck length is the minimum distance allowed by string theory. The black hole then begins to rebound in a huge explosion, causing the big bang. Since this process may repeat itself throughout the universe, this means that there may be other distant black holes/universes.

(The idea that our universe might be a black hole is not as far-fetched as it seems. We have the intuitive notion that a black hole must be extremely dense, with an enormous, crushing gravitational field, but this is not always the case. The size of a black hole's event horizon is proportional to its mass. The more massive a black hole is, the larger its event horizon. But a larger event horizon means that matter is spread out over a larger volume; as a result, the density actually decreases as the mass increases. In fact, if a black hole were to weigh as much as our universe, its size would be approximately the size of our universe, and its density would be quite low, comparable to the density of our universe.)

Some astrophysicists, however, are not impressed with the application of string theory and M-theory to cosmology. Joel Primack of the University of California at Santa Cruz is less charitable than others: "I think it's silly to make much of a production about this stuff . . . The ideas in these papers are essentially untestable." Only time will tell if Primack is right, but because the pace of string theory has been accelerating, we may find a resolution of this problem soon, and it may come from our space satellites. As we see in chapter 9, a new generation of gravity wave detectors to be sent into outer space by 2020, like LISA, may give us the ability to rule out or verify some of these theories. If the inflation theory is correct, LISA might possibly be able to detect violent gravity waves created by the original inflationary process. The ekpyrotic universe, however, predicts a slow collision between universes and hence much milder gravity waves. LISA might be able to rule out one of these theories experimentally. In other words, encoded within gravity waves created by the original big bang are the data necessary to determine which scenario is correct. LISA (or its successors) may give solid experimental results concerning inflation, string theory, and M-theory.

MINI–BLACK HOLES

Since string theory is really a theory of the entire universe, to test it directly requires creating a universe in the laboratory (see chapter 9). Normally, we expect quantum effects from gravity to occur at the Planck energy, which is a quadrillion times more powerful than our most powerful particle accelerator, making direct tests of string theory impossible. But if there really is a parallel universe that exists less than a millimeter from ours, then the energy at which unification and quantum effects occur may be quite low, within reach of the next generation of particle accelerators, such as the Large Hadron Collider (LHC). This, in turn, has sparked an avalanche of interest in black hole physics, the most exciting being the "mini–black hole." Mini–black holes, which act as if they are subatomic particles, are a "laboratory" in which one can test some of the predictions of string theory. Physicists are excited about the possibility of creating them with the LHC. (Mini–black holes are so small, comparable to an electron in size, that there is no threat that they will swallow up Earth. Cosmic rays routinely hit Earth with energies exceeding these mini–black holes, with no ill effect on the planet.)

As revolutionary as it may seem, a black hole masquerading as a subatomic particle is actually an old idea, first introduced by

Einstein in 1935. In Einstein's view, there must be a unified field theory in which matter, made of subatomic particles, could be viewed as some sort of distortion in the fabric of space-time. To him, subatomic particles like the electron were actually "kinks" or wormholes in curved space that, from a distance, looked like a particle. Einstein, with Nathan Rosen, toyed with the idea that an electron may actually be a mini–black hole in disguise. In his way, he tried to incorporate matter into this unified field theory, which would reduce subatomic particles to pure geometry.

Mini–black holes were introduced again by Stephen Hawking, who proved that black holes must evaporate and emit a faint glow of energy. Over many eons, a black hole would emit so much energy that it would gradually shrink, eventually becoming the size of a subatomic particle.

String theory is now reintroducing the concept of mini–black holes. Recall that black holes form when a large amount of matter is compressed to within its Schwarzschild radius. Because mass and energy can be converted into each other, black holes can also be created by compressing energy. There is considerable interest in whether the LHC may be able to produce mini–black holes among the debris created by smashing two protons together at 14 trillion electron volts of energy. These black holes would be very tiny, weighing perhaps only a thousand times the mass of an electron, and last for only 10^{-23} seconds. But they would be clearly visible among the tracks of subatomic particles created by the LHC.

Physicists also hope that cosmic rays from outer space may contain mini–black holes. The Pierre Auger Cosmic Ray Observatory in Argentina is so sensitive that it can detect some of the largest bursts of cosmic rays ever recorded by science. The hope is that mini–black holes may be found naturally among cosmic rays, which would create a characteristic shower of radiation when they hit Earth's upper atmosphere. One calculation shows that the Auger Cosmic Ray detector might be able to see up to ten cosmic ray showers per year triggered by a mini–black hole.

The detection of a mini–black hole either at the LHC in Switzerland or the Auger Cosmic Ray detector in Argentina, perhaps within this decade, would provide perhaps good evidence for the existence of parallel universes. Although it would not conclusively prove the correctness of string theory, it would convince the entire physics community that string theory is consistent with all experimental results and is in the right direction.

BLACK HOLES AND THE INFORMATION PARADOX

String theory may also shed light on some of the deepest paradoxes of black hole physics, such as the information paradox. As you will recall, black holes are not perfectly black but emit small amounts of radiation via tunneling. Because of the quantum theory, there is always the small chance that radiation can escape the viselike grip of a black hole's gravity. This leads to a slow leakage of radiation from a black hole, called Hawking radiation.

This radiation, in turn, has a temperature associated with it (which is proportional to the surface area of the black hole's event horizon). Hawking gave a general derivation of this equation that involved a lot of hand-waving. However, a rigorous derivation of this result would require using the full power of statistical mechanics (based on counting the quantum states of a black hole). Usually, statistical mechanical calculations are done by counting the number of states that an atom or molecule can occupy. But how do you count the quantum states of a black hole? In Einstein's theory, black holes are perfectly smooth, so counting their quantum states was problematic.

String theorists were anxious to close this gap, so Andrew Strominger and Cumrum Vafa of Harvard decided to analyze a black hole using M-theory. Since the black hole itself was too difficult to work with, they took a different approach and asked a clever question: what is the dual to a black hole? (We recall that an electron is dual to a magnetic monopole, such as a single north pole. Hence, by examining an electron in a weak electric field, which is easy to do, we can analyze a much more difficult experiment: a monopole placed in a very large magnetic field.) The hope was that the dual of the black hole would be easier to analyze than the black hole itself, although they would ultimately have the same final result. By a series of mathematical manipulations, Strominger and Vafa were able to show that the black hole was dual to a collection of one-branes and five-branes. This was a tremendous relief, since counting the quantum states of these branes was known. When Strominger and Vafa then calculated the number of quantum states, they found that the answer precisely reproduced Hawking's result.

This was a piece of welcome news. String theory, which is sometimes ridiculed for not connecting with the real world, gave perhaps the most elegant solution for black hole thermodynamics.

Now, string theorists are trying to tackle the most difficult problem in black hole physics, the "information paradox." Hawking has argued that if you throw something into a black hole, the information it carries is lost forever, never to return again. (This would be a clever way to commit the perfect crime. A criminal could use a black hole to destroy all incriminating evidence.) From a distance, the only parameters that we can measure for a black hole are its mass, spin, and charge. No matter what you throw into a black hole, you lose all its information. (This goes by the statement that "black holes have no hair"—that is, they have lost all information, all hair, except for these three parameters.)

The loss of information from our universe seems to be an inevitable consequence of Einstein's theory, but this violates the principles of quantum mechanics, which state that information can never really be lost. Somewhere, the information must be floating in our universe, even if the original object was sent down the throat of a black hole.

"Most physicists want to believe that information is not lost," Hawking has written, "as this would make the world safe and predictable. But I believe that if one takes Einstein's general relativity seriously, one must allow for the possibility that spacetime ties itself in knots and that information gets lost in the folds. Determining whether or not information actually does get lost is one of the major questions in theoretical physics today."

This paradox, which pits Hawking against most string theorists, still has not been resolved. But the betting among string theorists is that we will eventually find where the missing information went. (For example, if you throw a book into a black hole, it is conceivable that the information contained in the book will gently seep back out into our universe in the form of tiny vibrations contained within the Hawking radiation of an evaporating black hole. Or perhaps it reemerges from a white hole on the other side of the black hole.) That is why I personally feel that when someone finally calculates what happens to information when it disappears into a black hole in string theory, he or she will find that information is not really lost but subtly reappears somewhere else.

In 2004, in a stunning reversal, Hawking made the front page of the *New York Times* when he announced before TV cameras that he was wrong about the information problem. (Thirty years ago, he bet other physicists that information could never leak out of a black hole. The loser of the bet was to give the winner an encyclopedia, from which information can be easily retrieved.) Redoing some of his earlier calculations, he concluded that if an object such as a book fell into a black hole, it might disturb the radiation field it emits, allowing information to leak back into the universe. The information contained within the book would be encoded in the radiation slowly seeping out of the black hole, but in mangled form.

On one hand, this put Hawking in line with the majority of quantum physicists, who believe that information cannot be lost. But it also raised the question: can information pass to a parallel universe? On the surface, his result seemed to cast doubt on the idea that information may pass through a wormhole into a parallel universe. However, no one believes that this is the last word on the subject. Until string theory is fully developed, or a complete quantum gravitational calculation is made, no one will believe that the information paradox is fully resolved.

THE HOLOGRAPHIC UNIVERSE

Last, there is a rather mysterious prediction of M-theory that is still not understood but may have deep physical and philosophical consequences. This result forces us to ask the question: is the universe a hologram? Is there a "shadow universe" in which our bodies exist in a compressed two-dimensional form? This also raises another, equally disturbing question: is the universe a computer program? Can the universe be placed on a CD, to be played at our leisure?

Holograms are now found on credit cards, in children's museums, and in amusement parks. They are remarkable because they can capture a complete three-dimensional image on a two-dimensional surface. Normally, if you glance at a photograph and then move your head, the image on the photograph does not change. But a hologram is different. When you glance at a holographic picture and then move your head, you find the picture changing, as if you were looking at the image through a window or a keyhole. (Holograms may eventually lead to three-dimensional TV and movies. In the future, perhaps we will relax in our living room and glance at a wall screen that gives us the complete three-dimensional image of distant locations, as if the TV wall screen were actually a window peering out over a new landscape. Furthermore, if the wall screen were shaped like a large cylinder with our living room placed in the center, it would appear as if we were transported to a new world. Everywhere we looked, we would see the three-dimensional image of a new reality, indistinguishable from the real thing.)

The essence of the hologram is that the two-dimensional surface of the hologram encodes all the information necessary to reproduce a three-dimensional image. (Holograms are made in the laboratory by shining laser light onto a sensitive photographic plate and allowing the light to interfere with laser light from the original source. The interference of the two light sources creates an interference pattern that "freezes" the image onto the two-dimensional plate.)

Some cosmologists have conjectured that this may also apply to the universe itself—that perhaps we *live* in a hologram. The origin of this strange speculation arises from black hole physics. Bekenstein and Hawking conjecture that the total amount of information contained in a black hole is proportional to the surface area of its event horizon (which is a sphere). This is a strange result, because usually the information stored in an object is proportional to its volume. For example, the amount of information stored in a book is proportional to its size, not to the surface area of its cover. We know this instinctively, when we say that we cannot judge a book by its cover. But this intuition fails for black holes: we can completely judge a black hole by its cover.

We may dismiss this curious hypothesis because black holes are strange oddities in themselves, where normal intuition breaks down. However, this result also applies to M-theory, which may give us the best description of the entire universe. In 1997, Juan Maldacena, at the Institute for Advanced Study at Princeton, created quite a sensation when he showed that string theory leads to a new type of holographic universe.

He started with a five-dimensional "anti–de Sitter universe" which often appears in string theory and supergravity theory. A de Sitter universe is one with a positive cosmological constant that creates an accelerating universe. (We recall that our universe is currently best represented as a de Sitter universe, with a cosmological constant pushing the galaxies away at faster and faster velocities. An anti–de Sitter universe has a negative cosmological constant and hence can implode.) Maldacena showed that there is a duality between this five-dimensional universe and its "boundary," which is a four-dimensional universe. Strangely enough, any beings living in this five-dimensional space would be mathematically equivalent to beings living in this four-dimensional space. There is no way to tell them apart.

By crude analogy, think of fish swimming inside a goldfish bowl. These fish think that their fish bowl corresponds to reality. Now imagine a two-dimensional holographic image of these fish that is projected onto the surface of the fish bowl. This image contains an exact replica of the original fish, except they are flattened. Any movement the fish make in the fish bowl is mirrored by the flat image on the surface of the fish bowl. Both the fish swimming in the bowl and the flattened fish living on the surface of the bowl think that they are the real fish, that the other is an illusion. Both fish are alive and act as if they are the true fish. Which description is correct? Actually, both are, since they are mathematically equivalent and indistinguishable.

What excited string theorists is the fact that five-dimensional anti–de Sitter space is relatively easy to calculate with, while four-dimensional field theories are notoriously difficult to handle. (Even today, after decades of hard work, our most powerful computers cannot solve the four-dimensional quark model and derive the masses of the proton and neutron. The equations for the quarks themselves are fairly well understood, but solving them in four dimensions to obtain the properties of protons and neutrons has proved to be more difficult than previously thought.) One goal is to calculate the masses and properties of the proton and neutron, using this strange duality.

This holographic duality may also have practical applications, such as solving the information problem in black hole physics. In four dimensions, it is extremely difficult to prove that information isn't lost when we throw objects through a black hole. But such a space is dual to a five-dimensional world, in which information is perhaps never lost. The hope is that problems that are intractable in four dimensions (such as the information problem, calculating the masses of the quark model, and so forth) may eventually be solved in five dimensions, where the mathematics is simpler. And it is always possible that this analogy is actually a reflection of the real world—that we really exist as holograms.

IS THE UNIVERSE A COMPUTER PROGRAM?

John Wheeler, as we saw earlier, believed that all physical reality could be reduced to pure information. Bekenstein takes the idea of black hole information one step further into uncharted waters by asking the question: is the entire universe a computer program? Are we just bits on a cosmic CD?

The question of whether we are living in a computer program was brought brilliantly to the silver screen in the movie *The Matrix*, where aliens have reduced all physical reality to a computer program. Billions of humans think that they are leading everyday lives, oblivious of the fact that all this is a computer-generated fantasy, while their real bodies are asleep in pods, where the aliens use them as a power source.

In the movie, it is possible to run smaller computer programs that can create artificial minirealities. If one wants to become a kung fu master or a helicopter pilot, one just inserts a CD into a computer, the program is fed into our brain, and presto! one instantly learns these complicated skills. As the CD is run, a whole new subreality is created. But it raises an intriguing question: can all of reality be placed on a CD? The computer power necessary to simulate reality for billions of sleeping humans is truly staggering. But in theory: can the entire universe be digitalized in a finite computer program?

The roots of this question go back to Newton's laws of motion, with very practical applications for commerce and our lives. Mark Twain was famous for stating, "Everyone complains about the weather, but no one ever does anything about it." Modern civilization cannot change the course of even a single thunderstorm, but physicists have asked a more modest question: can we predict the weather? Can a computer program be devised that will predict the course of complex weather patterns on Earth? This has very practical applications for everyone concerned about the weather, from farmers wanting to know when to harvest their crops to meteorologists wanting to know the course of global warming in this century.

In principle, computers can use Newton's laws of motion to compute with almost arbitrary accuracy the course of molecules that make up the weather. But in practice, computer programs are extremely crude and are not reliable at predicting the weather beyond a few days or so, at best. To predict the weather, one would need to determine the motion of every air molecule—something that is magnitudes beyond our most powerful computer; there is also the problem of chaos theory and the "butterfly effect," where even the tiniest vibration from a butterfly's wing can cause a ripple effect that, at key junctures, may decisively change the weather hundreds of miles away.

Mathematicians summarize this situation by stating that the smallest model that can accurately describe the weather is the weather itself. Rather than microanalyzing each molecule, the best we can do is to look for estimates of tomorrow's weather and also larger trends and patterns (such as the greenhouse effect).

So it is exceedingly difficult for a Newtonian world to be reduced to a computer program, since there are too many variables and too many "butterflies." But in the quantum world, strange things happen.

Bekenstein, as we saw, showed that the total information content of a black hole is proportional to the surface area of its event horizon. There is an intuitive way of seeing this. Many physicists believe that the smallest possible distance is the Planck length of 10^{-33} cm. At this incredibly small distance, space-time is no longer smooth but becomes "foamy," resembling a froth of bubbles. We can divide up the spherical surface of the horizon into tiny squares, each one the size of the Planck length. If each of these squares contains one bit of information, and we add up all the squares, we find roughly the total information content of the black hole. This seems to indicate that each of these "Planck squares" is the smallest unit of information. If this is true, then Bekenstein claims that perhaps information is the true language of physics, not field theory. As he puts it, "Field theory, with its infinity, cannot be the final story."

Ever since the work of Michael Faraday in the nineteenth century, physics has been formulated in the language of fields, which are smooth and continuous, and which measure the strength of magnetism, electricity, gravity, and so on at any point in space-time. But field theory is based on continuous structures, not digitalized ones. A field can occupy any value, while a digitalized number can only represent discrete numbers based on 0s and 1s. This is the difference, for example, between a smooth rubber sheet found in Einstein's theory and a fine wire mesh. The rubber sheet can be divided up into an infinite number of points, while a wire mesh has a smallest distance, the mesh length.

Bekenstein suggests that "a final theory must be concerned not with fields, not even with spacetime, but rather with information exchange among physical processes."

If the universe can be digitalized and reduced to 0s and 1s, then what is the total information content of the universe? Bekenstein estimates that a black hole about a centimeter across could contain 10^{66} bits of information. But if an object a centimeter in size can hold that many bits of information, then he estimates that the visible universe probably contains much more information, no less than 10^{100} bits of information (which can in principle be squeezed into a sphere a tenth of a light-year across. This colossal number, 1 followed by 100 zeros, is called a google.)

If this picture is correct, we have a strange situation. It might mean that while a Newtonian world cannot be simulated by computers (or can only be simulated by a system as large as itself), in a quantum world, perhaps the universe itself *can* be put onto a CD! In theory, if we can put 10^{100} bits of information on a CD, we can watch any event in our universe unfold in our living room. In principle, one could arrange or reprogram the bits on this CD, so that physical reality proceeds in a different fashion. In some sense, one would have a God-like ability to rewrite the script.

(Bekenstein also admits that the total information content of the universe could be much larger than that. In fact, the smallest volume that can contain the information of the universe may be the size of the universe itself. If this is true, then we are back to where we started: the smallest system that can model the universe is the universe itself.)

String theory, however, offers a slightly different interpretation of the "smallest distance" and whether we can digitalize the universe on a CD. M-theory possesses what is called T-duality. Recall that the Greek philosopher Zeno thought that a line could be divided into an infinite number of points, without limit. Today, quantum physicists like Bekenstein believe that the smallest distance may be the Planck distance of 10^{-33} centimeters, where the fabric of space-time becomes foamy and bubbly. But M-theory gives us a new twist to this. Let's say that we take a string theory and wrap up one dimension into a circle of radius R. Then we take another string and wrap up one dimension into a circle of radius $1/R$. By comparing these two quite different theories, we find that they are exactly the same.

Now let R become extremely small, much smaller than the Planck length. This means that the physics within the Planck length is identical to the physics outside the Planck length. At the Planck length, space-time may become lumpy and foamy, but the physics inside the Planck length and the physics at very large distances can be smooth and in fact are identical.

This duality was first found in 1984 by my old colleague Keiji Kikkawa and his student Masami Yamasaki, of Osaka University. Although string theory apparently concludes that there is a "smallest distance," the Planck length, physics does not abruptly end at the Planck length. The new twist is that physics smaller than the Planck length is equivalent to physics larger than the Planck length.

If this rather topsy-turvy interpretation is correct, then it means that even within the "smallest distance" of string theory, an entire universe can exist. In other words, we can still use field theory, with its continuous (not digitalized) structures to describe the universe even to distances well inside the Planck energy. So perhaps the universe is not a computer program at all. In any event, since this is a well-defined problem, time will tell.

(This T-duality is the justification for the "pre–big bang" scenario of Veneziano I mentioned earlier. In that model, a black hole collapses down to the Planck length and then "bounces" back into the big bang. This bounce is not an abrupt event but the smooth T-duality between a black hole smaller than the Planck length and an expanding universe larger than the Planck length.)

THE END?

If M-theory is successful, if it is indeed a theory of everything, is it the end of physics as we know it?

The answer is no. Let me give you an example. Even if we know the rules of chess, knowing the rules does not make us a grand master. Similarly, knowing the laws of the universe does not mean that we are grand masters in terms of understanding its rich variety of solutions.

Personally, I think it still might be a bit premature to apply M-theory to cosmology, although it gives a startling new picture of the way the universe might have begun. The main problem, I think, is that the model is not in its final form. M-theory may very well be the theory of everything, but I believe that it is far from finished. The theory has been evolving backward since 1968, and its final equations have still not been found. (For example, string theory can be formulated via string field theory, as Kikkawa and I showed years ago. The counterpart of these equations for M-theory is unknown.)

Several problems confront M-theory. One is that physicists are now drowning in p-branes. A series of papers has been written trying to catalog the bewildering variety of membranes that can exist in different dimensions. There are membranes shaped like a doughnut with a hole, a doughnut with multiple holes, intersecting membranes, and so forth.

One is reminded of what happens when the fabled blind wise men confront an elephant. Touching the elephant in different places, each comes up with his own theory. One wise man, touching the tail, says that the elephant is a one-brane (a string). Another wise man, touching the ear, says that the elephant is a two-brane (a membrane). Finally, the last says that the other two wise men are wrong. Touching the legs, which feel like tree trunks, the third wise man says that the elephant is really a three-brane. Because they are blind, they cannot see the big picture, that the sum total of a one-brane, two-brane, and three-brane is nothing but a single animal, an elephant.

Similarly, it's hard to believe that the hundreds of membranes found in M-theory are somehow fundamental. At present, we have no comprehensive understanding of M-theory. My own point of view, which has guided my current research, is that these membranes and strings represent the "condensation" of space. Einstein tried to describe matter in purely geometrical terms, as some kind of kink in the fabric of space-time. If we have a bed sheet, for example, and a kink develops, the kink acts as if it has a life of its own. Einstein tried to model the electron and other elementary particles as some kind of disturbance in the geometry of space-time. Although he ultimately failed, this idea may be resurrected on a much higher level in M-theory.

I believe Einstein was on the right track. His idea was to generate subatomic physics via geometry. Instead of trying to find a geometric analog to point particles, which was Einstein's strategy, one could revise it and try to construct a geometric analog of strings and membranes made of pure space-time.

One way to see the logic of this approach is to look at physics historically. In the past, whenever physicists were confronted with a spectrum of objects, we realized that there was something more fundamental at the root. For example, when we discovered the spectral lines emitted from hydrogen gas, we eventually realized that they originated from the atom, from quantum leaps made by the electron as it circled the nucleus. Similarly, when confronted by the proliferation of strong particles in the 1950s, physicists eventually realized that they were nothing but bound states of quarks. And when confronted with the proliferation of quarks and other "elementary" particles of the Standard Model, most physicists now believe that they arise out of vibrations of the string.

With M-theory, we are confronted with the proliferation of p-branes of all type and varieties. It's hard to believe that these can be fundamental, because there are simply too many p-branes, and because they are inherently unstable and divergent. A simpler solution, which agrees with the historical approach, is to assume that M-theory originates from an even simpler paradigm, perhaps geometry itself.

In order to settle this fundamental question, we need to know the physical principle underlying the theory, not just its arcane mathematics. As physicist Brian Greene says, "Currently, string theorists are in a position analogous to an Einstein bereft of the equivalence principle. Since Veneziano's insightful guess in 1968, the theory has been pieced together, discovery by discovery, revolution by revolution. But a central organizing principle that embraces these discoveries and all other features of the theory within one overarching and systematic framework—a framework that makes the existence of each individual ingredient absolutely inevitable—is still missing. The discovery of this principle would mark a pivotal moment in the development of string theory, as it would likely expose the theory's inner workings with unforeseen clarity."

It would also make sense of the millions of solutions so far found for string theory, each one representing a fully self-consistent universe. In the past, it was thought that, of this forest of solutions, only one represented the true solution of string theory. Today, our thinking is shifting. So far, there is no way to select out one universe out of the millions that have been discovered so far. There is a growing body of opinion that states that if we cannot find the unique solution to string theory, it's probably because there is none. All solutions are equal. There is a multiverse of universes, each one consistent with all the laws of physics. This then leads us to what is called the anthropic principle and the possibility of a "designer universe."

CHAPTER EIGHT

A Designer Universe?

> Numerous universes might have been botched and bun-
> gled throughout an eternity, ere this system was struck
> out; much labor lost, many fruitless trials made, and a
> slow but continual improvement carried out during infi-
> nite ages in the art of world-making.
>
> —David Hume

WHEN I WAS A CHILD in second grade, my teacher made a casual
remark that I will never forget. She said, "God so loved the
earth, that He put the earth just right from the sun." As a child of
six, I was shocked by the simplicity and power of this argument. If
God had put Earth too far from the Sun, then the oceans would have
frozen. If He had put Earth too close, then the oceans would have
boiled off. To her, this meant that not only did God exist, but that He
was also benevolent, so loving Earth that He put it just right from
the Sun. It made a deep impact on me.

Today, scientists say that Earth lives in the "Goldilocks zone"
from the Sun, just far enough so that liquid water, the "universal sol-
vent," can exist to create the chemicals of life. If Earth were farther
from the Sun, it might become like Mars, a "frozen desert," where
temperatures have created a harsh, barren surface where water and
even carbon dioxide are often frozen solid. Even beneath the soil of
Mars one finds permafrost, a permanent layer of frozen water.

If Earth were closer to the Sun, then it might become more like the planet Venus, which is nearly identical to Earth in size but is known as the "greenhouse planet." Because Venus is so close to the Sun, and its atmosphere is made of carbon dioxide, the energy of sunlight is captured by Venus, sending temperatures soaring to 900 degrees Fahrenheit. Because of this, Venus is the hottest planet, on average, in the solar system. With rains of sulfuric acid, atmospheric pressures a hundred times greater than those found on Earth, and scorching temperatures, Venus is perhaps the most hellish planet in the solar system, largely because it is closer to the Sun than is Earth.

Analyzing my second grade teacher's argument, scientists would say that her statement is an example of the anthropic principle, which states that the laws of nature are arranged so that life and consciousness are possible. Whether these laws are arranged by some greater design or by accident has been the subject of much debate, especially in recent years, because of the overwhelming number of "accidents" or coincidences that have been found which make life and consciousness possible. To some, this is evidence of a deity who has deliberately arranged the laws of nature to make life, and us, possible. But to other scientists, it means we are the by-products of a series of lucky accidents. Or perhaps, if one believes the ramifications of inflation and M-theory, there is a multiverse of universes.

To appreciate the complexity of these arguments, consider first the coincidences that make life on Earth possible. We live not just within the Goldilocks zone of the Sun, we also live within a series of other Goldilocks zones. For example, our Moon is just the right size to stabilize Earth's orbit. If the Moon were much smaller, even tiny perturbations in Earth's spin would slowly accumulate over hundreds of millions of years, causing Earth to wobble disastrously and creating drastic changes in the climate so as to make life impossible. Computer programs show that without a large Moon (about a third the size of Earth), Earth's axis might have shifted by as much as 90 degrees over a period of many millions of years. Since scientists believe the creation of DNA required hundreds of millions of years of climactic stability, an Earth that periodically tips on its axis would create catastrophic changes in the weather, making the creation of

DNA impossible. Fortunately, our Moon is "just right" in size to stabilize the orbit of Earth, so that such a disaster will not happen. (The moons of Mars are not large enough to stabilize its spin. As a result, Mars is slowly beginning to enter another era of instability. In the past, astronomers believe, Mars might have wobbled on its axis by as much as 45 degrees.)

Due to small tidal forces, the Moon is also moving away from Earth at the rate of about 4 centimeters per year; in about 2 billion years, it will be too far to stabilize Earth's spin. This could be disastrous for life on Earth. Billions of years from now, not only will the night sky be moonless, we might see an entirely different set of constellations, as Earth tumbles in its orbit. The weather on Earth will become unrecognizable, making life impossible.

Geologist Peter Ward and astronomer Donald Brownlee of the University of Washington write, "Without the Moon there would be no moonbeams, no month, no lunacy, no Apollo program, less poetry, and a world where every night was dark and gloomy. Without the Moon it is also likely that no birds, redwoods, whales, trilobite, or other advanced life would ever grace the earth."

Similarly, computer models of our solar system show that the presence of the planet Jupiter in our solar system is a fortuitous one for life on Earth, because its immense gravity helps to fling asteroids into outer space. It took almost a billion years, during the "age of meteors," which extended from 3.5 billion to 4.5 billion years ago, to "clean out" our solar system of the debris of asteroids and comets left over from its creation. If Jupiter were much smaller and its gravity much weaker, then our solar system would still be full of asteroids, making life on Earth impossible, as asteroids plunged into our oceans and destroyed life. Hence, Jupiter, too, is just the right size.

We also live in the Goldilocks zone of planetary masses. If Earth were a bit smaller, its gravity would be so weak that it could not keep its oxygen. If it were too large, it would retain many of its primordial, poisonous gases, making life impossible. Earth has "just the right" weight to keep an atmospheric composition beneficial to life.

We also live in the Goldilocks zone of permissible planetary orbits. Remarkably, the orbits of the other planets, except for Pluto, are all nearly circular, meaning that planetary impacts are quite rare in the solar system. This means that Earth won't come close to any gas giants, whose gravity could easily disrupt Earth's orbit. This is again good for life, which requires hundreds of millions of years of stability.

Likewise, Earth also exists within the Goldilocks zone of the Milky Way galaxy, about two-thirds of the way from the center. If the solar system were too close to the galactic center, where a black hole lurks, the radiation field would be so intense that life would be impossible. And if the solar system were too far away, there would not be enough higher elements to create the necessary elements of life.

Scientists can provide scores of examples where Earth lies within myriad Goldilocks zones. Astronomers Ward and Brownlee argue that we live within so many narrow bands or Goldilocks zones that perhaps intelligent life on earth is unique to the galaxy, maybe even to the universe. They recite a remarkable list of ways that Earth has "just the right" amount of oceans, plate tectonics, oxygen content, heat content, tilt of its axis, and so on to create intelligent life. If Earth were outside just one these very narrow bands, we would not be here to discuss the question.

Was Earth placed in the middle of all these Goldilocks zones because God loved it? Perhaps. We can, however, arrive at a conclusion that does not rely on a deity. Perhaps there are millions of dead planets in space that *are* too close to their suns, whose moons are too small, whose Jupiters are too small, or that are too close to their galactic center. The existence of Goldilocks zones with respect to Earth does not necessarily mean that God has bestowed a special blessing on us; it might simply be a coincidence, one rare example among millions of dead planets in space that lie outside Goldilocks zones.

The Greek philosopher Democritus, who hypothesized the existence of atoms, wrote, "There are worlds infinite in number and different in size. In some there is neither sun nor moon. In others, there are more than one sun and moon. The distances between the worlds are unequal, in some directions there are more of them . . .

Their destruction comes about through collision with one another. Some worlds are destitute of animal and plant life and of all moisture."

By 2002, in fact, astronomers had discovered one hundred extrasolar planets that were orbiting other stars. Extrasolar planets are being discovered at the rate of one every two weeks or so. Since extrasolar planets do not give off any light of their own, astronomers identify them via various indirect means. The most reliable is to look for the wobbling of the mother star, which moves back and forth as its Jupiter-sized planet circles around it. By analyzing the Doppler shift of the light emitted from the wobbling star, one can calculate how fast it is moving and use Newton's laws to calculate the mass of its planet.

"You can think of the star and the large planet as dance partners, spinning around while clasping their outstretched hands. The smaller partner on the outside is moving greater distances in a larger circle, while the larger inside partner only moves his or her feet in a very small circle—the movement around the very small inner circle is the 'wobble' that we see in these stars," says Chris McCarthy of the Carnegie Institution. This process is now so accurate that we can detect tiny variations in velocity of 3 meters per second (the speed of a brisk walk) in a star hundreds of light-years away.

Other, more ingenious methods are being proposed to find even more planets. One is to look for a planet when it eclipses the mother star, which leads to a slight decrease in its brightness as the planet passes in front of the star. And within fifteen to twenty years, NASA will send its interferometry space satellite into orbit, which will be able to find smaller, Earth-like planets in outer space. (Since the brightness of the mother star overwhelms the planet, this satellite will use light interference to cancel out the mother star's intense halo, leaving the Earth-like planet unobscured.)

So far, none of the Jupiter-sized extrasolar planets we've discovered resembles our Earth, and all are probably dead. Astronomers have discovered them in highly eccentric orbits or in orbits extremely close to their mother star; in either case, an Earth-like planet within a Goldilocks zone will be impossible. In these solar systems, the Jupiter-sized planet would cross the Goldilocks zone and fling any small Earth-sized planet into outer space, preventing life as we know it from forming.

Highly eccentric orbits are common in space—so common, in fact, that when a "normal" solar system was discovered in space, it made headlines in 2003. Astronomers in the United States and Australia alike heralded the discovery of a Jupiter-sized planet orbiting the star HD 70642. What was so unusual about this planet (about twice the size of our Jupiter) was that it was in a circular orbit in roughly the same ratio as Jupiter is to our sun.

In the future, however, astronomers should be able to catalog all the nearby stars for potential solar systems. "We are working to place all 2,000 of the nearest sun-like stars under survey, all the sun-like stars out to 150 light-years," says Paul Butler of the Carnegie Institution of Washington, who was involved in the first discovery of an extrasolar planet in 1995. "Our goal is two-fold—to provide a reconnaissance—a first census—of our nearest neighbors in space, and to provide the first data to address the fundamental question, how common or how rare is our own solar system," he says.

COSMIC ACCIDENTS

In order to create life, our planet must have been relatively stable for hundreds of millions of years. But a world that is stable for hundreds of millions of years is astonishingly difficult to make.

Start with the way atoms are made, with the fact that a proton weighs slightly less than a neutron. This means that neutrons eventually decay into protons, which occupy a lower energy state. If the proton were just 1 percent heavier, it would decay into a neutron, and all nuclei would become unstable and disintegrate. Atoms would fly apart, making life impossible.

Another cosmic accident that makes life possible is that the proton is stable and does not decay into an antielectron. Experiments have shown that the proton lifetime is truly astronomical, much longer than the lifetime of the universe. For the purpose of creating stable DNA, protons must be stable for at least hundreds of millions of years.

If the strong nuclear force were a bit weaker, nuclei like deuterium would fly apart, and none of the elements of the universe could have been successively built up in the interior of stars via nucleosynthesis. If the nuclear force were a bit stronger, stars would burn their nuclear fuel too quickly, and life could not evolve.

If we vary the strength of the weak force, we also find that life once again is impossible. Neutrinos, which act via the weak nuclear force, are crucial to carry the energy outward from an exploding supernova. This energy, in turn, is responsible for the creation of the higher elements beyond iron. If the weak force were a bit weaker, neutrinos would interact hardly at all, meaning that supernovae could not create the elements beyond iron. If the weak force were a bit stronger, neutrinos might not escape properly from a star's core, again preventing the creation of the higher elements that make up our bodies and our world.

Scientists have, in fact, assembled long lists of scores of such "happy cosmic accidents." When faced with this imposing list, it's shocking to find how many of the familiar constants of the universe lie within a very narrow band that makes life possible. If a single one of these accidents were altered, stars would never form, the universe would fly apart, DNA would not exist, life as we know it would be impossible, Earth would flip over or freeze, and so on.

Astronomer Hugh Ross, to emphasize how truly remarkable this situation is, has compared it to a Boeing 747 aircraft being completely assembled as a result of a tornado striking a junkyard.

THE ANTHROPIC PRINCIPLE

Again, all the arguments presented above are lumped together under the anthropic principle. There are several points of view one can take concerning this controversial principle. My second-grade teacher felt that these happy coincidences implied the existence of a grand design or plan. As physicist Freeman Dyson once said, "It's as if the universe knew we were coming." This is an example of the strong anthropic principle, the idea that the fine-tuning of the physical constants was not an accident but implies a design of some sort. (The weak anthropic principle simply states that the physical constants of the universe are such that they make life and consciousness possible.)

Physicist Don Page has summarized the various forms of the anthropic principle that have been proposed over the years:

weak anthropic principle: "What we observe about the universe is restricted by the requirement of our existence as observers."

strong-weak anthropic principle: "In at least one world . . . of the many-worlds universe, life must develop."

strong anthropic principle: "The universe must have the properties for life to develop at some time within it."

final anthropic principle: "Intelligence must develop within the universe and then never die out."

One physicist who takes the strong anthropic principle seriously, and claims that it is a sign of a God, is Vera Kistiakowsky, a physicist at MIT. She says, "The exquisite order displayed by our scientific understanding of the physical world calls for the divine." A scientist who seconds that opinion is John Polkinghorne, a particle physicist who gave up his position at Cambridge University and became a priest of the Church of England. He writes that the universe is "not just 'any old world,' but it's special and finely tuned for life because it is the creation of a Creator who wills that it should be so." Indeed, Isaac Newton himself, who introduced the concept of immutable laws which guided the planets and stars without divine intervention, believed that the elegance of these laws pointed to the existence of God.

But the physicist and Nobel laureate Steven Weinberg is not convinced. He acknowledges the appeal of the anthropic principle: "It is almost irresistible for humans to believe that we have some special relation to the universe, that human life is not just a more-or-less farcical outcome of a chain of accidents reaching back to the first three minutes, but that we were somehow built in from the beginning." However, he concludes that the strong anthropic principle is "little more than mystical mumbo jumbo."

Others are also less convinced about the anthropic principle's power. The late physicist Heinz Pagels was once impressed with the anthropic principle but eventually lost interest because it had no predictive power. The theory is not testable, nor is there any way to extract new information from it. Instead, it yields an endless stream of empty tautologies—that we are here because we are here.

Guth, too, dismisses the anthropic principle, stating that, "I find it hard to believe that anybody would ever use the anthropic principle if he had a better explanation for something. I've yet, for example, to hear an anthropic principle of world history . . . The anthropic principle is something that people do if they can't think of anything better to do."

MULTIVERSE

Other scientists, like Sir Martin Rees of Cambridge University, think that these cosmic accidents give evidence for the existence of the multiverse. Rees believes that the only way to resolve the fact that we live within an incredibly tiny band of hundreds of "coincidences" is to postulate the existence of millions of parallel universes. In this multiverse of universes, most universes are dead. The proton is not stable. Atoms never condense. DNA never forms. The universe collapses prematurely or freezes almost immediately. But in our universe, a series of cosmic accidents has happened, not necessarily because of the hand of God but because of the law of averages.

In some sense, Sir Martin Rees is the last person one might expect to advance the idea of parallel universes. He is the Astronomer Royal of England and bears much responsibility for representing the establishment viewpoint toward the universe. Silver-haired, distinguished, impeccably dressed, Rees is equally fluent speaking about the marvels of the cosmos as about the concerns of the general public.

It is no accident, he believes, that the universe is fine-tuned to allow life to exist. There are simply too many accidents for the universe to be in such a narrow band that allows for life. "The apparent fine-tuning on which our existence depends could be a coincidence," writes Rees. "I once thought so. But that view now seems too narrow . . . Once we accept this, various apparently special features of our universe—those that some theologians once adduced as evidence for Providence or design—occasion no surprise."

Rees has tried to give substance to his arguments by quantifying some of these concepts. He claims that the universe seems to be governed by six numbers, each of which is measurable and finely tuned. These six numbers must satisfy the conditions for life, or else they create dead universes.

First is Epsilon, which equals 0.007, which is the relative amount of hydrogen that converts to helium via fusion in the big bang. If this number were 0.006 instead of 0.007, this would weaken the nuclear force, and protons and neutrons would not bind together. Deuterium (with one proton and one neutron) could not form, hence the heavier elements would never have been created in the stars, the atoms of our body could not have formed, and the entire universe would have dissolved into hydrogen. Even a small reduction in the nuclear force would create instability in the periodic chart of the elements, and there would be fewer stable elements out of which to create life.

If Epsilon were 0.008, then fusion would have been so rapid that no hydrogen would have survived from the big bang, and there would be no stars today to give energy to the planets. Or perhaps two protons would have bound together, also making fusion in the stars impossible. Rees points to the fact that Fred Hoyle found that even a shift as small as 4 percent in the nuclear force would have made the formation of carbon impossible in the stars, making the higher elements and hence life impossible. Hoyle found that if one changed the nuclear force slightly, then beryllium would be so unstable that it could never be a "bridge" to form carbon atoms.

Second is N, equal to 10^{36}, which is the strength of the electric force divided by the strength of gravity, which shows how weak gravity is. If gravity were even weaker, then stars could not condense and create the enormous temperatures necessary for fusion. Hence, stars would not shine, and the planets would plunge into freezing darkness.

But if gravity were a bit stronger, this would cause stars to heat up too fast, and they would burn up their fuel so quickly that life could never get started. Also, a stronger gravity would mean that galaxies would form earlier and would be quite small. The stars would be more densely packed, making disastrous collisions between various stars and planets.

Third is Omega, the relative density of the universe. If Omega were too small, then the universe would have expanded and cooled too fast. But if Omega were too large, then the universe would have collapsed before life could start. Rees writes, "At one second after the big bang, Omega cannot have differed from unity by more than one part in a million billion (one in 10^{15}) in order that the universe should now, after 10 billion years, be still expanding and with a value of Omega that has certainly not departed wildly from unity."

Fourth is Lambda, the cosmological constant, which determines the acceleration of the universe. If it were just a few times larger, the antigravity it would create would blow the universe apart, sending it into an immediate big freeze, making life impossible. But if the cosmological constant were negative, the universe would have contracted violently into a big crunch, too soon for life to form. In other words, the cosmological constant, like Omega, must also be within a certain narrow band to make life possible.

Fifth is Q, the amplitude of the irregularities in the cosmic microwave background, which equals 10^{-5}. If this number were a bit smaller, then the universe would be extremely uniform, a lifeless mass of gas and dust, which would never condense into the stars and galaxies of today. The universe would be dark, uniform, featureless, and lifeless. If Q were larger, then matter would have condensed earlier in the history of the universe into huge supergalactic structures. These "great gobs of matter would have condensed into huge black holes," says Rees. These black holes would be heavier than an entire cluster of galaxies. Whatever stars can form in these huge cluster of gas would be so tightly packed that planetary systems would be impossible.

Last is D, the number of spatial dimensions. Due to interest in M-theory, physicists have returned to the question of whether life is possible in higher or lower dimensions. If space is one-dimensional, then life probably cannot exist because the universe is trivial. Usually, when physicists try to apply the quantum theory to one-dimensional universes, we find that particles pass through one other without interacting. So it's possible that universes existing in one dimension cannot support life because particles cannot "stick" together to form increasingly complex objects.

In two space dimensions, we also have a problem because life forms would probably disintegrate. Imagine a two-dimensional race of flat beings, called Flatlanders, living on a tabletop. Imagine them trying to eat. The passage extending from its mouth to its rear would split the Flatlander in half, and he would fall apart. Thus, it's difficult to imagine how a Flatlander could exist as a complex being without disintegrating or falling into separate pieces.

Another argument from biology indicates that intelligence cannot exist in fewer than three dimensions. Our brain consists of a large number of overlapping neurons connected by a vast electrical network. If the universe were one- or two-dimensional, then it would be difficult to construct complex neural networks, especially if they short-circuit by being placed on top of each other. In lower dimensions, we are severely limited by the number of complex logic circuits and neurons we can place in a small area. Our own brain, for example, consists of about 100 billion neurons, about as many stars as in the Milky Way galaxy, with each neuron connected to about 10,000 other neurons. Such complexity would be hard to duplicate in lower dimensions.

In four space dimensions, one has another problem: planets are not stable in their orbits around the Sun. Newton's inverse square law is replaced by an inverse cube law, and in 1917, Paul Ehrenfest, a close colleague of Einstein, speculated about what physics might look like in other dimensions. He analyzed what is called the Poisson-Laplace equation (which governs the motion of planetary objects as well as electric charges in atoms) and found that orbits are not stable in four or higher spatial dimensions. Since electrons in atoms as well as planets experience random collisions, this means that atoms and solar systems probably cannot exist in higher dimensions. In other words, three dimensions are special.

To Rees, the anthropic principle is one of the most compelling arguments for the multiverse. In the same way that the existence of Goldilocks zones for Earth implies extrasolar planets, the existence of Goldilocks zones for the universe implies there are parallel universes. Rees comments, "If there is a large stock of clothing, you're not surprised to find a suit that fits. If there are many universes, each governed by a differing set of numbers, there will be one where there is a particular set of numbers suitable to life. We are in that one." In other words, our universe is the way it is because of the law of averages over many universes in the multiverse, not because of a grand design.

Weinberg seems to agree on this point. Weinberg, in fact, finds the idea of a multiverse intellectually pleasing. He never did like the idea that time could suddenly spring into existence at the big bang, and that time could not exist before that. In a multiverse, we have the eternal creation of universes.

There is another, quirky reason why Rees prefers the multiverse idea. The universe, he finds, contains a small amount of "ugliness." For example, Earth's orbit is slightly elliptical. If it were perfectly spherical, then one might argue, as theologians have, that it was a by-product of divine intervention. But it is not, indicating a certain amount of randomness within the narrow Goldilocks band. Similarly, the cosmological constant is not perfectly zero but is small, which indicates that our universe is "no more special than our presence requires." This is all consistent with our universe being randomly generated by accident.

EVOLUTION OF UNIVERSES

Being an astronomer, rather than a philosopher, Rees says that the bottom line is that all these theories have to be testable. In fact, that is the reason why he favors the multiverse idea rather than competing, mystical theories. The multiverse theory, he believes, can be tested in the next twenty years.

One variation of the multiverse idea is actually testable today. Physicist Lee Smolin goes even further than Rees and assumes that an "evolution" of universes took place, analogous to Darwinian evolution, ultimately leading to universes like ours. In the chaotic inflationary theory, for example, the physical constants of the "daughter" universes have slightly different physical constants than the mother universe. If universes can sprout from black holes, as some physicists believe, then the universes that dominate the multiverse are those that have the most black holes. This means that, as in the animal kingdom, the universes that give rise to the most "children" eventually dominate to spread their "genetic information"— the physical constants of nature. If true, then our universe might have had an infinite number of ancestor universes in the past, and our universe is a by-product of trillions of years of natural selection. In other words, our universe is the by-product of survival of the fittest, meaning it is the child of universes with the maximum number of black holes.

Although a Darwinian evolution among universes is a strange and novel idea, Smolin believes that it can be tested by simply counting the number of black holes. Our universe should be maximally favorable to the creation of black holes. (However, one still has to prove that universes with the most black holes are the ones that favor life, like ours.)

Because this idea is testable, counterexamples can be considered. For example, perhaps it can be shown, by hypothetically adjusting the physical parameters of the universe, that black holes are most readily produced in universes that are lifeless. For example, perhaps one might be able to show that a universe with a much stronger nuclear force has stars that burn out extremely quickly, creating large numbers of supernovae that then collapse into black holes. In such a universe, a larger value for the nuclear force means that stars live for brief periods, and hence life cannot get started. But this universe might also have more black holes, thereby disproving Smolin's idea. The advantage of this idea is that can be tested, reproduced, or falsified (the hallmark of any true scientific theory). Time will tell whether it holds up or not.

Although any theory involving wormholes, superstrings, and higher dimensions is beyond our current experimental ability, new experiments are now being conducted and future ones planned that may determine whether these theories are correct or not. We are in the midst of a revolution in experimental science, with the full power of satellites, space telescopes, gravity wave detectors, and lasers being brought to bear on these questions. The bountiful harvest from these experiments could very well resolve some of the deepest questions in cosmology.

CHAPTER NINE

Searching for Echoes from the Eleventh Dimension

Remarkable claims require remarkable proof.

—Carl Sagan

PARALLEL UNIVERSES, dimensional portals, and higher dimensions, as spectacular as they are, require airtight proof of their existence. As the astronomer Ken Croswell remarks, "Other universes can get intoxicating: you can say anything you want about them and never be proven wrong, as long as astronomers never see them." Previously, it seemed hopeless to test many of these predictions, given the primitiveness of our experimental equipment. However, recent advances in computers, lasers, and satellite technology have put many of these theories tantalizingly close to experimental verification.

Direct verification of these ideas may prove to be exceedingly difficult, but indirect verification may be within reach. We sometimes forget that much of astronomical science is done indirectly. For example, no one has ever visited the Sun or the stars, yet we know what the stars are made of by analyzing the light given off by these luminous objects. By analyzing the spectrum of light within starlight, we know indirectly that the stars are made primarily of hydrogen and some helium. Likewise, no one has ever seen a black hole, and in fact black holes are invisible and cannot be directly seen. However, we see indirect evidence of their existence by looking for accretion disks and computing the mass of these dead stars.

In all these experiments, we look for "echoes" from the stars and black holes to determine their nature. Likewise, the eleventh dimension may be beyond our direct reach, but there are ways in which inflation and superstring theory may be verified, in light of the new revolutionary instruments now at our disposal.

GPS AND RELATIVITY

The simplest example of the way satellites have revolutionized research in relativity is the Global Positioning System (GPS), in which twenty-four satellites continually orbit Earth, emitting precise, synchronized pulses which allow one to triangulate one's position on the planet to remarkable accuracy. The GPS has become an essential feature of navigation, commerce, as well as warfare. Everything from computerized maps inside cars to cruise missiles depends on the ability to synchronize signals to within 50 billionths of a second to locate an object on Earth to within 15 yards. But in order to guarantee such incredible accuracy, scientists must calculate slight corrections to Newton's laws due to relativity, which states that radio waves will be slightly shifted in frequency as satellites soar in outer space. In fact, if we foolishly discard the corrections due to relativity, then the GPS clocks will run faster each day by 40,000 billions of a second, and the entire system will become unreliable. Relativity theory is thus absolutely essential for commerce and the military. Physicist Clifford Will, who once briefed a U.S. Air Force general about the crucial corrections to the GPS coming from Einstein's theory of relativity, once commented that he knew that relativity theory had come of age when even senior Pentagon officials had to be briefed on it.

GRAVITY WAVE DETECTORS

So far, almost everything we know about astronomy has come in the form of electromagnetic radiation, whether it's starlight or radio or microwave signals from deep space. Now scientists are introducing the first new medium for scientific discovery, gravity itself. "Every time we have looked at the sky in a new way, we have seen a new universe," says Gary Sanders of Cal Tech and deputy director of the gravity wave project.

It was Einstein, in 1916, who first proposed the existence of gravity waves. Consider what would happen if the Sun disappeared. Recall the analogy of a bowling ball sinking into a mattress? Or better, a trampoline net? If the ball is suddenly removed, the trampoline net will immediately spring back into its original position, creating shock waves that ripple outward along the trampoline net. If the bowling ball is replaced by the Sun, then we see that shock waves of gravity travel at a specific speed, the speed of light.

Although Einstein later found an exact solution of his equations that allowed for gravity waves, he despaired of ever seeing his prediction verified in his lifetime. Gravity waves are extremely weak. Even the shock waves of colliding stars are not strong enough to be measured by current experiments.

At present, gravity waves have only been detected indirectly. Two physicists, Russell Hulse and Joseph Taylor, Jr., conjectured that if you analyze circling binary neutron stars that chase each other in space, then each star would emit a stream of gravity waves, similar to the wake created by stirring molasses, as their orbit slowly decays. They analyzed the death spiral of two neutron stars as they slowly spiraled toward each other. The focus of their investigation was the double neutron star PSR 1913+16, located about 16,000 light-years from Earth, which orbit around each other every 7 hours, 45 minutes, in the process emitting gravity waves into outer space.

Using Einstein's theory, they found that the two stars should come closer by a millimeter every revolution. Although this is a fantastically small distance, it increases to a yard over a year, as the orbit of 435,000 miles slowly decreases in size. Their pioneering work showed that the orbit decayed precisely as Einstein's theory predicted on the basis of gravity waves. (Einstein's equations, in fact, predict that the stars will eventually plunge into each other within 240 million years, due to the loss of energy radiated into space in the form of gravity waves.) For their work, they won the Nobel Prize in physics in 1993.

We can also go backward and use this precision experiment to measure the accuracy of general relativity itself. When the calculations are done backward, we find that general relativity is at least 99.7 percent accurate.

LIGO GRAVITY WAVE DETECTOR

But to extract usable information about the early universe, one must observe gravity waves directly, not indirectly. In 2003, the first operational gravity wave detector, LIGO (Laser Interferometer Gravitational-Wave Observatory), finally came online, realizing a decades-old dream of probing the mysteries of the universe with gravity waves. The goal of LIGO is to detect cosmic events that are too distant or tiny to be observed by Earth telescopes, such as colliding black holes or neutron stars.

LIGO consists of two gigantic laser facilities, one in Hanford, Washington, and the other in Livingston Parish, Louisiana. Each facility has two pipes, each 2.5 miles long, creating a gigantic L-shaped tubing. Within each tube a laser is fired. At the joint of the L, both laser beams collide, and their waves interfere with each other. Normally, if there are no disturbances, then the two waves are synchronized so that they cancel each other out. But when even the tiniest gravity wave emitted from colliding black holes or neutron stars hits the apparatus, it causes one arm to contract and expand differently than the other arm. This disturbance is sufficient to disrupt the delicate cancellation of the two laser beams. As a result, the two beams, instead of canceling each other out, create a characteristic wavelike interference pattern that can be computer-analyzed in detail. The larger the gravity wave, the greater the mismatch between the two laser beams, and the larger the interference pattern.

LIGO is an engineering marvel. Since air molecules may absorb the laser light, the tube containing the light has to be evacuated down to a trillionth of atmospheric pressure. Each detector takes up 300,000 cubic feet of space, meaning that LIGO has the largest artificial vacuum in the world. What gives LIGO such sensitivity, in part, is the design of the mirrors, which are controlled by tiny magnets, six in all, each the size of an ant. The mirrors are so polished that they are accurate to one part in 30 billionths of an inch. "Imagine the earth were that smooth. Then the average mountain wouldn't rise more than an inch," says GariLynn Billingsley, who monitors the mirrors. They are so delicate that they can be moved by less than a millionth of a meter, which makes the LIGO mirrors perhaps the most sensitive in the world. "Most control systems engineers' jaws drop when they hear what we're trying to do," says LIGO scientist Michael Zucker.

Because LIGO is so exquisitely balanced, it is sometimes plagued by slight, unwanted vibrations from the most unlikely sources. The detector in Louisiana, for example, cannot be run during the day because of loggers who are cutting trees 1,500 feet from the site. (LIGO is so sensitive that even if the logging were to take place a mile away, it still could not be run during the daytime.) Even at night, vibrations from passing freight trains at midnight and 6 a.m. bracket how much continuous time the LIGO can operate.

Even something as faint as ocean waves striking the coastline miles away can affect the results. Ocean waves breaking on North American beaches wash ashore every six seconds, on average, and this creates a low growl that can actually be picked up by the lasers. The noise is so low in frequency, in fact, that it actually penetrates right through the earth. "It feels like a rumble," says Zucker, commenting about this tidal noise. "It's a huge headache during the Louisiana hurricane season." LIGO is also affected by the tides created by the Moon's and Sun's gravity tugging on Earth, creating a disturbance of several millionths of an inch.

In order to eliminate these incredibly tiny disturbances, LIGO engineers have gone to extraordinary lengths to isolate much of the apparatus. Each laser system rests on top of four huge stainless steel platforms, each stacked on top of each other; each level is separated by springs to damp any vibration. Sensitive optical instruments each have their own seismic isolation system; the floor is a slab of 30-inch-thick concrete that is not coupled to the walls.

LIGO is actually part of an international consortium, including the French-Italian detector called VIRGO in Pisa, Italy, a Japanese detector called TAMA outside Tokyo, and a British-German detector called GEO600 in Hanover, Germany. Altogether, LIGO's final construction cost will be $292 million (plus $80 million for commissioning and upgrades), making it the most expensive project ever funded by the National Science Foundation.

But even with this sensitivity, many scientists concede that LIGO may not be sensitive enough to detect truly interesting events in its lifetime. The next upgrade of the facility, LIGO II, is scheduled to occur in 2007 if funding is granted. If LIGO does not detect gravity waves, the betting is that LIGO II will. LIGO scientist Kenneth Libbrecht claims that LIGO II will improve the sensitivity of the equipment a thousandfold: "You go from [detecting] one event every 10 years, which is pretty painful, to an event every three days, which is very nice."

For LIGO to detect the collision of two black holes (within a distance of 300 million light-years), a scientist could wait anywhere from a year to a thousand years. Many astronomers may have second thoughts about investigating such an event with LIGO if it means that their great-great-great . . . grandchildren will be the ones to witness the event. But as LIGO scientist Peter Saulson has said, "People take pleasure in solving these technical challenges, much the way medieval cathedral builders continued working knowing they might not see the finished church. But if there wasn't a fighting chance to see a gravity wave during my life career, I wouldn't be in this field. It's not just Nobel fever . . . The levels of precision we are striving for mark our business; if you do this, you have 'the right stuff.' " With LIGO II, the chances are much better of finding a truly interesting event in our lifetime. LIGO II might detect colliding black holes within a much larger distance of 6 billion light-years at a rate of ten per day to ten per year.

Even LIGO II, however, will not be powerful enough to detect gravity waves emitted from the instant of creation. For that, we must wait another fifteen to twenty years for LISA.

LISA GRAVITY WAVE DETECTOR

LISA (Laser Interferometry Space Antenna) represents the next generation in gravity wave detectors. Unlike LIGO, it will be based in outer space. Around 2010, NASA and the European Space Agency plan to launch three satellites into space; they will orbit around the Sun at approximately 30 million miles from Earth. The three laser detectors will form an equilateral triangle in space (5 million kilometers on a side). Each satellite will have two lasers that allow it to be in continual contact with the other two satellites. Although each laser will fire a beam with only half a watt of power, the optics are so sensitive that they will be able to detect vibrations coming from gravity waves with an accuracy of one part in a billion trillion (corresponding to a shift that is one hundredth the width of a single atom). LISA should be able to detect gravity waves from a distance of 9 billion light-years, which cuts across most of the visible universe.

LISA will be so accurate that it might detect the original shock waves from the big bang itself. This will give us by far the most accurate look at the instant of creation. If all goes according to plan, LISA should be able to peer to within the first trillionth of a second after the big bang, making it perhaps the most powerful of all cosmological tools. It is believed that LISA may be able to find the first experimental data on the precise nature of the unified field theory, the theory of everything.

LISA (or its successors) may provide the "smoking gun" for the inflationary theory. So far, inflation is consistent with all cosmological data (flatness, fluctuations in the cosmic background, and so forth). But that doesn't mean the theory is correct. To clinch the theory, scientists want to examine the gravity waves that were set off by the inflationary process itself. The "fingerprint" of gravity waves created at the instant of the big bang should tell the difference between inflation and any rival theory. Kip Thorne of Cal Tech believes that LISA (or its successors) may determine whether some version of string theory is correct. As I explain in chapter 7, the inflationary universe theory predicts that gravity waves emerging from the big bang should be quite violent, corresponding to the rapid, exponential expansion of the early universe, while the ekpyrotic model predicts a much gentler expansion, accompanied by much smoother gravity waves. LISA should be able to rule out various rival theories of the big bang and make a crucial test of string theory.

EINSTEIN LENSES AND RINGS

Yet another powerful tool in exploring the cosmos is the use of gravitational lenses and "Einstein rings." As early as 1801, Berlin astronomer Johan Georg von Soldner was able to calculate the possible deflection of starlight by the Sun's gravity (although, because Soldner used strictly Newtonian arguments, he was off by a crucial factor of 2. Einstein wrote, "Half of this deflection is produced by the Newtonian field of attraction of the sun, the other half by the geometrical modification ['curvature'] of space caused by the sun.")

In 1912, even before he completed the final version of general relativity, Einstein contemplated the possibility of using this deflection as a "lens," in the same way that your glasses bend light before it reaches your eye. In 1936, a Czech engineer, Rudi Mandl, wrote to Einstein asking whether a gravity lens could magnify light from a nearby star. The answer was yes, but it would be beyond their technology to detect this.

In particular, Einstein realized that you would see optical illusions, such as double images of the same object, or a ringlike distortion of light. Light from a very distant galaxy passing near our Sun, for example, would travel both to the left and right of our Sun before the beams rejoined and reached our eye. When we gaze at the distant galaxy, we see a ringlike pattern, an optical illusion caused by general relativity. Einstein concluded that there was "not much hope of observing this phenomenon directly." In fact, he wrote that this work "is of little value, but it makes the poor guy [Mandl] happy."

Over forty years later, in 1979, the first partial evidence of lensing was found by Dennis Walsh of the Jordell Bank Observatory in England, who discovered the double quasar Q0957+561. In 1988, the first Einstein ring was observed from the radio source MG1131+0456. In 1997, the Hubble space telescope and the UK's MERLIN radio telescope array caught the first completely circular Einstein ring by analyzing the distant galaxy 1938+666, vindicating Einstein's theory once again. (The ring is tiny, only a second of arc, or roughly the size of a penny viewed from two miles away.) The astronomers described the excitement they felt witnessing this historic event: "At first sight, it looked artificial and we thought it was some sort of defect in the image, but then we realized we were looking at a perfect Einstein ring!" said Ian Brown of the University of Manchester. Today, Einstein's rings are an essential weapon in the arsenal of astrophysicists. About sixty-four double, triple, and multiple quasars (illusions caused by Einstein lensing) have been seen in outer space, or roughly one in every five hundred observed quasars.

Even invisible forms of matter, like dark matter, can be "seen" by analyzing the distortion of light waves they create. In this way, one can obtain "maps" showing the distribution of dark matter in the universe. Since Einstein lensing distorts galactic clusters by creating large arcs (rather than rings), it is possible to estimate the concentration of dark matter in these clusters. In 1986, the first giant galactic arcs were discovered by astronomers at the National Optical Astronomy Observatory, Stanford University, and Midi-Pyrenees Observatory in France. Since then, about a hundred galactic arcs have been discovered, the most dramatic in the galactic cluster Abell 2218.

Einstein lenses can also be used as an independent method to measure the amount of MACHOs in the universe (which consist of ordinary matter like dead stars, brown dwarfs, and dust clouds). In 1986, Bohdan Paczynski of Princeton realized that if MACHOs passed in front of a star, they would magnify its brightness and create a second image.

In the early 1990s, several teams of scientists (such as the French EROS, the American-Australian MACHO, and the Polish-American OGLE) applied this method to the center of the Milky Way galaxy and found more than five hundred microlensing events (more than expected, because some of this matter consisted of low-mass stars and not true MACHOs). This same method can be used to find extrasolar planets orbiting other stars. Since a planet would exert a tiny but noticeable gravitational effect on the mother star's light, Einstein lensing can in principle detect them. Already, this method has identified a handful of candidates for extrasolar planets, some of them near the center of the Milky Way.

Even Hubble's constant and the cosmological constant can be measured using Einstein lenses. Hubble's constant is measured by making a subtle observation. Quasars brighten and dim with time; one might expect that double quasars, being images of the same object, would oscillate at the same rate. Actually, these twin quasars do not quite oscillate in unison. Using the known distribution of matter, astronomers can calculate the time delay divided by the total time it took light to reach Earth. By measuring the time delay in the brightening of the double quasars, one can then calculate its distance from Earth. Knowing its redshift, one can then calculate the Hubble constant. (This method was applied to the quasar Q0957+561, which was found to be roughly 14 billion light-years from Earth. Since then, the Hubble constant has been computed by analyzing seven other quasars. Within error bars, these calculations agree with known results. What is interesting is that this method is totally independent of the brightness of stars, such as Cepheids and type Ia supernovae, which gives an independent check on the results.)

The cosmological constant, which may hold the key to the future of our universe, can also be measured by this method. The calculation is a bit crude, but it is also in agreement with other methods. Since the total volume of the universe was smaller billions of years ago, the probability of finding quasars that will form an Einstein lens was also greater in the past. Thus, by measuring the number of double quasars at different times in the evolution for the universe, one can roughly calculate the total volume of the universe and hence the cosmological constant, which is helping to drive the universe's expansion. In 1998, astronomers at the Harvard-Smithsonian Center for Astrophysics made the first crude estimate of the cosmological constant and concluded that it probably made up no more than 62 percent of the total matter/energy content of the universe. (The actual WMAP result is 73 percent.)

DARK MATTER IN YOUR LIVING ROOM

Dark matter, if it does pervade the universe, does not solely exist in the cold vacuum of space. In fact, it should also be found in your living room. Today, a number of research teams are racing to see who will be the first to snare the first particle of dark matter in the laboratory. The stakes are high; the team that is capable of capturing a particle of dark matter darting through their detectors will be first to detect a new form of matter in two thousand years.

The central idea behind these experiments is to have a large block of pure material (such as sodium iodide, aluminum oxide, freon, germanium, or silicon), in which particles of dark matter may interact. Occasionally, a particle of dark matter may collide with the nucleus of an atom and cause a characteristic decay pattern. By photographing the tracks of the particles involved in this decay, scientists can then confirm the presence of dark matter.

Experimenters are cautiously optimistic, since the sensitivity of their equipment gives them the best opportunity yet to observe dark matter. Our solar system orbits around the black hole at the center of the Milky Way galaxy at 220 kilometers per second. As a result, our planet is passing through a considerable amount of dark matter. Physicists estimate that a billion dark matter particles flow through every square meter of our world every second, including through our bodies.

Although we live in a "dark matter wind" that blows through our solar system, experiments to detect dark matter in the laboratory have been exceedingly difficult to perform because dark matter particles interact so weakly with ordinary matter. For example, scientists would expect to find anywhere from 0.01 to 10 events per year occurring within a single kilogram of material in the lab. In other words, you would have to carefully watch large quantities of this material over a period of many years to see events consistent with dark matter collisions.

So far, experiments with acronyms like UKDMC in the United Kingdom; ROSEBUD in Canfranc, Spain; SIMPLE in Rustrel, France; and Edelweiss in Frejus, France, have not yet detected any such events. An experiment called DAMA, outside Rome, created a stir in 1999 when scientists reportedly sighted dark matter particles. Because DAMA uses 100 kilograms of sodium iodide, it is the largest detector in the world. However, when the other detectors tried to reproduce DAMA's result, they found nothing, casting doubt on the DAMA findings.

Physicist David B. Cline notes, "If the detectors do register and verify a signal, it would go down as one of the great accomplishments of the twenty-first century . . . The greatest mystery in modern astrophysics may soon be solved."

If dark matter is found soon, as many physicists hope, it might give support to supersymmetry (and possibly, over time, to superstring theory) without the use of atom smashers.

SUSY (SUPERSYMMETRIC) DARK MATTER

A quick look at the particles predicted by supersymmetry shows that there are several likely candidates that can explain dark matter. One is the neutralino, a family of particles which contains the superpartner of the photon. Theoretically, the neutralino seems to fit the data. Not only is it neutral in charge, and hence invisible, and also massive (so it is affected only by gravity) but it is also stable. (This is because it has the lowest mass of any particle in its family and hence cannot decay to any lower state.) Last, and perhaps most important, the universe should be full of neutralinos, which would make them ideal candidates for dark matter.

Neutralinos have one great advantage: they might solve the mystery of why dark matter makes up 23 percent of the matter/energy content of the universe while hydrogen and helium make up only a paltry 4 percent.

Recall that when the universe was 380,000 years old, the temperature dropped until atoms were no longer ripped apart by collisions caused by the intense heat of the big bang. At that time, the expanding fireball began to cool, condense, and form stable, whole atoms. The abundance of atoms today dates back roughly to that time period. The lesson is that the abundance of matter in the universe dates back to the time when the universe had cooled enough so that matter could be stable.

This same argument can be used to calculate the abundance of neutralinos. Shortly after the big bang, the temperature was so blistering hot that even neutralinos were destroyed by collisions. But as the universe cooled, at a certain time the temperature dropped enough so that neutralinos could form without being destroyed. The abundance of neutralinos dates back to this early era. When we do this calculation, we find that the abundance of neutralinos is much larger than atoms, and in fact approximately corresponds to the actual abundance of dark matter today. Supersymmetric particles, therefore, can explain the reason why dark matter is overwhelmingly abundant throughout the universe.

SLOAN SKY SURVEY

Although many of the advances in the twenty-first century will be made in instrumentation involving satellites, this does not mean that research in earthbound optical and radio telescopes has been set aside. In fact, the impact of the digital revolution has changed the way optical and radio telescopes are utilized, making possible statistical analyses of hundreds of thousands of galaxies. Telescope technology is now having a sudden second lease on life as a result of this new technology.

Historically, astronomers have fought over the limited amount of time they were permitted to use the world's biggest telescopes. They jealously guarded their precious time on these instruments and spent many hours toiling in cold, damp rooms throughout the night. Such an antiquated observation method was highly inefficient and often sparked bitter feuds among astronomers who felt slighted by the "priesthood" monopolizing time on the telescope. All this is changing with the coming of the Internet and high-speed computing.

Today, many telescopes are fully automated and can be programmed thousands of miles away by astronomers located on different continents. The results of these massive star surveys can be digitized and then placed on the Internet, where powerful supercomputers can then analyze the data. One example of the power of this digital method is SETI@home, a project based at the University of California at Berkeley to analyze signals for signs of extraterrestrial intelligence. The massive data from the Aricebo radio telescope in Puerto Rico is chopped up into tiny digital pieces and then sent via the Internet to PCs around the world, mainly to amateurs. A screen saver software program analyzes the data for intelligent signals when the PC is not in use. Using this method, the research group has constructed the largest computer network in the world, linking about 5 millions PCs from all points of the globe.

The most prominent example of today's digital exploration of the universe is the Sloan Sky Survey, which is the most ambitious survey of the night sky ever undertaken. Like the earlier Palomar Sky Survey, which used old-fashioned photographic plates stored in bulky volumes, the Sloan Sky Survey will create an accurate map of the celestial objects in the sky. The survey has constructed three-dimensional maps of distant galaxies in five colors, including the redshift of over a million galaxies. The output of the Sloan Sky Survey is a map of the large-scale structure of the universe several hundred times larger than previous efforts. It will map in exquisite detail one quarter of the entire sky and determine the position and brightness of 100 million celestial objects. It will also determine the distance to more than a million galaxies and about 100,000 quasars. The total information generated by the survey will be 15 terabytes (a trillion bytes), which rivals the information stored within the Library of Congress.

The heart of the Sloan Survey is a 2.5-meter telescope based in southern New Mexico containing one of the most advanced cameras ever produced. It contains thirty delicate electronic light sensors, called CCDs (charge-coupled devices), each 2 inches square, sealed in a vacuum. Each sensor, which is cooled down to -80 degrees C by liquid nitrogen, contains 4 million picture elements. All the light collected by the telescope can therefore be instantly digitized by the CCDs and then fed directly into a computer for processing. For less than $20 million, the survey creates a stunning picture of the universe at a cost of a hundredth of the Hubble space telescope.

The survey then puts some of this digitized data on the Internet, where astronomers all over the world can pore over it. In this way, we can also harness the intellectual potential of the world's scientists. In the past, all too often scientists in the Third World were unable to get access to the latest telescopic data and the latest journals. This was a tremendous waste of scientific talent. Now, because of the Internet, they can download the data from sky surveys, read articles as they appear on the Internet, and also publish articles on the Web with the speed of light.

The Sloan Survey is already changing the way astronomy is conducted, with new results based on analyses of hundreds of thousands of galaxies, which would have been prohibitive just a few years ago. For example, in May 2003, a team of scientists from Spain, Germany, and the United States announced that they had analyzed 250,000 galaxies for evidence of dark matter. Out of this huge number, they focused on three thousand galaxies with star clusters orbiting around them. By using Newton's laws of motion to analyze the motion of these satellites, they calculated the amount of dark matter that must surround the central galaxy. Already, these scientists have ruled out a rival theory. (An alternative theory, first proposed in 1983, tried to explain the anomalous orbits of stars in the galaxies by modifying Newton's laws themselves. Perhaps dark matter did not really exist at all but was due to an error within Newton's laws. The survey data cast doubt on this theory.)

In July 2003, another team of scientists from Germany and the United States announced that they had analyzed 120,000 nearby galaxies using the Sloan Survey to unravel the relationship between galaxies and the black holes inside them. The question is: which came first, the black hole or the galaxy that harbors them? The result of this investigation indicates that galaxy and black hole formation are intimately tied together, and that they probably were formed together. It showed that, of the 120,000 galaxies analyzed in the survey, fully 20,000 of them contain black holes that are still growing (unlike the black hole in the Milky Way galaxy, which seems to be quiescent). The results show that galaxies containing black holes that are still growing in size are much larger than the Milky Way galaxy, and that they grow by swallowing up relatively cold gas from the galaxy.

COMPENSATING FOR THERMAL FLUCTUATIONS

Yet another way that optical telescopes have been revitalized is through lasers to compensate for the distortion of the atmosphere. Stars do not twinkle because they vibrate; stars twinkle mainly because of tiny thermal fluctuations in the atmosphere. This means that in outer space, far from the atmosphere, the stars glare down on our astronauts continuously. Although this twinkling gives much of the beauty of the night sky, to an astronomer it is a nightmare, resulting in blurry pictures of celestial bodies. (As a child, I remember staring at the fuzzy pictures of the planet Mars, wishing there was some way to obtain crystal clear pictures of the red planet. If only the disturbances from the atmosphere could be eliminated by rearranging the light beams, I thought, maybe the secret of extraterrestrial life could be solved.)

One way to compensate for this blurriness is to use lasers and high-speed computers to subtract out the distortion. This method uses "adaptive optics," pioneered by a classmate of mine from Harvard, Claire Max of the Lawrence Livermore National Laboratory, and others, using the huge W. M. Keck telescope in Hawaii (the largest in the world) and also the smaller 3-meter Shane telescope at the Lick Observatory in California. For example, by shooting a laser beam into outer space, one can measure tiny temperature fluctuations in the atmosphere. This information is analyzed by computer, which then makes tiny adjustments in the mirror of a telescope which compensate for the distortion of starlight. In this way, one can approximately subtract out the disturbance from the atmosphere.

This method was successfully tested in 1996 and since then has produced crystal-sharp pictures of planets, stars, and galaxies. The system fires light from a tunable dye laser with 18 watts of power into the sky. The laser is attached to the 3-meter telescope, whose deformable mirrors are adjusted to make up for the atmospheric distortion. The image itself is caught on a CCD camera and digitalized. With a modest budget, this system has obtained pictures almost comparable to the Hubble space telescope. One can see fine details in the outer planets and even peer into the heart of a quasar using this method, which breathes new life into optical telescopes.

This method has also increased the resolution of the Keck telescope by a factor of 10. The Keck Observatory, located at the summit of Hawaii's dormant volcano Mauna Kea, almost 14,000 feet above sea level, consists of twin telescopes that weigh 270 tons each. Each mirror, measuring 10 meters (394 inches) across, is composed of thirty-six hexagonal pieces, each of which can be independently manipulated by computer. In 1999, an adaptive optics system was installed into Keck II, consisting of a small, deformable mirror that can change shape 670 times per second. Already, this system has captured the image of stars orbiting around the black hole at the center of our Milky Way galaxy, the surface of Neptune and Titan (a moon of Saturn), and even an extrasolar planet which eclipsed the mother star 153 light-years from Earth. Light from the star HD 209458 dimmed exactly as predicted, as the planet moved in front of the star.

LASHING RADIO TELESCOPES TOGETHER

Radio telescopes have also been revitalized by the computer revolution. In the past, radio telescopes were limited by the size of their dish. The larger the dish, the more radio signals could be gathered from space and analyzed. However, the larger the dish, the more expensive it becomes. One way to overcome this problem is to lash several dishes together to mimic the radio-gathering capability of a super radio telescope. (The largest radio telescope that can be lashed together on Earth is the size of Earth itself.) Previous efforts to lash together radio telescopes in Germany, Italy, and the United States proved partially successful.

One problem with this method is that signals from all the various radio telescopes must be combined precisely and then fed into a computer. In the past, this was prohibitively difficult. However, with the coming of the Internet and cheap high-speed computers, costs have dropped considerably. Today, creating radio telescopes with the effective size of the planet Earth is no longer a fantasy.

In the United States, the most advanced device employing this interference technology is the VLBA (very long baseline array), which is a collection of ten radio antennas located at different sites, including New Mexico, Arizona, New Hampshire, Washington, Texas, the Virgin Islands, and Hawaii. Each VLBA station contains a huge, 82-foot-diameter dish which weighs 240 tons and stands as tall as a ten-story building. Radio signals are carefully recorded at each site on tape, which is then shipped to the Socorro Operations Center, New Mexico, where they are correlated and analyzed. The system went online in 1993 at a cost of $85 million.

Correlating the data from these ten sites creates an effective, giant radio telescope that is 5,000 miles wide and can produce some of the sharpest images on Earth. It is equivalent to standing in New York City and reading a newspaper in Los Angeles. Already, the VLBA has produced "movies" of cosmic jets and supernova explosions and the most accurate distance measurement ever made of an object outside the Milky Way galaxy.

In the future, even optical telescopes may use the power of interferometry, although this is quite difficult because of the short wavelength of light. There is a plan to bring the optical data from the two telescopes at the Keck Observatory in Hawaii and interfere them, essentially creating a giant telescope much larger than either one.

MEASURING THE ELEVENTH DIMENSION

In addition to the search for dark matter and black holes, what is most intriguing to physicists is the search for higher dimensions of space and time. One of the more ambitious attempts to verify the existence of a nearby universe was done at the University of Colorado at Boulder. Scientists there tried to measure deviations from Newton's famous inverse square law.

According to Newton's theory of gravity, the force of attraction between any two bodies diminishes with the square of the distance separating them. If you double the distance from Earth to the Sun, then the force of gravity goes down by 2 squared, or 4. This, in turn, measures the dimensionality of space.

So far, Newton's law of gravity holds at cosmological distances involving large clusters of galaxies. But no one has adequately tested his law of gravity down to tiny length scales because it was prohibitively difficult. Because gravity is such a weak force, even the tiniest disturbance can destroy the experiment. Even passing trucks create vibrations large enough to nullify experiments trying to measure the gravity between two small objects.

The physicists in Colorado built a delicate instrument, called a high-frequency resonator, that was able to test the law of gravity down to a 10th of a millimeter, the first time this had ever been done on such a tiny scale. The experiment consisted of two very thin tungsten reeds suspended in a vacuum. One of the reeds vibrated at a frequency of 1,000 cycles per second, looking somewhat like a vibrating diving board. Physicists then looked for any vibrations that were transmitted across the vacuum to the second reed. The apparatus was so sensitive that it could detect motion in the second reed caused by the force of a billionth of the weight of a grain of sand. If there was a deviation in Newton's law of gravity, then there should have been slight disturbances recorded in the second reed. However, after analyzing distances down to 108 millionths of a meter, the physicists found no such deviation. "So far, Newton is holding his ground," said C. D. Hoyle of the University of Trento in Italy, who analyzed the experiment for *Nature* magazine.

This result was negative, but this has only whetted the appetite of other physicists who want to test deviations to Newton's law down to the microscopic level.

Yet another experiment is being planned at Purdue University. Physicists there want to measure tiny deviations in Newton's gravity not at the millimeter level but at the atomic level. They plan to do this by using nanotechnology to measure the difference between nickel 58 and nickel 64. These two isotopes have identical electrical and chemical properties, but one isotope has six more neutrons than the other. In principle, the only difference between these isotopes is their weight.

These scientists envision creating a Casimir device consisting of two sets of neutral plates made out of the two isotopes. Normally, when these plates are held closely together, nothing happens because they have no charge. But if they are brought extremely close to each other, the Casimir effect takes place, and the two plates are attracted slightly, an effect that has been measured in the laboratory. But because each set of parallel plates is made out of different isotopes of nickel, they will be attracted slightly differently, depending on their gravity.

In order to maximize the Casimir effect, the plates have to be brought extremely close together. (The effect is proportional to the inverse fourth power of the separation distance. Hence, the effect grows rapidly as the plates are brought together.) The Purdue physicists will use nanotechnology to make plates separated by atomic distances. They will use state-of-the-art microelectromechanical torsion oscillators to measure tiny oscillations in the plates. Any difference between the nickel 58 and nickel 64 plates can then be attributed to gravity. In this way, they hope to measure deviations to Newton's laws of motion down to atomic distances. If they find a deviation from Newton's famed inverse square law with this ingenious device, it may signal the presence of a higher-dimensional universe separated from our universe by the size of an atom.

LARGE HADRON COLLIDER

But the device that may decisively settle many of these questions is the LHC (Large Hadron Collider), now nearing completion near Geneva, Switzerland, at the famed CERN nuclear laboratory. Unlike previous experiments on strange forms of matter that naturally occur in our world, the LHC might have enough energy to create them directly in the laboratory. The LHC will be able to probe tiny distances, down to 10^{-19} meters, or 10,000 times smaller than a proton, and create temperatures not seen since the big bang. "Physicists are sure that nature has new tricks up her sleeve that must be revealed in those collisions—perhaps an exotic particle known as the Higgs boson, perhaps evidence of a miraculous effect called supersymmetry, or perhaps something unexpected that will turn theoretical particle physics on its head," writes Chris Llewellyn Smith, former director general of CERN and now president of the University College in London. Already, CERN has seven thousand users of its equipment, which amounts to more than half of all the experimental particle physicists on the planet. And many of them will be directly involved in the LHC experiments.

The LHC is a powerful circular machine, 27 kilometers in diameter, large enough to completely encircle many cities around the world. Its tunnel is so long that it actually straddles the French-Swiss border. The LHC is so expensive that it has taken a consortium of several European nations to build it. When it is finally turned on in 2007, powerful magnets arranged along the circular tubing will force a beam of protons to circulate at ever-increasing energies, until they reach about 14 trillion electron volts.

The machine consists of a large circular vacuum chamber with huge magnets placed strategically along its length to bend the powerful beam into a circle. As the particles circulate in the tubing, energy is injected into the chamber, increasing the velocity of the protons. When the beam finally hits a target, it releases a titanic burst of radiation. Fragments created by this collision are then photographed by batteries of detectors to look for evidence of new, exotic, subatomic particles.

The LHC is truly a mammoth machine. While LIGO and LISA push the envelope in terms of sensitivity, the LHC is the ultimate in sheer brute strength. Its powerful magnets, which bend the beam of protons into a graceful arc, generate a field of 8.3 teslas, which is 160,000 times greater than Earth's magnetic field. To generate such monstrous magnetic fields, physicists ram 12,000 amps of electrical current down a series of coils, which have to be cooled down to −271 degrees C, where the coils lose all resistance and become superconducting. In all, it has 1,232 15-meter-long magnets, which are placed along 85 percent of the entire circumference of the machine.

In the tunnel, protons are accelerated to 99.999999 percent of the speed of light until they hit a target, located at four places around the tube, thereby creating billions of collisions each second. Huge detectors are placed there (the largest is the size of a six-story building) to analyze the debris and hunt for elusive subatomic particles.

As Smith mentioned earlier, one of the goals of the LHC is to find the elusive Higgs boson, which is the last piece of the Standard Model that has still eluded capture. It is important because this particle is responsible for spontaneous symmetry breaking in particle theories and gives rise to the masses of the quantum world. Estimates of the mass of the Higgs boson place it somewhere between 115 and 200 billion electron volts (the proton, by contrast, weighs about 1 billion electron volts). (The Tevatron, a much smaller machine located at Fermilab outside Chicago, may actually be the first accelerator to bag the elusive Higgs boson, if the particle's mass is not too heavy. In principle, the Tevatron may produce up to 10,000 Higgs bosons if it operates as planned. The LHC, however, will gen-

erate particles with seven times more energy. With 14 trillion electron volts to play with, the LHC can conceivably become a "factory"
for Higgs bosons, creating millions of them in its proton collisions.)

Another goal of the LHC is to create conditions not seen since the
big bang itself. In particular, physicists believe that the big bang
originally consisted of a loose collection of extremely hot quarks and
gluons, called a quark-gluon plasma. The LHC will be able to produce
this kind of quark-gluon plasma, which dominated the universe in
the first 10 microseconds of its existence. In the LHC, one can collide
nuclei of lead with an energy of 1.1 trillion electron volts. With such
a colossal collision, the four hundred protons and neutrons can
"melt" and free the quarks into this hot plasma. In this way, cosmology may gradually become less an observational science and
more an experimental science, with precise experiments on quark-
gluon plasmas done right in the laboratory.

There is also the hope that the LHC might find mini–black holes
among the debris created by smashing protons together at fantastic
energy, as mentioned in chapter 7. Normally the creation of quantum black holes should take place at the Planck energy, which is a
quadrillion times beyond the energy of the LHC. But if a parallel universe exists within a millimeter of our universe, this reduces the energy at which quantum gravitational effects become measurable,
putting mini–black holes within reach of the LHC.

And last, there is still the hope that the LHC might be able to find
evidence of supersymmetry, which would be a historic breakthrough
in particle physics. These particles are believed to be partners of the
ordinary particles we see in nature. Although string theory and supersymmetry predict that each subatomic particle has a "twin" with
differing spin, supersymmetry has never been observed in nature,
probably because our machines are not powerful enough to detect it.

The existence of superparticles would help to answer two nagging
questions. First, is string theory correct? Although it is exceedingly
difficult to detect strings directly, it may be possible to detect the
lower octaves or resonances of string theory. If particles are discovered, it would go a long way toward giving string theory experimen-

tal justification (although this still would not be direct proof of its correctness).

Second, it would give perhaps the most plausible candidate for dark matter. If dark matter consists of subatomic particles, they must be stable and neutral in charge (otherwise they would be visible), and they must interact gravitationally. All three properties can be found among the particles predicted by string theory.

The LHC, which will be the most powerful particle accelerator when it is finally turned on, is actually a second choice for most physicists. Back in the 1980s, President Ronald Reagan approved the Superconducting Supercollider (SSC), a monstrous machine 50 miles in circumference which was to have been built outside Dallas, Texas; it would have dwarfed the LHC. While the LHC is capable of producing particle collisions with 14 trillion electron volts of energy, the SSC was designed to produce collisions with 40 trillion electron volts. The project was initially approved but, in the final days of hearings, the U.S. Congress abruptly canceled the project. It was a tremendous blow to high-energy physics and set the field back for an entire generation.

Primarily, the debate was about the $11 billion cost of the machine and greater scientific priorities. The scientific community itself was badly split on the SSC, with some physicists claiming that the SSC might drain funds from their own research. The controversy grew so heated that even the *New York Times* wrote a critical editorial about the dangers that "big science" would smother "small science." (These arguments were misleading, since the SSC budget came out of a different source than the budget for small science. The real competitor for funds was the Space Station, which many scientists feel is a true waste of money.)

But in retrospect, the controversy was also about learning to speak to the public in language they can understand. In some sense, the physics world was used to having its monster atom smashers approved by Congress because the Russians were building them as well. The Russians, in fact, were building their UNK accelerator to compete against the SSC. National prestige and honor were at stake. But the Soviet Union broke apart, their machine was canceled, and the wind gradually went out of the sails of the SSC program.

TABLETOP ACCELERATORS

With the LHC, physicists are gradually approaching the upper limit of energy attainable with the present generation of accelerators, which now dwarf many modern cities and cost tens of billions of dollars. They are so huge that only large consortiums of nations can afford them. New ideas and principles are necessary if we are to push the barriers facing conventional accelerators. The holy grail for particle physicists is to create a "tabletop" accelerator that can create beams with billions of electron volts of energy at a fraction of the size and cost of conventional accelerators.

To understand the problem, imagine a relay race, where the runners are distributed around a very large circular race track. The runners exchange a baton as they race around the track. Now imagine that every time the baton is passed from one runner to another, the runners get an extra burst of energy, so they run successively faster along the track.

This is similar to a particle accelerator, where the baton consists of a beam of subatomic particles moving around the circular track. Every time the beam passes from one runner to another, the beam receives an injection of radio frequency (RF) energy, accelerating it to faster and faster velocities. This is how particle accelerators have been built for the past half century. The problem with conventional particle accelerators is that we are hitting the limit of RF energy that can be used to drive the accelerator.

To solve this vexing problem, scientists are experimenting with radically different ways of pumping energy into the beam, such as with powerful laser beams, which are growing exponentially in power. One advantage of laser light is that it is "coherent"—that is, all the waves of light are vibrating in precise unison, making it possible to create enormously powerful beams. Today, laser beams can generate bursts of energy carrying trillions of watts (terrawatts) of power for a brief period of time. (By contrast, a nuclear power plant can generate only a paltry billion watts of power, but at a steady rate.) Lasers that generate up to a thousand trillion watts (a quadrillion watts, or a petawatt) are now becoming available.

Laser accelerators work by the following principle. Laser light is hot enough to create a gas of plasma (a collection of ionized atoms), which then moves in wavelike oscillations at high velocities, like a tidal wave. Then a beam of subatomic particles "surfs" in the wake created by this wave of plasma. By injecting more laser energy, the plasma wave travels at faster velocity, boosting the energy of the particle beam surfing on it. Recently, by blasting a 50-terrawatt laser at a solid target, the scientists at the Rutherford Appleton Laboratory in England produced a beam of protons emerging from the target carrying up to 400 million electron volts (MeV) of energy in a collimated beam. At École Polytechnique in Paris, physicists have accelerated electrons to 200 MeV over a distance of a millimeter.

The laser accelerators created so far have been tiny and not very powerful. But assume for a moment that this accelerator could be scaled up so that it operates not just over a millimeter but over a full meter. Then it would be able to accelerate electrons to 200 giga electron volts over a distance of a meter, fulfilling the goal of a tabletop accelerator. Another milestone was reached in 2001, when the physicists at SLAC (Stanford Linear Accelerator Center) were able to accelerate electrons over a distance of 1.4 meters. Instead of using a laser beam, they created a plasma wave by injecting a beam of charged particles. Although the energy they attained was low, it demonstrated that plasma waves can accelerate particles over distances of a meter.

Progress in this promising area of research is extremely rapid: the energy attained by these accelerators is growing by a factor of 10 every five years. At this rate, a prototype tabletop accelerator may be within reach. If successful, it may make the LHC look like the last of the dinosaurs. Although promising, there are, of course, still many hurdles facing such a tabletop accelerator. Like a surfer who "wipes out" riding a treacherous ocean wave, maintaining the beam so that it properly rides the plasma wave is difficult (problems include focusing the beam and maintaining its stability and intensity). But none of these problems seems insurmountable.

THE FUTURE

There are some long shots in proving string theory. Edward Witten holds out the hope that, at the instant of the big bang, the universe expanded so rapidly that maybe a string was expanded along with it, leaving a huge string of astronomical proportions drifting in space. He muses, "Although somewhat fanciful, this is my favorite scenario for confirming string theory, as nothing would settle the issue quite as dramatically as seeing a string in a telescope."

Brian Greene lists five possible examples of experimental data that could confirm string theory or at least give it credibility:

1. The tiny mass of the elusive, ghostlike neutrino could be experimentally determined, and string theory might explain it.
2. Small violations of the Standard Model could be found that violate point-particle physics, such as the decays of certain subatomic particles.
3. New long-range forces (other than gravity and electromagnetism) could be found experimentally that would signal a certain choice of a Calabi-Yau manifold.
4. Dark matter particles could be found in the laboratory and compared to predictions of string theory.
5. String theory might be able to calculate the amount of dark energy in the universe.

My own view is that verification of string theory might come entirely from pure mathematics, rather than from experiment. Since string theory is supposed to be a theory of everything, it should be a theory of everyday energies as well as cosmic ones. Thus, if we can finally solve the theory completely, we should be able to calculate the properties of ordinary objects, not just exotic ones found in outer space. For example, if string theory can calculate the masses of the proton, neutron, and electron from first principles, this would be an accomplishment of first magnitude. In all models of physics (except string theory), the masses of these familiar particles are put in by hand. We do not need an LHC, in some sense, to verify the theory, since we already know the masses of scores of subatomic particles, all of which should be determined by string theory with no adjustable parameters.

As Einstein said, "I am convinced that we can discover by means of purely mathematical construction the concepts and the laws . . . which furnish the key to the understanding of natural phenomena. Experience may suggest the appropriate mathematical concepts, but they most certainly cannot be deduced from it . . . In a certain sense, therefore, I hold it true that pure thought can grasp reality, as the ancients dreamed."

If true, then perhaps M-theory (or whatever theory finally leads us to a quantum theory of gravity) will make possible the final journey for all intelligent life in the universe, the escape from our dying universe trillions upon trillions of years from now to a new home.

PART THREE

ESCAPE INTO HYPERSPACE

CHAPTER TEN

The End of Everything

[Consider] the view now held by most physicists, namely
that the sun with all the planets will in time grow too
cold for life, unless indeed some great body dashes into
the sun and thus gives it fresh life—believing as I do
that man in the distant future will be a far more perfect
creature than he now is, it is an intolerable thought that
he and all other sentient beings are doomed to complete
annihilation after such long-continued slow progress.

—Charles Darwin

ACCORDING TO NORSE LEGEND, the final day of reckoning, or
Ragnarok, the Twilight of the Gods, will be accompanied by cataclysmic upheavals. Midgard (Middle Earth) as well as the heavens
will be caught in the viselike grip of a bone-chilling frost. Piercing
winds, blinding blizzards, ruinous earthquakes, and famine will
stalk the land, as men and women perish helplessly in great numbers. Three such winters will paralyze the earth, without any relief,
while the ravenous wolves eat up the sun and the moon, plunging
the world into total darkness. The stars in the heaven will fall, the
earth will tremble, and the mountains will disintegrate. Monsters
will break free, as the god of chaos, Loki, escapes, spreading war,
confusion, and discord across the bleak land.

Odin, the father of the gods, will assemble his brave warriors for the last time in Valhalla for the final conflict. Eventually, as the gods die one by one, the evil god Surtur will breathe fire and brimstone, igniting a gigantic inferno that will engulf both heaven and earth. As the entire universe is plunged into flames, the earth sinks into the oceans, and time itself stops.

But out of the great ash, a new beginning stirs. A new earth, unlike the old, gradually rises out of the sea, as new fruits and exotic plants spring forth copiously from the fertile soil, giving birth to a new race of humans.

The Viking legend of a gigantic freeze followed by flames and a final battle presents a grim tale of the end of the world. In mythologies around the world, similar themes can be found. The end of the world is accompanied by great climactic catastrophes, usually a great fire, earthquakes, or a blizzard, followed by the final battle between good and evil. But there is also a message of hope. Out of the ashes comes renewal.

Scientists, facing the cold laws of physics, must now confront similar themes. Hard data, rather than mythology whispered around campfires, dictates how scientists view the final end of the universe. But similar themes may prevail in the scientific world. Among the solutions of Einstein's equations we also see possible futures involving freezing cold, fire, catastrophe, and an end to the universe. But will there be a final rebirth?

According to the picture emerging from the WMAP satellite, a mysterious antigravity force is accelerating the expansion of the universe. If it continues for billions or trillions of years, the universe will inevitably reach a big freeze similar to the blizzard foretelling the twilight of the gods, ending all life as we know it. This antigravity force pushing the universe apart is proportional to the volume of the universe. Thus, the larger the universe becomes, the more antigravity there is to push the galaxies apart, which in turn increases the volume of the universe. This vicious cycle repeats itself endlessly, until the universe enters a runaway mode and grows exponentially fast.

Eventually, this will mean that thirty-six galaxies in the local group of galaxies will make up the entire visible universe, as billions of neighboring galaxies speed past our event horizon. With the space between galaxies expanding faster than the speed of light, the universe will become terribly lonely. Temperatures will plunge, as the remaining energy is spread thinner and thinner across space. As temperatures drop to near absolute zero, intelligent species will have to face their ultimate fate: freezing to death.

THREE LAWS OF THERMODYNAMICS

If all the world is a stage, as Shakespeare said, then ultimately there must be an act III. In act 1, we had the big bang and the rise of life and consciousness on Earth. In act 2, perhaps we will live to explore the stars and galaxies. Finally, in act 3, we face the final death of the universe in the big freeze.

Ultimately, we find that the script must follow the laws of thermodynamics. In the nineteenth century, physicists formulated the three laws of thermodynamics which govern the physics of heat and began contemplating the eventual death of the universe. In 1854, the great German physicist Hermann von Helmholtz realized that the laws of thermodynamics could be applied to the universe as a whole, meaning that everything around us, including the stars and galaxies, would eventually have to run down.

The first law states that the total amount of matter and energy is conserved. Although energy and matter may turn into each other (via Einstein's celebrated equation $E = mc^2$), the total amount of matter and energy can never be created or destroyed.

The second law is the most mysterious and most profound. It states that the total amount of entropy (chaos or disorder) in the universe always increases. In other words, everything must eventually age and run down. The burning of forests, the rusting of machines, the fall of empires, and the aging of the human body all represent the increase of entropy in the universe. It is easy, for example, to burn a piece of paper. This represents a net increase in total chaos. However, it is impossible to reassemble the smoke back into paper. (Entropy can be made to decrease with the addition of mechanical work, as in a refrigerator, but only in a small local neighborhood; the total entropy for the entire system—the refrigerator plus all its surroundings—always increases.)

Arthur Eddington once said about the second law: "The law that entropy always increases—the Second Law of Thermodynamics—holds, I think, the supreme position among the laws of Nature . . . If your theory is found to be against the Second Law of Thermodynamics, I can give you no hope; there is nothing for it but to collapse in deepest humiliation."

(At first, it seems as if the existence of complex life forms on Earth violates the second law. It seems remarkable that out of the chaos of the early Earth emerged an incredible diversity of intricate life forms, even harboring intelligence and consciousness, lowering the amount of entropy. Some have taken this miracle to imply the hand of a benevolent creator. But remember that life is driven by the natural laws of evolution, and that total entropy still increases, because additional energy fueling life is constantly being added by the Sun. If we include the Sun and Earth, then the total entropy still increases.)

The third law states that no refrigerator can reach absolute zero. One may come within a tiny fraction of a degree above absolute zero, but you can never reach a state of zero motion. (And if we incorporate the quantum principle, this implies that molecules will always have a small amount of energy, since zero energy implies that we know the exact location and velocity of each molecule, which would violate the uncertainty principle.)

If the second law is applied to the entire universe, it means that the universe will eventually run down. The stars will exhaust their nuclear fuel, galaxies will cease to illuminate the heavens, and the universe will be left as a lifeless collection of dead dwarf stars, neutron stars, and black holes. The universe will be plunged in eternal darkness.

Some cosmologists have tried to evade this "heat death" by appealing to an oscillating universe. Entropy would increase continually as the universe expanded and eventually contracted. But after the big crunch, it is not clear what would become of the entropy in the universe. Some have entertained the idea that perhaps the universe might simply repeat itself exactly in the next cycle. More realistic is the possibility that the entropy would be carried over to the next cycle, which means that the lifetime of the universe would gradually lengthen for each cycle. But no matter how one looks at the question, the oscillating universe, like the open and closed universes, will eventually result in the destruction of all intelligent life.

THE BIG CRUNCH

One of the first attempts to apply physics to explain the end of the universe was a paper written in 1969 by Sir Martin Rees entitled, "The Collapse of the Universe: An Eschatological Study." Back then, the value of Omega was still largely unknown, so he assumed it was two, meaning that the universe would eventually stop expanding and die in a big crunch rather than a big freeze.

He calculated that the expansion of the universe will eventually grind to a halt, when the galaxies are twice as far away as they are today, when gravity finally overcomes the original expansion of the universe. The redshift we see in the heavens will become a blueshift, as the galaxies begin to race toward us.

In this version, about 50 billion years from now, catastrophic events will take place, signaling the final death throes of the universe. One hundred million years before the final crunch, the galaxies in the universe, including our own Milky Way galaxy, will begin to collide with each other and eventually merge. Oddly, Rees discovered that individual stars will dissolve even before they began to collide with each other, for two reasons. First, the radiation from the other stars in the heavens will gain energy as the universe contracts; thus, the stars will be bathed in the blistering blueshifted light of other stars. Second, the temperature of the background microwave radiation will be vastly increased as the temperature of the universe skyrockets. The combination of these two effects will create temperatures that exceed the surface temperature of the stars, which will absorb heat faster than they can get rid of it. In other words, the stars will probably disintegrate and disperse into super-hot gas clouds.

Intelligent life, under these circumstances, would inevitably perish, seared by the cosmic heat pouring in from the nearby stars and galaxies. There is no escape. As Freeman Dyson has written, "Regrettably I have to concur that in this case we have no escape from frying. No matter how deep we burrow into the Earth to shield ourselves from blue-shifted background radiation, we can only postpone by a few million years our miserable end."

If the universe is headed for a big crunch, then the remaining question is whether the universe might collapse and then rebound, as in the oscillating universe. This is the scenario adopted in Poul Anderson's novel *Tau Zero*. If the universe were Newtonian, this might be possible, if there was sufficient sideways motion as the galaxies were compressed into each other. In this case, the stars might not be squeezed into a single point but might miss each other at the point of maximum compression and then rebound, without colliding with each other.

The universe, however, is not Newtonian; it obeys Einstein's equations. Roger Penrose and Stephen Hawking have shown that, under very general circumstances, a collapsing collection of galaxies will necessarily be squeezed down to a singularity. (This is because the sideways motion of the galaxies contains energy and hence interacts with gravity. Thus, the gravitational pull in Einstein's theory is much greater than that found in Newtonian theory for collapsing universes, and the universe collapses into a single point.)

FIVE STAGES OF THE UNIVERSE

Recent data from the WMAP satellite, however, favors the big freeze. To analyze the life history of the universe, scientists like Fred Adams and Greg Laughlin of the University of Michigan have tried to divide up the age of the universe into five distinct states. Since we are discussing truly astronomical time scales, we will adopt a logarithmic time frame. Thus, 10^{20} years will be represented as 20. (This timetable was drawn up before the implications of an accelerating universe were fully appreciated. But the general breakdown of the stages of the universe remains the same.)

The question that haunts us is: can intelligent life use its ingenuity to survive in some form through these stages, through a series of natural catastrophes and even the death of the universe?

Stage 1: Primordial Era

In the first stage (between -50 and 5, or between 10^{-50} and 10^5 seconds), the universe underwent rapid expansion but also rapid cooling. As it cooled, the various forces, which were once united into a master "superforce," gradually broke apart, yielding the familiar four forces of today. Gravity broke off first, then the strong nuclear force, and finally the weak nuclear force. At first, the universe was opaque and the sky was white, since light was absorbed soon after it was created. But 380,000 years after the big bang, the universe cooled enough for atoms to form without being smashed apart by the intense heat. The sky turned black. The microwave background radiation dates back to this period.

During this era, primordial hydrogen fused into helium, creating the current mixture of stellar fuel that has spread throughout the universe. At this stage of the evolution of the universe, life as we know it was impossible. The heat was too intense; any DNA or other autocatalytic molecules that were formed would have been burst apart by random collisions with other atoms, making the stable chemicals of life impossible.

Stage 2: Stelliferous Era

Today, we live in stage 2 (between 6 and 14, or between 10^6 and 10^{14} seconds), when hydrogen gas has been compressed and stars have ignited, lighting up the heavens. In this era, we find hydrogen-rich stars that blaze away for billions of years until they exhaust their nuclear fuels. The Hubble space telescope has photographed stars in all their stages of evolution, including young stars surrounded by a swirling disk of dust and debris, probably the predecessor to planets and a solar system.

In this stage, the conditions are ideal for the creation of DNA and life. Given the enormous number of stars in the visible universe, astronomers have tried to give plausible arguments, based on the known laws of science, for the rise of intelligent life on other planetary systems. But any intelligent life form will have to face a number of cosmic hurdles, many of its own making, such as environmental pollution, global warming, and nuclear weapons. Assuming that intelligent life has not destroyed itself, then it must face a daunting series of natural disasters, any one of which may end in catastrophe.

On a time scale of tens of thousands of years, there may be an ice age, similar to the one that buried North America under almost a mile of ice, making human civilization impossible. Before ten thousand years ago, humans lived like wolves in packs, foraging for scraps of food in small, isolated tribes. There was no accumulation of knowledge or science. There was no written word. Humanity was preoccupied with one goal: survival. Then, for reasons we still do not understand, the Ice Age ended, and humans began the rapid rise from the ice to the stars. However, this brief interglacial period cannot last forever. Perhaps in another ten thousand years, another Ice Age will blanket most of the world. Geologists believe that the effects of tiny variations in Earth's spin around its axis eventually build up, allowing the jet stream from the ice caps to descend to lower latitudes, blanketing Earth in freezing ice. At that point, we might have to go underground to keep warm. Earth was once completely covered in ice. This might happen again.

On a time scale of thousands to millions of years, we must prepare for meteor and comet impacts. Most likely a meteor or comet impact destroyed the dinosaurs 65 million years ago. Scientists believe that an extraterrestrial object, perhaps less than 10 miles across, plowed into the Yucatan Peninsula of Mexico, gouging out a crater 180 miles across and shooting enough debris into the atmosphere to cut off sunlight and darken Earth, causing freezing temperatures that killed off vegetation and the dominant life form on Earth at that time, the dinosaurs. Within less than a year, the dinosaurs and most of the species on Earth perished.

Judging by the rate of past impacts, there is a 1 in 100,000 chance over the next fifty years of an asteroid impact that would cause worldwide damage. The chance of a major impact over millions of years probably grows to nearly 100 percent.

(In the inner solar system, where Earth resides, there are perhaps 1,000 to 1,500 asteroids that are a kilometer across or greater, and a million asteroids 50 meters across or larger. Asteroid observations pour into the Smithsonian Astrophysical Observatory in Cambridge at the rate of about fifteen thousand per day. Fortunately, only forty-two known asteroids have a small but finite probability of impacting with Earth. In the past, there have been a number of false alarms concerning these asteroids, the most famous involving the asteroid 1997XF11, which astronomers mistakenly said might hit Earth in thirty years, generating worldwide headlines. But by carefully examining the orbit of one asteroid called 1950DA, scientists have calculated that there is only a tiny—but nonzero—probability that it may hit Earth on March 16, 2880. Computer simulations done at the University of California at Santa Cruz show that, if this asteroid hits the oceans, it will create a tidal wave 400 feet tall, which would swamp most of the coastal areas in devastating floods.)

On a scale of billions of years, we have to worry about the Sun swallowing up Earth. The Sun is already 30 percent hotter today than it was in its infancy. Computer studies have shown that, in 3.5 billion years, the Sun will be 40 percent brighter than it is today, meaning that Earth will gradually heat up. The Sun will appear larger and larger in the day sky, until it fills up most of the sky from horizon to horizon. In the short term, living creatures, desperately trying to escape the scorching heat of the Sun, may be forced back into the oceans, reversing the historic march of evolution on this planet. Eventually, the oceans themselves will boil, making life as we know it impossible. In about 5 billion years, the Sun's core will exhaust its supply of hydrogen gas and mutate into a red giant star. Some red giants are so large that they could gobble up Mars if they were located at the position of our Sun. However, our Sun will probably expand only to the size of the orbit of Earth, devouring Mercury and Venus and melting the mountains of Earth. So it is likely our Earth will die in fire, rather than ice, leaving a burnt-out cinder orbiting the Sun.

Some physicists have argued that before this occurs, we should be able to use advanced technology to move Earth to a larger orbit around the Sun, if we haven't already migrated from Earth to other planets in gigantic space arks. "As long as people get smarter faster than the Sun gets brighter, the Earth should thrive," remarks astronomer and writer Ken Croswell.

Scientists have proposed several ways to move Earth from its current orbit around the Sun. One simple way would be to carefully divert a series of asteroids from the asteroid belt so that they whip around Earth. This slingshot effect would give a boost to Earth's orbit, increasing its distance from the Sun. Each boost would move Earth only incrementally, but there would be plenty of time to divert hundreds of asteroids to accomplish this feat. "During the several billion years before the Sun bloats into a red giant, our descendants could snare a passing star into an orbit around the Sun, then jettison the Earth from its solar orbit into an orbit around the new star," adds Croswell.

Our Sun will suffer a different fate from Earth; it will die in ice, rather than fire. Eventually, after burning helium for 700 million years as a red giant, the Sun will exhaust most of its nuclear fuel, and gravity will compress it into a white dwarf about the size of Earth. Our Sun is too small to undergo the catastrophe called a supernova and turn into a black hole. After our Sun turns into a white dwarf star, eventually it will cool down, thereby glowing a faint red color, then brown, and finally black. It will drift in the cosmic void as a piece of dead nuclear ash. The future of almost all the atoms we see around us, including the atoms of our bodies and our loved ones, is to wind up on a burnt-out cinder orbiting a black dwarf star. Because this dwarf star will weigh only 0.55 solar masses, what's left of Earth will settle into an orbit about 70 percent farther out than it is today.

On this scale, we see that the blossoming of plants and animals on Earth will only last a mere billion years (and we are halfway through this golden era today). "Mother Nature wasn't designed to make us happy," says astronomer Donald Brownlee. Compared to the life span of the entire universe, the flowering of life lasts only the briefest instant of time.

Stage 3: Degenerate Era

In stage 3 (between 15 and 39), the energy of the stars in the universe will finally be exhausted. The seemingly eternal process of burning hydrogen and then helium finally comes to a halt, leaving behind lifeless hunks of dead nuclear matter in the form of dwarf stars, neutron stars, and black holes. The stars in the sky cease to shine; the universe is gradually plunged into darkness.

Temperatures will fall dramatically in stage 3, as stars lose their nuclear engines. Any planet circling around a dead star will freeze. Assuming that Earth is still intact, what is left of its surface will become a frozen sheet of ice, forcing intelligent life forms to seek a new home.

While giant stars may last for a few million years and hydrogen-burning stars like our Sun for billions of years, tiny red dwarf stars may actually burn for trillions of years. This is why attempting to relocate the orbit of Earth around a red dwarf star in theory makes sense. The closest stellar neighbor to Earth, Promixa Centauri, is a red dwarf star that is only 4.3 light-years from Earth. Our closest neighbor weighs only 15 percent of the Sun's mass and is four hundred times dimmer than the Sun, so any planet orbiting it would have to be extremely close to benefit from its faint starlight. Earth would have to orbit this star twenty times closer than it currently is from the Sun to receive the same amount of sunlight. But once in orbit around a red dwarf star, a planet would have energy to last for trillions of years.

Eventually, the only stars that will continue to burn nuclear fuel will be the red dwarfs. In time, however, even they will turn dark. In a hundred trillion years, the remaining red dwarfs will finally expire.

Stage 4: Black Hole Era

In stage 4 (between 40 to 100), the only source of energy will be the slow evaporation of energy from black holes. As shown by Jacob Bekenstein and Stephen Hawking, black holes are not really black; they actually radiate a faint amount of energy, called evaporation. (In practice, this black hole evaporation is too small to be observed experimentally, but on long time scales evaporation ultimately determines the fate of a black hole.)

Evaporating black holes can have various lifetimes. A mini–black hole the size of a proton might radiate 10 billion watts of power for the lifetime of the solar system. A black hole weighing as much as the Sun will evaporate in 10^{66} years. A black hole weighing as much as a galactic cluster will evaporate in 10^{117} years. However, as a black hole nears the end of its lifespan, after slowly oozing out radiation it suddenly explodes. It's possible that intelligent life, like homeless people huddled next to the dying embers of dim fires, will congregate around the faint heat emitted from evaporating black holes to extract a bit of warmth from them, until they evaporate.

Stage 5: Dark Era

In stage 5 (beyond 101), we enter the dark era of the universe, when all heat sources are finally exhausted. In this stage, the universe drifts slowly toward the ultimate heat death, as the temperature approaches absolute zero. At this point, the atoms themselves almost come to a halt. Perhaps even the protons themselves will have decayed, leaving a drifting sea of photons and a thin soup of weakly interacting particles (neutrinos, electrons, and their antiparticle, the positron). The universe may consist of a new type of "atom" called positronium, consisting of electrons and positrons that circulate around each other.

Some physicists have speculated that these "atoms" of electrons and antielectrons might be able to form new building blocks for intelligent life in this dark era. However, the difficulties facing this idea are formidable. An atom of positronium is comparable in size to an ordinary atom. But an atom of positronium in the dark era would be about 10^{12} megaparsecs across, millions of times larger than the observable universe of today. So in this dark era, while these "atoms" may form, they would be the size of an entire universe. Since the universe during the dark era will have expanded to enormous distances, it would easily be able to accommodate these gigantic atoms of positronium. But since these positronium atoms are so large, it means that any "chemistry" involving these "atoms" would be on colossal time scales totally different from anything we know.

As cosmologist Tony Rothman writes, "And so, finally, after 10^{117} years, the cosmos will consist of a few electrons and positrons locked in their ponderous orbits, neutrinos and photons left over from baryon decay, and stray protons remaining from positronium annihilation and black holes. For this too is written in the Book of Destiny."

CAN INTELLIGENCE SURVIVE?

Given the mind-numbing conditions found at the end of the big freeze, scientists have debated whether any intelligent life form can possibly survive. At first, it seems pointless to discuss intelligent life surviving in stage 5, when temperatures plunge to near absolute zero. However, there is actually a spirited debate among physicists about whether intelligent life can survive.

The debate centers upon two key questions. The first is: can intelligent beings operate their machines when temperatures approach absolute zero? By the laws of thermodynamics, because energy flows from a higher temperature to a lower temperature, this movement of energy can be used to do usable mechanical work. For example, mechanical work can be extracted by a heat engine that connects two regions at different temperatures. The greater the difference in temperature, the greater the efficiency of the engine. This is the basis of the machines that powered the Industrial Revolution, such as the steam engine and the locomotive. At first, it seems impossible to extract any work from a heat engine in stage 5, since all temperatures will be the same.

The second question is: can an intelligent life form send and receive information? According to information theory, the smallest unit that can be sent and received is proportional to the temperature. As the temperature drops to near absolute zero, the ability to process information is also severely impaired. Bits of information that can be transmitted as the universe cools will have to be smaller and smaller.

Physicist Freeman Dyson and others have reanalyzed the physics of intelligent life coping in a dying universe. Can ingenious ways, they ask, be found for intelligent life to survive even as temperatures drop near absolute zero?

As the temperature begins to drop throughout the universe, at first creatures may try to lower their body temperature using genetic engineering. This way, they could be much more efficient in using the dwindling energy supply. But eventually, body temperatures will reach the freezing point of water. At this time, intelligent beings may have to abandon their frail bodies of flesh and blood and assume robotic bodies. Mechanical bodies can withstand the cold much better than flesh. But machines also must obey the laws of information theory and thermodynamics, making life extremely difficult, even for robots.

Even if intelligent creatures abandon their robotic bodies and transform themselves into pure consciousness, there is still the problem of information processing. As the temperature continues to fall, the only way to survive will be to "think" slower. Dyson concludes that an ingenious life form would still be able to think for an indefinite amount of time by spreading out the time required for information processing and also by hibernating to conserve energy. Although the physical time necessary to think and process information may be spread out over billions of years, the "subjective time," as seen by the intelligent creatures themselves, will remain the same. They will never notice the difference. They will still be able to think deep thoughts but only on a much, much slower time scale. Dyson concludes, on a strange but optimistic note, that in this manner, intelligent life will be able to process information and "think" indefinitely. Processing a single thought may take trillions of years, but with respect to "subjective time," thinking will proceed normally.

But if intelligent creatures think slower, perhaps they might witness cosmic quantum transitions taking place in the universe. Normally, such cosmic transitions, such as the creation of baby universes or the transition to another quantum universe, take place over trillions of years and hence are purely theoretical. In stage 5, however, trillions of years in "subjective time" will be compressed and may appear to be only a few seconds to these creatures; they will think so slowly that they might see bizarre quantum events happen all the time. They might regularly see bubble universes appearing out of nowhere or quantum leaps into alternate universes.

But in light of the recent discovery that the universe is accelerating, physicists have reexamined the work of Dyson and have ignited a new debate, reaching the opposite conclusions—intelligent life will necessarily perish in an accelerating universe. Physicists Lawrence Krauss and Glenn Starkman have concluded, "Billions of years ago the universe was too hot for life to exist. Countless eons hence, it will become so cold and empty that life, no matter how ingenious, will perish."

In Dyson's original work, he assumed that the 2.7-degree microwave radiation in the universe would continue to drop indefinitely, so intelligent beings might extract usable work from these tiny temperature differences. As long as the temperature continued to drop, usable work could always be extracted. However, Krauss and Stackman point out that if the universe has a cosmological constant, then temperatures will not drop forever, as Dyson had assumed, but will eventually hit a lower limit, the Gibbons-Hawking temperature (about 10^{-29} degrees). Once this temperature is reached, the temperature throughout the universe will be the same, and hence intelligent beings will not be able to extract usable energy by exploiting temperature differences. Once the entire universe reaches a uniform temperature, all information processing will cease.

(In the 1980s, it was found that certain quantum systems, such as the Browning motion in a fluid, can serve as the basis of a computer, regardless of how cold the temperature is outside. So even as temperatures plunge, these computers can still compute by using less and less energy. This was good news to Dyson. But there was a catch. The system must satisfy two conditions: it must remain in equilibrium with its environment, and it must never discard information. But if the universe expands, equilibrium is impossible, because radiation gets diluted and stretched in its wavelength. An accelerating universe changes too rapidly for the system to reach equilibrium. And second, the requirement that it never discard information means that an intelligent being must never forget. Eventually, an intelligent being, unable to discard old memories, might find itself reliving old memories over and over again. "Eternity would be a prison, rather than an endlessly receding horizon of creativity and exploration. It might be nirvana, but would it be living?" Krauss and Starkman ask.)

In summary, we see that if the cosmological constant is close to zero, intelligent life can "think" indefinitely as the universe cools by hibernating and thinking slower. But in an accelerating universe such as ours, this is impossible. All intelligent life is doomed to perish, according the laws of physics.

From the vantage point of this cosmic perspective, we see therefore that the conditions for life as we know it are but a fleeting episode in a much larger tapestry. There is only a tiny window where the temperatures are "just right" to support life, neither too hot nor too cold.

LEAVING THE UNIVERSE

Death can be defined as the final cessation of all information processing. Any intelligent species in the universe, as it begins to understand the fundamental laws of physics, will be forced to confront the ultimate death of the universe and any intelligent life it may contain.

Fortunately, there is ample time to assemble the energy for such a journey, and there are alternatives, as we will see in the next chapter. The question we will explore is: do the laws of physics allow for our escape into a parallel universe?

CHAPTER ELEVEN

Escaping the Universe

Any sufficiently advanced technology is indistinguishable from magic.

—Arthur C. Clarke

I n THE NOVEL *Eon*, the science fiction author Greg Bear writes a harrowing tale about fleeing a devastated world into a parallel universe. A colossal, menacing asteroid from space has approached the planet Earth, causing mass panic and hysteria. However, instead of striking Earth, it strangely settles into an orbit around the planet. Teams of scientists are sent into space to investigate. However, instead of finding a desolate, lifeless surface, they find that the asteroid is actually hollow; it's a huge spaceship abandoned by a superior technological race. Inside the deserted spaceship, the book's heroine, a theoretical physicist named Patricia Vasquez, finds seven vast chambers, entrances to different worlds, with lakes, forests, trees, even entire cities. Next, she stumbles upon huge libraries containing the complete history of these strange people.

Picking up an old book, she finds that it is *Tom Sawyer*, by Mark Twain, but republished in 2110. She realizes that the asteroid is not from an alien civilization at all, but from Earth itself, 1,300 years in the future. She realizes the sickening truth: these old records tell of an ancient nuclear war that erupted in the distant past, killing billions of people, unleashing a nuclear winter that killed billions more. When she determines the date of this nuclear war, she is shocked to find that it is only two weeks into the future! She is helpless to stop the inevitable war that will soon consume the entire planet, killing her loved ones.

Eerily, she locates her own personal history in these old records, and finds that her future research in space-time will help to lay the groundwork for a vast tunnel in the asteroid, called the Way, which will allow the people to leave the asteroid and enter other universes. Her theories have proved that there are an infinite number of quantum universes, representing all possible realities. Moreover, her theories make possible the building of gateways located along the Way for entering these universes, each with a different alternate history. Eventually, she enters the tunnel, travels down the Way, and meets the people who fled in the asteroid, her descendants.

It is a strange world. Centuries before, people had abandoned strictly human form and can now assume various shapes and bodies. Even people long dead have their memories and personalities stored in computer banks and can be brought back to life. They can be resurrected and downloaded several times into new bodies. Implants placed in their bodies give them access to nearly infinite information. Although these people can have almost anything they wish, nonetheless our heroine is miserable and lonely in this technological paradise. She misses her family, her boyfriend, her Earth, all of which were destroyed in the nuclear war. She is eventually granted permission to scan the multiple universes that lie along the Way to find a parallel Earth in which nuclear war was averted and her loved ones are still alive. She eventually finds one and leaps into it. (Unfortunately, she makes a tiny mathematical error; she winds up in a universe in which the Egyptian empire never fell. She spends the rest of her days trying to leave this parallel Earth to find her true home.)

Although the dimensional gateway discussed in *Eon* is purely fictional, it raises an interesting question that relates to us: could one find haven in a parallel universe if conditions in our own universe became intolerable?

The eventual disintegration of our universe into a lifeless mist of electrons, neutrinos, and photons seems to foretell the ultimate doom of all intelligent life. On a cosmic scale, we see how fragile and transitory life is. The era when life is able to flourish is concentrated in a very narrow band, a fleeting period in the life of the stars that light up the night sky. It seems impossible for life to continue as the universe ages and cools. The laws of physics and thermodynamics are quite clear: if the expansion of the universe continues to accelerate in a runaway mode, intelligence as we know it cannot ultimately survive. But as the temperature of the universe continues to drop over the eons, can an advanced civilization try to save itself? By marshaling all its technology, and the technology of any other civilizations that may exist in the universe, can it escape the inevitability of the big freeze?

Because the rate at which the stages of the universe evolve is measured in billions to trillions of years, there is plenty of time for an industrious, clever civilization to attempt to meet these challenges. Although it is sheer speculation to imagine what kinds of technologies an advanced civilization may devise to prolong its existence, one can use the known laws of physics to discuss the broad options that may be available to them billions of years from now. Physics cannot tell us what specific plans an advanced civilization may adopt, but it might tell us what the range of parameters are for such an escape.

To an engineer, the main problem in leaving the universe is whether we have sufficient resources to build a machine that can perform such a difficult feat. But to a physicist, the main problem is different: whether the laws of physics allow for the existence of these machines in the first place. Physicists want a "proof of principle"—we want to show that, if you had sufficiently advanced technology, an escape into another universe would be possible according to the laws of physics. Whether we have sufficient resources is a lesser, practical detail that has to be left for civilizations billions of years in the future that are facing the big freeze.

According to Astronomer Royal Sir Martin Rees, "Wormholes, extra dimensions, and quantum computers open up speculative scenarios that could transform our entire universe eventually into a 'living cosmos.' "

TYPE I, II, AND III CIVILIZATIONS

To understand the technology of civilizations thousands to millions of years ahead of ours, physicists sometimes classify civilizations depending on their consumption of energy and the laws of thermo-dynamics. When scanning the heavens for signs of intelligent life, physicists do not look for little green men but for civilizations with the energy output of type I, II, and III civilizations. The ranking was introduced by Russian physicist Nikolai Kardashev in the 1960s for classifying the radio signals from possible civilizations in outer space. Each civilization type emits a characteristic form of radiation that can be measured and cataloged. (Even an advanced civilization that tries to conceal its presence can be detected by our instruments. By the second law of thermodynamics, any advanced civilization will create entropy in the form of waste heat that will inevitably drift into outer space. Even if they try to mask their presence, it is impossible to hide the faint glow created by their entropy.)

A type I civilization is one that has harnessed planetary forms of energy. Their energy consumption can be precisely measured: by def-inition, they are able to utilize the entire amount of solar energy striking their planet, or 10^{16} watts. With this planetary energy, they might control or modify the weather, change the course of hurri-canes, or build cities on the ocean. Such civilizations are truly mas-ters of their planet and have created a planetary civilization.

A type II civilization has exhausted the power of a single planet and has harnessed the power of an entire star, or approximately 10^{26} watts. They are able to consume the entire energy output of their star and might conceivably control solar flares and ignite other stars.

A type III civilization has exhausted the power of a single solar system and has colonized large portions of its home galaxy. Such a civilization is able to utilize the energy from 10 billion stars, or approximately 10^{36} watts.

Each type of civilization differs from the next lower type by a factor of 10 billion. Hence, a type III civilization, harnessing the power of billions of star systems, can use 10 billion times the energy output of a type II civilization, which in turn harnesses 10 billion times the output of a type I civilization. Although the gap separating these civilizations may seem astronomical, it is possible to estimate the time it might take to achieve a type III civilization. Assume that a civilization grows at a modest rate of 2 to 3 percent in its energy output per year. (This is a plausible assumption, since economic growth, which can be reasonably calculated, is directly related to energy consumption. The larger the economy, the greater its energy demands. Since the growth of the gross domestic product, or GDP, of many nations lies within 1 to 2 percent per year, we can expect its energy consumption to grow at roughly the same rate.)

At this modest rate, we can estimate that our current civilization is approximately 100 to 200 years from attaining type I status. It will take us roughly 1,000 to 5,000 years to achieve type II status, and perhaps 100,000 to 1,000,000 years to achieve type III status. On such a scale, our civilization today may be classified as a type 0 civilization, because we obtain our energy from dead plants (oil and coal). Even controlling a hurricane, which can unleash the power of hundreds of nuclear weapons, is beyond our technology.

To describe our present-day civilization, astronomer Carl Sagan advocated creating finer gradations between the civilization types. Type I, II, and III civilizations, we have seen, generate a total energy output of roughly 10^{16}, 10^{26}, and 10^{36} watts, respectively. Sagan introduced a type I.1 civilization, for example, which generates 10^{17} watts of power, a type I.2 civilization, which generates 10^{18} watts of power, and so on. By dividing each type I into ten smaller subtypes, we can begin to classify our own civilization. On this scale, our present civilization is more like a type 0.7 civilization—within striking distance of being truly planetary. (A type 0.7 civilization is still a thousand times smaller than a type I, in terms of energy production.)

Although our civilization is still quite primitive, we already see signs of a transition taking place. When I gaze at the headlines, I constantly see reminders of this historic evolution. In fact, I feel privileged to be alive to witness it:

* The Internet is an emerging type I telephone system. It has the capability of becoming the basis of a universal planetary communication network.

* The economy of the type I society will be dominated not by nations but by large trading blocs resembling the European Union, which itself was formed because of competition from NAFTA (the countries of North America).

* The language of our type I society will probably be English, which is already the dominant second language on Earth. In many third-world countries today, the upper classes and college educated tend to speak both English and the local language. The entire population of a type I civilization may be bilingual in this fashion, speaking both a local language and a planetary language.

* Nations, although they will probably exist in some form for centuries to come, will become less important, as trade barriers fall and as the world becomes more economically interdependent. (Modern nations, in part, were originally carved out by capitalists and those who wanted a uniform currency, borders, taxes, and laws with which to conduct business. As business itself becomes more international, national borders should become less relevant.) No single nation is powerful enough to stop this march to a type I civilization.

* Wars will probably always be with us, but the nature of war will change with the emergence of a planetary middle class more interested in tourism and the accumulation of wealth and resources than in overpowering other peoples and controlling markets or geographical regions.

* Pollution will increasingly be tackled on a planetary scale. Greenhouse gases, acid rain, burning rain forests, and such respect no national boundaries, and there will be pressure from

neighboring nations for offending entities to clean up their act. Global environmental problems will help to accelerate global solutions.

* As resources (such as fish harvests, grain harvests, water resources) gradually flatten out due to overcultivation and overconsumption, there will be increased pressure to manage our resources on a global scale or else face famine and collapse.

* Information will be almost free, encouraging society to be much more democratic, allowing the disenfranchised to gain a new voice, and putting pressure on dictatorships.

These forces are beyond the control of any single individual or nation. The Internet cannot be outlawed. In fact, any such move would be met more with laughter than with horror, because the Internet is the road to economic prosperity and science as well as culture and entertainment.

But the transition from type 0 to type I is also the most perilous, because we still demonstrate the savagery that typified our rise from the forest. In some sense, the advancement of our civilization is a race against time. On one hand, the march toward a type I planetary civilization may promise us an era of unparalleled peace and prosperity. On the other hand, the forces of entropy (the greenhouse effect, pollution, nuclear war, fundamentalism, disease) may yet tear us apart. Sir Martin Rees sees these threats, as well as those due to terrorism, bioengineered germs, and other technological nightmares, as some of the greatest challenges facing humanity. It is sobering that he gives us only a fifty-fifty chance of successfully negotiating this challenge.

This may be one of the reasons we don't see extraterrestrial civilizations in space. If they indeed exist, perhaps they are so advanced that they see little interest in our primitive type 0.7 society. Alternatively, perhaps they were devoured by war or killed off by their own pollution, as they strived to reach type I status. (In this sense, the generation now alive may be one of the most important generations ever to walk the surface of Earth; it may well decide if we safely make the transition to a type I civilization.)

But as Friedrich Nietzsche once said, what does not kill us makes us stronger. Our painful transition from type 0 to type I will surely be a trial by fire, with a number of harrowing close calls. If we can emerge from this challenge successfully, we will be stronger, in the same way that hammering molten steel serves to temper it.

TYPE I CIVILIZATION

When a civilization reaches type I status, it is unlikely to immediately reach for the stars; it is more likely to stay on the home planet for centuries, long enough to resolve the remaining nationalistic, fundamentalist, racial, and sectarian passions of its past. Science fiction writers frequently underestimate the difficulty of space travel and space colonization. Today, it costs $10,000 to $40,000 per pound to put anything into near-Earth orbit. (Imagine John Glenn made out of solid gold, and you begin to appreciate the steep cost of space travel.) Each space shuttle mission costs upward of $800 million (if we take the total cost for the space shuttle program and divide by the number of missions). It is likely that the cost of space travel will go down, but only by a factor of 10 in the next several decades, with the arrival of reusable launch vehicles (RLVs) which can be reused immediately after a mission is complete. Through most of the twenty-first century, space travel will remain a prohibitively expensive proposition except for the wealthiest individuals and nations.

(There is one possible exception to this: the development of "space elevators." Recent advances in nanotechnology make possible the production of threads made of superstrong and superlightweight carbon nanotubes. In principle, it is possible that these threads of carbon atoms could prove strong enough to connect Earth with a geosynchronous satellite orbiting more than 20,000 miles above Earth. Like Jack and the Beanstalk, one might be able to ascend this carbon nanotube to reach outer space for a fraction of the usual cost. Historically, space scientists dismissed space elevators because the tension on the string would be enough to break any known fiber.

However, carbon nanotube technology may change this. NASA is funding preliminary studies on this technology, and the situation will be closely analyzed over the years. But should such a technology prove possible, a space elevator could at best only take us into orbit around Earth, not to the other planets.)

The dream of space colonies must be tempered by the fact that the cost of manned missions to the Moon and the planets is many times the cost of near-Earth missions. Unlike the Earth-bound voyages of Columbus and the early Spanish explorers centuries ago, where the cost of a ship was a tiny fraction of the gross domestic product of Spain and where the potential economic rewards were huge, the establishment of colonies on the Moon and Mars would bankrupt most nations, while conferring almost no direct economic benefits. A simple manned mission to Mars could cost anywhere from $100 billion to $500 billion, with little to show for it financially in return.

Similarly, one also has to consider the danger to the human passengers. After half a century of experience with liquid-fueled rockets, the chances of a catastrophic failure involving rocket missions are about one in seventy. (In fact, the two tragic losses of the space shuttle fall within this ratio.) Space travel, we often forget, is different from tourism. With so much volatile fuel and so many hostile threats to human life, space travel will continue to be a risky proposition for decades to come.

On a scale of several centuries, however, the situation may gradually change. As the cost of space travel continues its slow decline, a few space colonies may gradually take hold on Mars. On this time scale, some scientists have even proposed ingenious mechanisms to terraform Mars, such as deflecting a comet and letting it vaporize in the atmosphere, thereby adding water vapor to the atmosphere. Others have advocated injecting methane gas into the atmosphere to create an artificial greenhouse effect on the red planet, raising temperatures and gradually melting the permafrost under the surface of Mars, thereby filling its lakes and streams for the first time in billions of years. Some have proposed more extreme, dangerous measures, such as detonating an underground nuclear warhead beneath the ice caps to melt the ice (which could pose a health hazard for space colonists of the future). But these suggestions are still wildly speculative.

More likely, a type I civilization will find space colonies a distant priority in the next few centuries. But for long-distance interplanetary missions, where time is not so pressing, the development of a solar/ion engine may offer a new form of propulsion between the stars. Such slow-moving engines would generate little thrust, but they can maintain that thrust for years at a time. These engines concentrate solar energy from the sun, heat up a gas like cesium, and then hurl the gas out the exhaust, giving a mild thrust that can be maintained almost indefinitely. Vehicles powered by such engines might be ideal for creating an interplanetary "interstate highway system" connecting the planets.

Eventually, type I civilizations might send a few experimental probes to nearby stars. Since the speed of chemical rockets is ultimately limited by the maximum speed of the gases in the rocket exhaust, physicists will have to find more exotic forms of propulsion if they hope to reach distances that are hundreds of light-years away. One possible design would be to create a fusion ramjet, a rocket that scoops hydrogen from interstellar space and fuses it, releasing unlimited amounts of energy in the process. However, proton-proton fusion is quite difficult to attain even on Earth, let alone in outer space in a starship. Such technology is at best another century in the future.

TYPE II CIVILIZATION

A type II civilization able to harness the power of an entire star might resemble a version of the Federation of Planets in the *Star Trek* series, without the warp drive. They have colonized a tiny fraction of the Milky Way galaxy and can ignite stars, and hence they qualify for an emerging type II status.

To fully utilize the output of the Sun, physicist Freeman Dyson has speculated that a type II civilization might build a gigantic sphere around the Sun to absorb its rays. This civilization might, for example, be able to deconstruct a planet the size of Jupiter and distribute the mass in a sphere around the Sun. Because of the second law of thermodynamics, the sphere would eventually heat up, giving off a characteristic infrared radiation that could be seen from outer space. Jun Jugaku of the Research Institute of Civilization in Japan and his colleagues have searched the heavens out to 80 light-years to try to locate other such civilizations and have found no evidence of these infrared emissions (although remember that our galaxy is 100,000 light-years across).

A type II civilization might colonize some of the planets in their solar system and even embark upon a program to develop interstellar travel. Because of the vast resources available to a type II civilization, they potentially might have developed such exotic forms of propulsion as an antimatter/matter drive for their starships, making possible travel near the speed of light. In principle, this form of energy is 100 percent energy-efficient. It is also experimentally possible but prohibitively expensive by type I standards (it takes an atom smasher to create beams of antiprotons that can be used to create antiatoms).

We can only speculate about how a type II society might function. However, it will have millennia to sort out disputes over property, resources, and power. A type II civilization could potentially be immortal. It is likely that nothing known to science could destroy such a civilization, except perhaps the folly of the inhabitants themselves. Comets and meteors could be deflected, ice ages could be diverted by changing the weather patterns, even the threat posed by a nearby supernova explosion could be avoided simply by abandoning the home planet and transporting the civilization out of harm's way—or even potentially by tampering with the thermonuclear engine of the dying star itself.

TYPE III CIVILIZATION

By the time a society reaches the level of a type III civilization, it may begin to contemplate the fantastic energies at which space and time become unstable. We recall that the Planck energy is the energy at which quantum effects dominate, and space-time becomes "foamy" with tiny bubbles and wormholes. The Planck energy is well beyond our reach today, but that is only because we judge energy from the point of view of a type 0.7 civilization. By the time a civilization has reached type III status, it will have access (by definition) to energies 10 billion times 10 billion (or 10^{20}) those found on Earth today.

Astronomer Ian Crawford of the University College in London, writes about type III civilizations, "Assuming a typical colony spacing of 10 light-years, a ship speed of 10 percent that of light, and a period of 400 years between the foundation of a colony and its sending out colonies of its own, the colonization wave front will expand at an average speed of 0.02 light-year a year. As the galaxy is 100,000 light-years across, it takes no more than about 5 million years to colonize it completely. Though a long time in human terms, this is only 0.05 percent of the age of the galaxy."

Scientists have made serious attempts to detect radio emissions from a type III civilization within our own galaxy. The giant Aricebo radio telescope in Puerto Rico has scanned much of the galaxy for radio emissions at 1.42 gigahertz, near the emission line of hydrogen gas. It has found no evidence of any radio emissions in that band from any civilization radiating between 10^{18} to 10^{30} watts of power (that is, from type I.2 to type II.4). However, this does not rule out civilizations that are just beyond us in technology, from type 0.8 to type I.1, or considerably ahead of us, such as type II.5 and beyond.

It also does not rule out other forms of communication. An advanced civilization, for example, might send signals by laser rather than radio. And if they use radio, they may use frequencies other than 1.42 gigahertz. For example, they might spread their signal out across many frequencies and then reassemble them at the receiving end. This way, a passing star or cosmic storm would not interfere with the entire message. Anyone listening in on this spread signal may hear only gibberish. (Our own e-mails are broken up into many pieces, with each piece sent through a different city, and then reassembled at the end for your PC. Similarly, advanced civilizations may decide to use sophisticated methods to break down a signal and reassemble it at the other end.)

If a type III civilization exists in the universe, then one of their most pressing concerns would be establishing a communication system connecting the galaxy. This, of course, depends on whether they can somehow master faster-than-light technology, such as via wormholes. If we assume that they cannot, then their growth will be stunted considerably. Physicist Freeman Dyson, quoting from the work of Jean-Marc Levy-Leblond, speculates that such a society may live in a "Carroll" universe, named after Lewis Carroll. In the past, Dyson writes, human society was based on small tribes in which space was absolute but time was relative. This meant that communication between scattered tribes was impossible, and we could only venture a short distance from our birthplace within a human lifetime. Each tribe was separated by the vastness of absolute space. With the coming of the Industrial Revolution, we entered the Newtonian universe, in which space and time became absolute, and we had ships and wheels that linked the scattered tribes into nations. In the twentieth century, we entered the Einsteinian universe, in which space and time were both relative, and we developed the telegraph, telephone, radio, and TV, resulting in instantaneous communication. A type III civilization may drift back to a Carroll universe once again, with pockets of space colonies separated by vast interstellar distances, unable to communicate because of the light barrier. To prevent the fragmentation of such a Carroll universe, a type III civilization might need to develop wormholes that allow for faster-than-light communication at the subatomic level.

TYPE IV CIVILIZATION

Once I was giving a talk at the London Planetarium, and a little boy of ten came up to me and insisted that there must be a type IV civilization. When I reminded him that there are only planets, stars, and galaxies, and that these are the only platforms that allow for the germination of intelligent life, he claimed that a type IV civilization could utilize the power of the continuum.

He was right, I realized. If a type IV civilization could exist, its energy source might be extragalactic, such as the dark energy we see around us, which makes up 73 percent of the matter/energy content of the universe. Although potentially an enormous reservoir of energy—by far the largest in the universe—this antigravity field is spread out over the vast empty reaches of the universe and is hence extremely weak at any point in space.

Nikola Tesla, the genius of electricity and rival to Thomas Edison, wrote extensively about harvesting the energy of the vacuum. He believed that the vacuum hid untold reservoirs of energy. If we could somehow tap into this source, it would revolutionize all of human society, he thought. However, extracting this fabulous energy would be extremely difficult. Think of searching for gold in the oceans. There is probably more gold dispersed in the oceans than all the gold at Fort Knox and the other treasuries of the world. However, the expense of extracting this gold over such a large area is prohibitive. Hence, the gold lying in the oceans has never been harvested.

Likewise, the energy hidden within dark energy exceeds the entire energy content of the stars and galaxies. However, it is spread out over billions of light-years and would be difficult to concentrate. But by the laws of physics, it is still conceivable that an advanced type III civilization, having exhausted the power of the stars in the galaxy, may somehow try to tap into this energy to make the transition to type IV.

INFORMATION CLASSIFICATION

Further refinements to the classification of civilizations can be made based on new technologies. Kardashev wrote down the original classification in the 1960s, before the explosion in computer miniaturization, advances in nanotechnology, and awareness of the problems of environmental degradation. In light of these developments, an advanced civilization might progress in a slightly different fashion, taking full advantage of the information revolution we are witnessing today.

As an advanced civilization develops exponentially, the copious production of waste heat could dangerously raise the temperature of the atmosphere of the planet and pose climactic problems. Colonies of bacteria grow exponentially in a petri dish until they exhaust the food supply and literally drown in their own waste. Similarly, because space travel will remain prohibitively expensive for centuries, and terraforming nearby planets, if possible, will be such an economic and scientific challenge, an evolving type I civilization could potentially suffocate in its own waste heat, or it could miniaturize and streamline its information production.

To see the effectiveness of such miniaturization, consider the human brain, which contains about 100 billion neurons (as many as there are galaxies in the visible universe) yet produces almost no heat. By rights, if a computer engineer today were to design an electronic computer capable of computing quadrillions of bytes per second, as the brain can apparently do effortlessly, it would probably be several square blocks in size and would require a reservoir of water to cool it down. Yet our brains can contemplate the most sublime thoughts without working up a sweat.

The brain accomplishes this because of its molecular and cellular architecture. First of all, it is not a computer at all (in the sense of being a standard Turing machine, with input tape, output tape, and central processor). The brain has no operating system, no Windows, no CPU, no Pentium chip that we commonly associate with computers. Instead, it is a highly efficient neural network, a learning machine, where memory and thought patterns are distributed throughout the brain rather than concentrated in a central processing unit. The brain does not even compute very quickly, because the electrical messages sent down neurons are chemical in nature. But it more than makes up for this slowness because it can execute parallel processing and can learn new tasks at astronomically fast speeds.

To improve on the crude efficiency of electronic computers, scientists are trying to use novel ideas, many taken from nature, to create the next generation of miniaturized computers. Already, scientists at Princeton have been able to compute on DNA molecules (treating DNA as a piece of computer tape based not on binary os and 1s, but on the four nucleic acids A, T, C, G); their DNA computer solved the traveling salesman problem for several cities (that is, calculate the shortest route connecting N cities). Similarly, molecular transistors have been created in the laboratory, and even the first primitive quantum computers (which can compute on individual atoms) have been constructed.

Given the advances in nanotechnology, it is conceivable that an advanced civilization will find much more efficient ways to develop rather than to create copious quantities of waste heat that threaten their existence.

TYPES A TO Z

Sagan introduced yet another way of ranking advanced civilizations according to their information content, which would be essential to any civilization contemplating leaving the universe. A type A civilization, for example, is one that processes 10^6 bits of information. This would correspond to a primitive civilization without a written language but with a spoken language. To understand how much information is contained within a type A civilization, Sagan used the example of the game twenty questions, where you are supposed to identify a mysterious object by asking no more than twenty questions that can be answered by a yes or a no. One strategy is to ask questions that divide the world into two large pieces, such as, "Is it living?" After asking twenty such questions, we have divided the world into 2^{20} pieces, or 10^6 pieces, which is the total information content of a type A civilization.

Once a written language is discovered, the total information content rapidly explodes. Physicist Phillip Morrison of MIT estimates that the total written heritage that survived from ancient Greece is about 10^9 bits, or a type C civilization by Sagan's ranking.

Sagan estimated our present-day information content. By estimating the number of books contained in all the libraries of the world (measured in the tens of millions) and the number of pages there are on each book, he came up with about 10^{13} bits of information. If we include photographs, this might rise to 10^{15} bits. This would place us as a type H civilization. Given our low energy and information output, we can be classified as a type 0.7 H civilization.

He estimated that our first contact with an extraterrestrial civilization would involve a civilization of a least type 1.5 J or 1.8 K because they have already mastered the dynamics of interstellar travel. At the minimum, such a civilization would be several centuries to millennia more advanced than ours. Similarly, a galactic type III civilization may be typified by the information content of each planet multiplied by the number of planets in the galaxy capable of supporting life. Sagan estimated that such a type III civilization would be type Q. An advanced civilization that can harness the information content of a billion galaxies, representing a large portion of the visible universe, would qualify the civilization as type Z, he estimated.

This is not a trivial academic exercise. Any civilization about to leave the universe will necessarily have to compute the conditions on the other side of the universe. Einstein's equations are notoriously difficult because, to calculate the curvature of space at any point, you have to know the location of all objects in the universe, each of which contributes to the bending of space. You also have to know the quantum corrections to the black hole, which at present are impossible to calculate. Since this is vastly too difficult for our computers, today physicists usually approximate a black hole by studying a universe dominated by a single collapsed star. To arrive at a more realistic understanding of the dynamics within the event horizon of a black hole or near the mouth of a wormhole, we necessarily have to know the location and energy content of all the nearby stars and compute quantum fluctuations. Again, this is prohibitively difficult. It is hard enough to solve the equations for a single star in an empty universe, let alone billions of galaxies floating in an inflated universe.

That is why any civilization that attempts to make the journey through a wormhole would have to have computational power far beyond that available to a type 0.7 H civilization like ours. Perhaps the minimum civilization with the energy and information content to seriously consider making the jump would be a type III Q.

It is also conceivable that intelligence may spread beyond the confines of the Kardashev classification. As Sir Martin Rees says, "It's quite conceivable that, even if life now exists only here on Earth, it will eventually spread through the galaxy and beyond. So life may not forever be an unimportant trace contaminant of the universe, even though it now is. In fact, I find it a rather appealing view, and I think it could be salutary if it became widely shared." But he warns us, "If we snuffed ourselves out, we'd be destroying genuine cosmic potentialities. So even if one believes that life is unique to the earth now, then that doesn't mean that life is forever going to be a trivial piece of the universe."

How would an advanced civilization contemplate leaving their dying universe? It would have to overcome a series of large obstacles.

STEP ONE: CREATE AND TEST A THEORY OF EVERYTHING

The first hurdle for a civilization hoping to leave the universe would be to complete a theory of everything. Whether it is string theory or not, we must have a way to reliably calculate quantum corrections to Einstein's equations, or else none of our theories are useful.

Fortunately, because M-theory is rapidly advancing, with some of the best minds on the planet working on this question, we shall know if it is truly the theory of everything or a theory of nothing fairly rapidly, within a few decades or possibly less.

Once a theory of everything or a theory of quantum gravity has been found, we have to verify the consequences of this theory using advanced technology. Several possibilities exist, including building large atom smashers to create super particles, or even huge gravity wave detectors based in space or on different moons throughout the solar system. (Moons are quite stable for long periods of time, free of erosion and atmospheric disturbances, so a planetary system of gravity wave detectors should be able to probe the details of the big bang, resolving any questions we may have about quantum gravity and creating a new universe.)

Once a theory of quantum gravity is found, and huge atom smashers and gravity wave detectors have confirmed its correctness, then we can begin to answer some essential questions concerning Einstein's equations and wormholes:

1. Are wormholes stable?

When passing through a Kerr rotating black hole, the problem is that your very presence disturbs the black hole; it may collapse before you make a complete passage through the Einstein-Rosen bridge. This stability calculation has to be redone in light of quantum corrections, which may change the calculation entirely.

2. Are there divergences?

If we pass through a transversable wormhole connecting two time eras, then the buildup of radiation surrounding the wormhole entrance may become infinite, which would be disastrous. (This is because radiation can pass through the wormhole, go back in time, and return after many years to enter the wormhole a second time. This process can be repeated an infinite number of times, leading to an infinite buildup of radiation. This problem can be solved, however, if the many-worlds theory holds, so that the universe splits every time radiation passes through the wormhole, and there is no infinite buildup of radiation. We need a theory of everything to settle this delicate question.)

3. Can we find large quantities of negative energy?

Negative energy, a key ingredient that can open up and stabilize wormholes, is already known to exist but only in small quantities. Can we find sufficient quantities of it to open and stabilize a wormhole?

Assuming that the answers to these questions can be found, then an advanced civilization may begin to seriously contemplate how to leave the universe, or face certain extinction. Several alternatives exist.

STEP TWO: FIND NATURALLY OCCURRING WORMHOLES AND WHITE HOLES

Wormholes, dimensional gateways, and cosmic strings may exist naturally in outer space. At the instant of the big bang, when there was a huge amount of energy released into the universe, wormholes and cosmic strings may have formed naturally. The inflation of the early universe might then have expanded these wormholes to macroscopic size. In addition, there is the possibility that exotic matter or negative matter exists naturally in outer space. This would help enormously in any effort to leave a dying universe. However, there is no guarantee that such objects exist in nature. No one has ever seen any of these objects, and it is simply too risky to bet the fate of all intelligent life on this assumption.

Next, there is the possibility that "white holes" may be found by scanning the heavens. A white hole is a solution of Einstein's equations in which time is reversed, so that objects are ejected from a white hole in the same way that objects are sucked into a black hole. A white hole might be found at the other end of a black hole, so that matter entering a black hole eventually comes out the white hole. So far, all astronomical searches have found no evidence of white holes, but their existence might be confirmed or disproved with the next generation of space-based detectors.

STEP THREE: SEND PROBES THROUGH A BLACK HOLE

There are decided advantages to using such black holes as wormholes. Black holes, as we have come to discover, are quite plentiful in the universe; if one can solve the numerous technical problems, they will have to be seriously considered by any advanced civilization as an escape hatch from our universe. Also, in passing through a black hole, we are not bound by the limitation that we cannot go backward in time to a time before the creation of the time machine. The wormhole in the center of the Kerr ring may connect our universe to quite different universes or different points in the same universe. The only way to tell would be to experiment with probes and use a supercomputer to calculate the distribution of masses in the universes and calculate quantum corrections to Einstein's equations through the wormhole.

Currently, most physicists believe that a trip through a black hole would be fatal. However, our understanding of black hole physics is still in its infancy, and this conjecture has never been tested. Assume, for the sake of argument, that a trip through a black hole might be possible, especially a rotating Kerr black hole. Then any advanced civilization would give serious thought to probing the interior of black holes.

Since a trip through a black hole would be a one-way trip, and because of the enormous dangers found near a black hole, an advanced civilization would likely try to locate a nearby stellar black hole and first send a probe through it. Valuable information could be sent back from the probe until it finally crossed the event horizon and all contact was lost. (A trip past the event horizon is likely to be quite lethal because of the intense radiation field surrounding it. Light rays falling into a black hole will be blueshifted and thereby will gain in energy as they get close to the center.) Any probe passing near the event horizon would have to be properly shielded against this intense barrage of radiation. In addition, this may destabilize the black hole itself, so that the event horizon would become a singularity, thereby closing the wormhole. The probe would determine precisely how much radiation there is near the event horizon and whether the wormhole could remain stable in spite of all this energy flux.

The data from the probe before it entered the event horizon would have to be radioed back to nearby spaceships, but therein lies another problem. To an observer on one of those spaceships, the probe would seem to be slowing down in time as it got closer to the event horizon. At it entered the event horizon, the probe in fact would seem to be frozen in time. To avoid this problem, probes would have to radio their data a certain distance away from the event horizon, or else even the radio signals would be redshifted so badly that the data would be unrecognizable.

STEP FOUR: CONSTRUCT A BLACK HOLE IN SLOW MOTION

Once the characteristics near the event horizon of black holes are carefully ascertained by probes, the next step might be to actually create a black hole in slow motion for experimental purposes. A type III civilization might try to reproduce the results suggested in Einstein's paper—that black holes can never form from swirling masses of dust and particles. Einstein tried to show that a collection of revolving particles will not reach the Schwarzschild radius by itself (and as a result black holes were impossible).

Swirling masses, by themselves, might not contract to a black hole. But this leaves open the possibility that one may artificially inject new energy and matter slowly into the spinning system, forcing the masses to gradually pass within the Schwarzschild radius. In this way, a civilization could manipulate the formation of a black hole in a controlled way.

For example, one can imagine a type III civilization corralling neutron stars, which are about the size of Manhattan but weigh more than our Sun, and forming a swirling collection of these dead stars. Gravity would gradually bring these stars closer together. But they would never hit the Schwarzschild radius, as Einstein showed. At this point, scientists from this advanced civilization might carefully inject new neutron stars into the mix. This might be enough to tip the balance, causing this swirling mass of neutron material to collapse to within the Schwarzschild radius. As a result, the collection of stars would collapse into a spinning ring, the Kerr black hole. By controlling the speed and radii of the various neutron stars, such a civilization would make the Kerr black hole open up as slowly as it wished.

Or, an advanced civilization might try to assemble small neutron stars together into a single, stationary mass, until it reached 3 solar masses in size, which is roughly the Chandrasekhar limit for neutron stars. Beyond this limit, the star would implode into a black hole by its own gravity. (An advanced civilization would have to be careful that the creation of a black hole did not set off a supernova-like explosion. The contraction to the black hole would have to be done very gradually and very precisely.)

Of course, for anyone passing through an event horizon, it is guaranteed to be a one-way trip. But for an advanced civilization facing the certainty of extinction, a one-way trip might be the only alternative. Still, there is the problem of radiation as one passes the event horizon. Light beams that follow us through the event horizon become more energetic as they increase in frequency. This would likely cause a rain of radiation that would be deadly to any astronaut who passed through the event horizon. Any advanced civilization would have to calculate the precise amount of such radiation and build proper shielding to prevent being fried.

Last, there is the stability problem: will the wormhole at the center of the Kerr ring be sufficiently stable to fall completely through? The mathematics of this question are not totally clear, since we would have to use a quantum theory of gravity to do a proper calculation. It may turn out that the Kerr ring is stable under certain very restrictive conditions as matter falls through the wormhole. This issue would have to be carefully resolved using the mathematics of quantum gravity and experiments on the black hole itself.

In summary, passage through a black hole would doubtless be a very difficult and dangerous journey. Theoretically, it cannot be ruled out until extensive experimentation is performed and a proper calculation is made of all quantum corrections.

STEP FIVE: CREATE A BABY UNIVERSE

So far, we have assumed that it might be possible to pass through a black hole. Now let's assume the reverse, that black holes are too unstable and too full of lethal radiation. One might then try an even more difficult path: to create a baby universe. The concept of an advanced civilization creating an escape hatch to another universe has intrigued physicists like Alan Guth. Because the inflationary theory is so crucially dependent on the creation of the false vacuum, Guth has wondered if some advanced civilization might artificially create a false vacuum and create a baby universe in the laboratory.

At first, the idea of creating a universe seems preposterous. After all, as Guth points out, to create our universe, you would need 10^{89} photons, 10^{89} electrons, 10^{89} positrons, 10^{89} neutrinos, 10^{89} antineutrinos, 10^{89} protons, and 10^{89} neutrons. While this task sounds daunting, Guth reminds us that although the matter/energy content of a universe is quite large, it is balanced by the negative energy derived from gravitation. The total net matter/energy may be as little as an ounce. Guth cautions, "Does this mean that the laws of physics truly enable us to create a new universe at will? If we tried to carry out this recipe, unfortunately, we would immediately encounter an annoying snag: since a sphere of false vacuum 10^{-26} centimeters across has a mass of one ounce, its density is a phenomenal 10^{80} grams per cubic centimeter! . . . If the mass of the entire observed universe were compressed to false-vacuum density, it would fit in a volume smaller than an atom!" The false vacuum would be the tiny region of space-time where an instability arises and a rift occurs in space-time. It may only take a few ounces of matter within the false vacuum to create a baby universe, but this tiny amount of matter has to be compressed down to an astronomically small distance.

There may be still another way to create a baby universe. One might heat up a small region of space to 10^{29} degrees K, and then rapidly cool it down. At this temperature, it is conjectured that spacetime becomes unstable; tiny bubble-universes would begin to form, and a false vacuum might be created. These tiny baby universes, which form all the time but are short-lived, may become real universes at that temperature. This phenomenon is already familiar with ordinary electric fields. (For example, if we create a large enough electric field, the virtual electron-antielectron pairs that constantly pop out in and out of the vacuum can suddenly become real, allowing these particles to spring into existence. Thus, concentrated energy in empty space can transform virtual particles into real ones. Similarly, if we apply enough energy at a single point, it is theorized that virtual baby universes may spring into existence, appearing out of nowhere.)

Assuming that such an unimaginable density or temperature can be achieved, the formation of a baby universe might look as follows. In our universe, powerful laser beams and particle beams may be used to compress and heat a tiny amount of matter to fantastic energies and temperatures. We would never see the baby universe as it begins to form, since it expands on the "other side" of the singularity, rather than in our universe. This alternate baby universe would potentially inflate in hyperspace via its own antigravity force and "bud" off our universe. We will, therefore, never see a new universe is forming on the other side of the singularity. But a wormhole would, like an umbilical cord, connect us with the baby universe.

There is a certain amount of danger, however, in creating a universe in an oven. The umbilical cord connecting our universe with the baby universe would eventually evaporate and create Hawking radiation equivalent to a 500-kiloton nuclear explosion, roughly twenty-five times the energy of the Hiroshima bomb. So there would be a price to pay for creating a new universe in an oven.

One last problem with this scenario of creating a false vacuum is that it would be easy for the new universe to simply collapse into a black hole, which, we recall, we assumed would be lethal. The reason for this is Penrose's theorem, which states that, for a wide variety of scenarios, any large concentration of sufficiently large mass will inevitably collapse into a black hole. Since Einstein's equations are time-reversal invariant, that is, they can be run either forward or backward in time, this means that any matter that falls out of our baby universe can be run backward in time, resulting in a black hole.

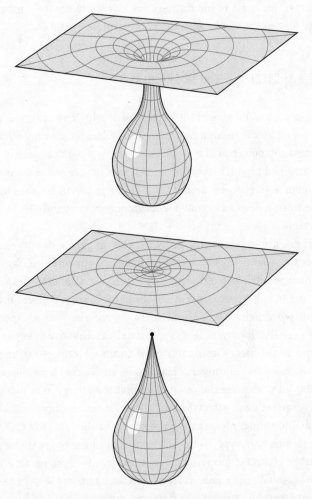

A baby universe could be artificially created by an advanced civilization in several ways. A few ounces of matter could be concentrated to enormous densities and energies, or matter could be heated to near the Planck temperature.

Thus, one would have to be very careful in constructing the baby universe to avoid the Penrose theorem.

Penrose's theorem rests on the assumption that the infalling matter is positive in energy (like the familiar world we see surrounding us). However, the theorem breaks down if we have negative energy or negative matter. Thus, even for the inflationary scenario, we need to obtain negative energy to create a baby universe, just as we would with the transversable wormhole.

STEP SIX: CREATE HUGE ATOM SMASHERS

How can we build a machine capable of leaving our universe, given unlimited access to high technology? At what point can we hope to harness the power of the Planck energy? By the time a civilization has attained type III status, it already has the power to manipulate the Planck energy, by definition. Scientists would be able to play with wormholes and assemble enough energy to open holes in space and time.

There are several ways in which this might be done by an advanced civilization. As I mentioned earlier, our universe may be a membrane with a parallel universe just a millimeter from ours, floating in hyperspace. If so, then the Large Hadron Collider may detect it within the next several years. By the time we advance to a type I civilization, we might even have the technology to explore the nature of this neighboring universe. So the concept of making contact with a parallel universe may not be such a farfetched idea.

But let us assume the worst case, that the energy at which quantum gravitational effects arise is the Planck energy, which is a quadrillion times greater than the energy of the LHC. To explore the Planck energy, a type III civilization would have to create an atom smasher of stellar proportions. In atom smashers, or particle accelerators, subatomic particles travel down a narrow tube. As energy is injected into the tubing, the particles are accelerated to high energies. If we use huge magnets to bend the particles' path into a large circle, then particles can be accelerated to trillions of electron volts of energy. The greater the radius of the circle, the greater the energy of the beam. The LHC has a diameter of 27 kilometers, which is pushing the limit of the energy available to a type 0.7 civilization.

But for a type III civilization, the possibility opens up of making an atom smasher the size of a solar system or even a star system. It is conceivable that an advanced civilization might fire a beam of subatomic particles into outer space and accelerate them to the Planck energy. As we recall, with the new generation of laser particle accelerators, within a few decades physicists might be able to create a tabletop accelerator capable of achieving 200 GeV (200 billion electron volts) over a distance of a meter. By stacking these tabletop accelerators one after the other, it is conceivable that one could attain energies at which space-time becomes unstable.

If we assume that future accelerators can boost particles only by 200 GeV per meter, which is a conservative assumption, we would need a particle accelerator 10 light-years long to reach the Planck energy. Although this is prohibitively large for any type I or II civilization, it is well within the ability of a type III civilization. To build such a gargantuan atom smasher, a type III civilization might either bend the path of the beam into a circle, thereby saving considerable space, or leave the path stretched out in a line that extends well past the nearest star.

One might, for example, build an atom smasher that fires subatomic particles along a circular path inside the asteroid belt. You would not need to build an expensive circular piece of tubing, because the vacuum of outer space is better than any vacuum we can create on Earth. But you would have to build huge magnets, placed at regular intervals on distant moons and asteroids in the solar system or in various star systems, which would periodically bend the beam.

When the beam comes near a moon or asteroid, huge magnets based on the moon would then yank the beam, changing its direction very slightly. (The lunar or asteroid stations would also have to refocus the beam at regular intervals, because the beam would gradually diverge the farther it traveled.) As the beam traveled by several moons, it would gradually form the shape of an arc. Eventually, the beam would travel in the approximate shape of a circle. One could also imagine two beams, one traveling clockwise around the solar system, the other counterclockwise. When the two beams collided, the energy released by the matter/antimatter collision would create energies approaching the Planck energy. (One can calculate that the magnetic fields necessary to bend such a powerful beam far exceed the technology of today. However, it is conceivable that an advanced civilization could use explosives to send a powerful surge of energy through coils to create a huge magnetic pulse. This titanic burst of magnetic energy could only be released once, since it would likely destroy the coils, so the magnets would have to be rapidly replaced before the particle beam returned for the next pass.)

Besides the horrendous engineering problems of creating such an atom smasher, there is also the delicate question of whether there is a limit to the energy of a particle beam. Any energetic beam of particles would eventually collide with the photons that make up the 2.7-degree background radiation and hence lose energy. In theory, this might, in fact, bleed so much energy from the beam that there would be an ultimate ceiling for the energy one could attain in outer space. This result still has not been checked experimentally. (In fact, there are indications that energetic cosmic ray impacts have exceeded this maximum energy, casting doubt on the whole calculation.) However, if it is true, then a more expensive modification of the apparatus would be required. First, one might enclose the entire beam in a vacuum tubing with shielding to keep out the 2.7-degree background radiation. Or, if the experiment is done in the far future, it is possible that the background radiation will be small enough so that it no longer matters.

STEP SEVEN: CREATE IMPLOSION MECHANISMS

One could also imagine a second device, based on laser beams and an implosion mechanism. In nature, tremendous temperatures and pressures are attained by the implosion method, as when a dying star collapses suddenly under the force of gravity. This is possible because gravity is only attractive, not repulsive, and hence the collapse takes place uniformly, so the star is compressed evenly to incredible densities.

This implosion method is very difficult to re-create on planet Earth. Hydrogen bombs, for example, have to be designed like a Swiss watch so that lithium deuteride, the active ingredient of a hydrogen bomb, is compressed to tens of millions of degrees to attain Lawson's criteria, at which the fusion process kicks in. (This is done by detonating an atomic bomb next to the lithium deuteride, and then focusing the X-ray radiation evenly on the surface of a piece of lithium deuteride.) This process, however, can only release energy explosively, not in a controlled fashion.

On Earth, attempts to use magnetism to compress hydrogen-rich gas have failed, mainly because magnetism does not compress gas uniformly. Because we have never seen a monopole in nature, magnetic fields are dipolar, like Earth's magnetic field. As a result, they are horribly nonuniform. Using them to squeeze gas is like trying to squeeze a balloon. Whenever you squeeze one end, the other end of the balloon bulges out.

Another way of controlling fusion might be to use a battery of laser beams, arranged along the surface of a sphere, so that the beams are fired radially onto a tiny pellet of lithium deuteride at the center. For example, at the Livermore National Laboratory, there is a powerful laser/fusion device used to simulate nuclear weapons. It fires a series of laser beams horizontally down a tunnel. Then mirrors based at the end of the tunnel carefully reflect each beam, so that the beams are directed radially onto a tiny pellet. The surface of the pellet is immediately vaporized, causing the pellet to implode and creating huge temperatures. In this fashion, fusion has actually been seen inside the pellet (although the machine consumes more energy than it creates and hence is not commercially viable).

Similarly, one can envision a type III civilization building large banks of laser beams on asteroids and moons of various star systems. This battery of lasers would then fire at once, releasing a series of powerful beams that converge at a single point, creating temperatures at which space and time become unstable.

In principle, there is no theoretical limit to the amount of energy that one can place on a laser beam. However, there are practical problems with creating extremely high-powered lasers. One of the main problems is the stability of lasing material, which will often overheat and crack at high energies. (This can be remedied by driving the laser beam by an explosion that occurs only once, such as nuclear detonations.)

The purpose of firing this spherical bank of laser beams is to heat a chamber so that the false vacuum is created inside, or to implode and compress a set of plates to create negative energy via the Casimir effect. To create such a negative-energy device, one would need to compress a set of spherical plates to within the Planck length, which is 10^{-33} centimeters. Because the distance separating atoms is 10^{-8} centimeters, and the distance separating the protons and neutrons in the nucleus is 10^{-13} cm, you see that the compression of these plates must be enormous. Because the total wattage that one can amass on a laser beam is essentially unlimited, the main problem is to create an apparatus that is stable enough to survive this enormous compression. (Since the Casimir effect creates a net attraction between the plates, we will also have to add charges to the plates to prevent them from collapsing.) In principle, a wormhole will develop within the spherical shells connecting our dying universe with a much younger, much hotter universe.

STEP EIGHT: BUILD A WARP DRIVE MACHINE

One key element necessary to assemble the devices described above is the ability to travel across vast interstellar distances. One possible way to do so is to use the Alcubierre warp drive machine, a machine first proposed by physicist Miguel Alcubierre in 1994. A warp drive machine does not alter the topology of space by punching a hole and leaping into hyperspace. It simply shrinks the space in front of you while expanding the space behind you. Think of walking across a carpet to reach a table. Instead of walking on the carpet, you could lasso the table and slowly drag it toward you, making the carpet bunch up in front of you. Thus, you have moved little; instead, the space in front of you has shrunk.

Recall that space itself can expand faster than the speed of light (since no net information is being transferred by expanding empty space). Similarly, it may be possible to travel faster than the speed of light by shrinking space faster than the speed of light. In effect, when traveling to a nearby star, we may barely leave Earth at all; we would simply collapse the space in front of us and expand the space behind us. Instead of traveling to Alpha Centauri, the nearest star, we are bringing Alpha Centauri to us.

Alcubierre showed that this is a viable solution of Einstein's equations—meaning that it falls within the laws of physics. But there is a price to pay. You would have to employ large quantities of both negative and positive energy to power your starship. (Positive energy could be used to compress the space in front of you and negative energy to lengthen the distance behind you.) To use the Casimir effect to create this negative energy, the plates would have to be separated by the Planck distance, 10^{-33} centimeters—too small to be achieved by ordinary means. To build such a starship, you would need to construct a large sphere and place the passengers inside. On the sides of the bubble, you would put a band of negative energy along the equator. The passengers inside the bubble would never move, but the space in front of the bubble would shrink faster than light, so that when the passengers left the bubble, they would have reached a nearby star.

In his original article, Alcubierre mentioned that his solution might not only take us to the stars, it might make possible time travel as well. Two years later, physicist Allen E. Everett showed that if one had two such starships, time travel would be possible by applying warp drive in succession. As Princeton physicist Gott says, "Thus, it appears that Gene Roddenberry, the creator of *Star Trek*, was indeed right to include all those time-travel episodes!"

But a later analysis by the Russian physicist Sergei Krasnikov revealed a technical defect in the solution. He showed that the inside of the starship is disconnected from the space outside the ship, so that messages cannot cross the boundary—that is, once inside the ship, you cannot change the path of the starship. The path has to be laid out before the trip is made. This is disappointing. In other words, you simply cannot spin a dial and set a course for the nearest star. But it does mean that such a theoretical starship could be a railway to the stars, an interstellar system in which the starships leave at regular intervals. One could, for example, build this railway by first using conventional rockets that travel at sublight speed to build rail stations at regular intervals between stars. Then the starship would travel between these stations at super light speed according to a timetable, with fixed departures and arrivals.

Gott writes, "A future supercivilization might want to lay down warpdrive paths among stars for starships to traverse, just as it might establish wormhole links among stars. A network of warpdrive paths might even be easier to create than one made up of wormholes because warpdrives would require only an alteration of existing space rather than the establishment of new holes connecting distant regions."

But precisely because such a starship must travel within the existing universe, it cannot be used to leave the universe. Nevertheless, the Alcubierre drive could help to construct a device to escape the universe. Such a starship might be useful, for example, in creating the colliding cosmic strings mentioned by Gott, which might take an advanced civilization back into its own past, when its universe was much warmer.

STEP NINE: USE NEGATIVE ENERGY FROM SQUEEZED STATES

In chapter 5, I mention that laser beams can create "squeezed states" which can be used to generate negative matter, which in turn can be used to open up and stabilize wormholes. When a powerful laser pulse hits a special optical material, it creates pairs of photons in its wake. These photons alternately enhance and suppress the quantum fluctuations found in the vacuum, giving both positive and negative energy pulses. The sum of these two energy pulses always averages to a positive energy, so that we do not violate known laws of physics.

In 1978, physicist Lawrence Ford at Tufts University proved three laws that such negative energy must obey, and they have been the subject of intense research ever since. First, Ford found that the amount of negative energy in a pulse is inversely related to its spatial and temporal extent—that is, the stronger the negative energy pulse, the shorter its duration. So if we create a large burst of negative energy with a laser to open up a wormhole, it can only last for an extremely short period of time. Second, a negative pulse is always followed by a positive energy pulse of larger magnitude (so the sum is still positive). Third, the longer the interval between these two pulses, the larger the positive pulse must be.

Under these general laws, one can quantify the conditions under which a laser or Casimir plates can produce negative energy. First, one might try to separate the negative energy pulse from the subsequent positive energy pulse by shining a laser beam into a box and having a shutter close immediately after the negative energy pulse enters. As a result, only the negative energy pulse would have entered the box. In principle, huge amounts of negative energy can be extracted in this way, followed by an even larger positive energy pulse (which is kept out of the box by the shutter). The interval between the two pulses can be quite long, as long as the energy of the positive pulse is large. In theory, this seems to be an ideal way in which to generate unlimited quantities of negative energy for a time machine or wormhole.

Unfortunately, there is a catch. The very act of closing the shutter creates a second positive energy pulse inside the box. Unless extraordinary precautions are taken, the negative energy pulse is wiped out. This will remain a technological feat for an advanced civilization to solve—to split off a powerful negative energy pulse from the subsequent positive energy pulse without having a secondary pulse wipe out the negative energy one.

These three laws can be applied to the Casimir effect. If we produce a wormhole that is one meter in size, we must have negative energy concentrated in a band no more than 10^{-22} meters (a millionth of the size of a proton). Once again, only an extremely advanced civilization might be able to create the technology necessary to manipulate these incredibly small distances or incredibly tiny time intervals.

STEP TEN: WAIT FOR QUANTUM TRANSITIONS

As we saw in chapter 10, intelligent beings facing the gradual cooling of their universe may have to think more slowly and hibernate for long periods of time. This process of slowing the rate of thinking could continue for trillions upon trillions of years, enough time for quantum events to happen. Normally, we can dismiss the spontaneous creation of bubble universes and transitions to other quantum universes because they would be such extremely rare events. However, in stage 5, intelligent beings may think so slowly that such quantum events could become relatively commonplace. In their own subjective time, their rate of thinking might appear to them to be perfectly normal, even though the actual time scale would be so long that quantum events become a normal occurrence.

If so, such beings would only have to wait until wormholes appear and quantum transitions occur in order to escape into another universe. (Although such beings might see quantum transitions as commonplace, one problem here is that these quantum events are totally unpredictable; it would be difficult to make the transition to another universe when one doesn't know precisely when the gateway might open or where it would lead. These beings might have to seize the opportunity to leave the universe as soon as a wormhole opened up, before they had a chance to fully analyze its properties.)

STEP ELEVEN: THE LAST HOPE

Assume for the moment that all future experiments with wormholes and black holes face a seemingly insurmountable problem: that the only stable wormholes are microscopic to subatomic in size. Assume also that an actual trip through a wormhole may place unacceptable stresses on our bodies, even within a protective vessel. Any number of challenges, such as intense tidal forces, radiation fields, incoming falling debris, would prove lethal. If that is the case, future intelligent life in our universe would have but one remaining option: to inject enough information into a new universe to recreate our civilization on the other side of the wormhole.

In nature, when living organisms are faced with a hostile environment, they sometimes devise ingenious methods to survive. Some mammals hibernate. Some fish and frogs have antifreeze-like chemicals circulating in their bodily fluids that allow them to be frozen alive. Fungi evade extinction by transforming into spores. Similarly, human beings might have to find a way to alter their physical existence to survive the trip to another universe.

Think of the oak tree, which scatters tiny seeds in all directions. The seeds are (a) small, resilient, and compact; (b) they contain the entire DNA content of the tree; (c) they are designed to travel a certain distance away from the mother tree; (d) they contain enough food to begin the process of regeneration in a distant land; (e) they take root by consuming nutrients and energy from the soil and living off the new land. Similarly, a civilization could try to mimic nature by sending its "seed" through a wormhole, using the most advanced nanotechnology available billions years from now, to copy each of these important properties.

As Stephen Hawking has said, "It seems . . . that quantum theory allows time travel on a microscopic basis." If Hawking is right, members of an advanced civilization could decide to alter their physical being into something that would survive the arduous journey back in time or to another universe, merging carbon with silicon and reducing their consciousness down to pure information. In the final analysis, our carbon-based bodies may well be too fragile to endure the physical hardship of a journey of this magnitude. Far in the future, we may be able to merge our consciousness with our robot creations, using advanced DNA engineering, nanotechnology, and

robotics. This may sound bizarre by today's standards, but a civilization billions to trillions of years in the future might find it the only
way to survive.

They might need to merge their brains and personalities directly
into machines. This could be done in several ways. One could create
a sophisticated software program that was able to duplicate all our
thinking processes, so that it had a personality identical to ours.
More ambitious is the program advocated by Hans Moravec of
Carnegie-Mellon University. He claims that, in the far future, we
may be able to reproduce, neuron for neuron, the architecture of our
brains onto silicon transistors. Each neural connection in the brain
would be replaced by a corresponding transistor that would duplicate the neuron's function inside a robot.

Because the tidal forces and radiation fields would likely be intense, future civilizations would have to carry the absolute minimum of fuel, shielding, and nutrients necessary to re-create our
species on the other side of a wormhole. Using nanotechnology, it
might be possible to send microscopic chains across the wormhole inside a device no wider than a cell.

If the wormhole was very small, on the scale of an atom, scientists would have to send large nanotubes made of individual atoms,
encoded with vast quantities of information sufficient to re-create
the entire species on the other side. If the wormhole was only the
size of a subatomic particle, scientists would have to devise a way to
send nuclei across the wormhole that would grab electrons on the
other side and reconstruct themselves into atoms and molecules. If
a wormhole was even smaller than that, perhaps laser beams made
of X rays or gamma rays of small wavelength could be used to send
sophisticated codes through the wormhole, giving instructions on
how to reconstruct civilization on the other side.

The goal of such a transmission would be to construct a microscopic "nanobot" on the other side of the wormhole, whose mission
would be to find a suitable environment in which to regenerate our
civilization. Because it would be constructed on an atomic scale, it
would not need huge booster rockets or a large amount of fuel to find
a suitable planet. In fact, it could effortlessly approach light-speed

because it is relatively easy to send subatomic particles to near light-speed using electric fields. Also, it would not need life support or other clumsy pieces of hardware, since the main content of the nanobot is the pure information necessary to regenerate the race.

Once the nanobot had found a new planet, it would create a large factory using the raw materials already available on the planet to build many replicas of itself and make a large cloning laboratory. The necessary DNA sequences could be produced in this laboratory and then injected into cells to begin the process of regenerating whole organisms and eventually the entire species. These cells in the lab would then be grown into fully adult beings, with the memory and personality of the original human placed into the brain.

In some sense, this process would be similar to injecting our DNA (the total information content of a type III civilization or beyond) into an "egg cell," containing the genetic instructions capable of re-creating an embryo on the other side. The "fertilized egg" would be compact, sturdy, and mobile, yet would contain the entire body of information necessary to recreate a type III civilization. A typical human cell contains only 30,000 genes, arranged on 3 billion DNA base pairs, but this concise piece of information is sufficient to re-create an entire human being, utilizing resources found outside the sperm (the nourishment provided by the mother). Similarly, the "cosmic egg" would consist of the totality of information necessary to regenerate an advanced civilization; the resources to do this (raw materials, solvents, metals, and so forth) would be found on the other side. In this way, an advanced civilization, such as a type III Q, might be able to use their formidable technology to send enough information (about 10^{24} bits of information) across a wormhole sufficient to re-create their civilization on the other side.

Let me emphasize that every step I've mentioned in this process is so far beyond today's capability that it must read like science fiction. But billions of years into the future, for a type III Q civilization facing extinction, it may be the only possible path to salvation. Certainly, there is nothing in the laws of physics or biology to prevent this from occurring. My point is that the ultimate death of our universe may not necessarily mean the death of intelligence. Of course, if the ability to transfer intelligence from one universe to another is possible, it leaves open the possibility that a life form from another universe, facing its own big freeze, could try to burrow into some distant part of our own universe, where it is warmer and more hospitable.

In other words, the unified field theory, instead of being a useless but elegant curiosity, may ultimately provide the blueprint for the survival of intelligent life in the universe.

CHAPTER TWELVE

Beyond the Multiverse

The Bible teaches us how to go to heaven, not how the
heavens go.

—Cardinal Baronius,
repeated by Galileo during his trial

Why is there something, rather than nothing? The un-
rest which keeps the never-stopping clock of metaphysics going is the thought that the non-existence of
the world is just as possible as its existence.

—William James

The most beautiful experience we can have is the mysterious. It is the fundamental emotion which stands at the
cradle of true art and true science. Whosoever does not
know it and can no longer wonder, no longer marvel, is
as good as dead, and his eyes are dimmed.

—Albert Einstein

In 1863, THOMAS H. HUXLEY wrote, "The question of all questions
for humanity, the problem which lies behind all others and is more
interesting than any of them, is that of the determination of man's
place in Nature and his relation to the Cosmos."

Huxley was famous as "Darwin's bulldog," the man who fero-
ciously defended the theory of evolution to a deeply conservative

Victorian England. English society saw humanity standing proudly at the very center of creation; not only was the solar system the center of the universe, but humanity was the crowning achievement of God's creation, the pinnacle of God's divine handiwork. God had created us in His very image.

By openly challenging this religious orthodoxy, Huxley had to defend Darwin's theory against the salvos launched by the religious establishment, thereby helping to establish a more scientific understanding of our role in the tree of life. Today, we recognize that, among the giants of science, Newton, Einstein, and Darwin have done the yeoman's work in helping to define our rightful place in the cosmos.

Each of them grappled with the theological and philosophical implications of his work in determining our role in the universe. In the conclusion to *Principia*, Newton declares, "The most beautiful system of the sun, planes, and comets could only proceed from the counsel and dominion of an intelligent and powerful Being." If Newton discovered the laws of motion, then there must be a divine lawgiver.

Einstein, too, was convinced of the existence of what he called the Old One, but one who did not intervene in the affairs of men. His goal, instead of glorifying God, was to "read the Mind of God." He would say, "I want to know how God created this world. I am not interested in this phenomenon or that. I want to know God's thoughts. The rest are details." Einstein would justify his intense interest in these theological matters by concluding, "Science without religion is lame. But religion without science is blind."

But Darwin was hopelessly divided on the question of the role of humanity in the universe. Although he is credited as the one who dethroned humanity from the center of the biological universe, he confessed in his autobiography concerning "the extreme difficulty or rather impossibility of conceiving this immense and wonderful universe, including man with his capacity for looking far backwards and far into futurity, as the result of blind chance or necessity." He confided to a friend, "My theology is simply a muddle."

Unfortunately, the "determination of man's place in Nature and his relation to the Cosmos" has been fraught with danger, especially to those who dared to challenge the rigid dogma of the ruling orthodoxy. It was no accident that Nicolaus Copernicus wrote his pioneering book, *De Revolutionibus Orbium Coelestium* (*On the Revolutions of the Celestial Orbs*) on his deathbed in 1543, beyond the morbid reach of the Inquisition. It was also inevitable that Galileo, who had been protected for so long by his powerful patrons in the Medici family, would eventually suffer the wrath of the Vatican for popularizing an instrument that revealed a universe that so sharply contradicted church doctrine: the telescope.

The mixture of science, religion, and philosophy is indeed a potent brew, so volatile that the great philosopher Giordano Bruno was burned at the stake in 1600 in the streets of Rome for refusing to repudiate his belief that there were an infinite number of planets in the heavens, harboring an infinite number of living beings. He wrote, "Thus is the excellence of God magnified and the greatness of his kingdom made manifest; he is glorified not in one, but in countless suns; not in a single earth, a single world, but in a thousand thousand, I say in an infinity of worlds."

Galileo's and Bruno's sin was not that they dared to divine the laws of the heavens; their true sin was that they dethroned humanity from its exalted place at the center of the universe. It would take over 350 years, until 1992, for the Vatican to issue a belated apology to Galileo. No apology was ever issued to Bruno.

HISTORICAL PERSPECTIVE

Since Galileo, a series of revolutions have overturned our conception of the universe and our role in it. During the Middle Ages, the universe was seen as a dark, forbidding place. Earth was like a small, flat stage, full of corruption and sin, enclosed by a mysterious, celestial sphere where omens like comets would terrify kings and peasants alike. And if we were deficient in our praise of God and church, we would face the wrath of the theater critics, the self-righteous members of the Inquisition and their hideous instruments of persuasion.

Newton and Einstein freed us from the superstition and mysticism of the past. Newton gave us precise, mechanical laws that guided all celestial bodies, including our own. The laws were so precise, in fact, that human beings became mere parrots reciting their lines. Einstein revolutionized how we viewed the stage of life. Not only was it impossible to define a uniform measure of time and space, the stage itself was curved. Not only was the stage replaced by a stretched rubber sheet, it was expanding as well.

The quantum revolution gave us an even more bizarre picture of the world. On one hand, the downfall of determinism meant that the puppets were allowed to cut their strings and read their own lines. Free will was restored, but at the price of having multiple and uncertain outcomes. This meant that actors could be in two places at the same time and could disappear and reappear. It became impossible to tell for certain where an actor was on the stage or what time it was.

Now, the concept of the multiverse has given us another paradigm shift, where the word "universe" itself could become obsolete. With the multiverse, there are parallel stages, one above the other, with trapdoors and hidden tunnels connecting them. Stages, in fact, give rise to other stages, in a never-ending process of genesis. On each stage, new laws of physics emerge. On perhaps only a handful of these stages are the conditions for life and consciousness met.

Today, we are actors living in act 1, at the beginning of the exploration of the cosmic wonders of this stage. In act 2, if we don't destroy our planet through warfare or pollution, we may be able to leave Earth and explore the stars and other heavenly bodies. But we are now becoming aware that there is the final scene, act 3, when the play ends, and all the actors perish. In act 3, the stage becomes so cold that life becomes impossible. The only possible salvation is to leave the stage entirely via a trapdoor and start over again with a new play and a new stage.

COPERNICAN PRINCIPLE VS. ANTHROPIC PRINCIPLE

Clearly, in the transition from the mysticism of the Middle Ages to the quantum physics of today, our role, our place in the universe, has shifted dramatically with each scientific revolution. Our world has been expanding exponentially, forcing us to change our conception of ourselves. When I view this historic progression, I am sometimes overwhelmed by two contradictory emotions, as I gaze upon the seemingly limitless number of stars in the celestial firmament or contemplate the myriad forms of life on Earth. On one hand, I feel dwarfed by the immensity of the universe. When contemplating the vast, empty expanse of the universe, Blaise Pascal once wrote, "The eternal silence of those infinite spaces strikes me with terror." On the other hand, I cannot help but be mesmerized by the splendid diversity of life and the exquisite complexity of our biological existence.

Today, when approaching the question of scientifically ascertaining our role in the universe, there are in some sense two extreme philosophical points of view represented in the physics community: the Copernican principle and the anthropic principle.

The Copernican principle states that there is nothing special about our place in the universe. (Some wags have dubbed this the mediocrity principle.) So far, every astronomical discovery seems to vindicate this point of view. Not only did Copernicus banish Earth from the center of the universe, Hubble displaced the entire Milky Way galaxy from the center of the universe, giving us instead an expanding universe of billions of galaxies. The recent discovery of dark matter and dark energy underscores the fact that the higher chemical elements that make up our bodies comprise only 0.03 percent of the total matter/energy content of the universe. With the inflation theory, we must contemplate the fact that the visible universe is like a grain of sand embedded in a much larger, flat universe, and that this universe itself may be constantly sprouting new universes. And finally, if M-theory proves successful, we must face the possibility that even the familiar dimensionality of space and time must be expanded to eleven dimensions. Not only have we been banished from the center of the universe, we may find that even the visible universe is but a tiny fraction of a much larger multiverse.

Faced with the enormity of this realization, one is reminded of the poem by Stephen Crane, who once wrote,

> A man said to the universe:
> "Sir, I exist!"
> "However," replied the universe,
> "The fact has not created in me
> A sense of obligation."

(One is reminded of Douglas Adams's science fiction spoof *Hitchhiker's Guide to the Galaxy*, in which there is a device called the Total Perspective Vortex, which is guaranteed to transform any sane person into a raving lunatic. Inside the chamber is a map of the entire universe with a tiny arrow reading, "You are here.")

But at the other extreme, we have the anthropic principle, which makes us realize that a miraculous set of "accidents" makes consciousness possible in this three-dimensional universe of ours. There is a ridiculously narrow band of parameters that makes intelligent life a reality, and we happen to thrive in this band. The stability of the proton, the size of the stars, the existence of higher elements, and so on, all seem to be finely tuned to allow for complex forms of life and consciousness. One can debate whether this fortuitous circumstance is one of design or accident, but no one can dispute the intricate tuning necessary to make us possible.

Stephen Hawking remarks, "If the rate of expansion one second after the big bang had been smaller by even one part in a hundred thousand million, [the universe] would have recollapsed before it reached its present size . . . The odds against a universe like ours emerging out of something like the big bang are enormous. I think there are clearly religious implications."

We often fail to appreciate how precious life and consciousness really are. We forget that something as simple as liquid water is one of the most precious substances in the universe, that only Earth (and perhaps Europa, a moon of Jupiter) has liquid water in any quantity in the solar system, perhaps even in this sector of the galaxy. It is also likely that the human brain is the most complex object nature has created in the solar system, perhaps out to the nearest star. When we view the vivid pictures of the lifeless terrain of Mars or Venus, we are struck by the fact that those surfaces are totally barren of cities and lights or even the complex organic chemicals of life. Countless worlds exist in deep space devoid of life, much less of intelligence. It should make us appreciate how delicate life is, and what a miracle it is that it flourishes on Earth.

The Copernican principle and the anthropic principle are in some sense opposite perspectives which bracket the extremes of our existence and help us to understand our true role in the universe. While the Copernican principle forces us to confront the sheer enormity of the universe, and perhaps the multiverse, the anthropic principle forces us to realize how rare life and consciousness really are.

But ultimately, the debate between the Copernican principle and the anthropic principle cannot determine our role in the universe unless we view this question from an even larger perspective, from the point of view of the quantum theory.

QUANTUM MEANING

The world of quantum science sheds much light on the question of our role in the universe, but from a different point of view. If one subscribes to the Wigner interpretation of the Schrödinger cat problem, then we necessarily see the hand of consciousness everywhere. The infinite chain of observers, each one viewing the previous observer, ultimately leads to a cosmic observer, perhaps God himself. In this picture, the universe exists because there is a deity to observe it. And if Wheeler's interpretation is correct, then the entire universe is dominated by consciousness and information. In this picture, consciousness is the dominant force that determines the nature of existence.

Wigner's viewpoint, in turn, led Ronnie Knox to pen the following poem about an encounter between a skeptic and God, pondering if a tree exists in the courtyard when there is no one there to observe it:

> *There was once a man who said, "God*
> *Must think it exceedingly odd*
> *If he finds that this tree*
> *Continues to be*
> *When there's no one about in the Quad."*

An anonymous wag then penned the following reply:

> *Dear sir, Your astonishment's odd*
> *I am always about in the Quad*
> *And that's why the tree*
> *Will continue to be,*
> *Since observed by Yours faithfully—God*

In other words, trees exist in the courtyard because a quantum observer is always there to collapse the wave function—God himself.

Wigner's interpretation puts the question of consciousness at the very center of the foundation of physics. He echoes the words of the great astronomer James Jeans, who once wrote, "Fifty years ago, the universe was generally looked on as a machine . . . When we pass to extremes of size in either direction—whether to the cosmos as a whole, or to the inner recesses of the atom—the mechanical interpretation of Nature fails. We come to entities and phenomena which are in no sense mechanical. To me they seem less suggestive of mechanical than of mental processes; the universe seems to be nearer to a great thought than to a great machine."

This interpretation takes perhaps its most ambitious form in Wheeler's theory of it from bit. "It is not only that we are adapted to the universe. The universe is also adapted to us." In other words, in some sense we create our own reality by making observations. He calls this "Genesis by observership." Wheeler claims that we live in a "participatory universe."

These same words are echoed by Nobel laureate biologist George Wald, who wrote, "It would be a poor thing to be an atom in a universe without physicists. And physicists are made of atoms. A physicist is the atom's way of knowing about atoms." Unitarian minister Gary Kowalski summarizes this belief by saying, "The universe, it could be said, exists to celebrate itself and revel in its own beauty. And if the human race is one facet of the cosmos growing toward awareness of itself, our purpose must surely be to preserve and perpetuate our world as well as to study it, not to despoil or destroy what has taken so long to produce."

In this line of reasoning, the universe does have a point: *to produce sentient creatures like us who can observe it so that it exists.* According to this perspective, the very existence of the universe depends on its ability to create intelligent creatures who can observe it and hence collapse its wave function.

One may take comfort in the Wigner interpretation of the quantum theory. However, there is the alternate interpretation, the many-worlds interpretation, which gives us an entirely different conception of the role of humanity in the universe. In the many-worlds interpretations, Schrödinger's cat can be both dead and alive simultaneously, simply because the universe itself has split into two separate universes.

MEANING IN THE MULTIVERSE

It is easy to get lost in the infinite multitude of universes in the many-worlds theory. The moral implications of these parallel quantum universes are explored in a short story by Larry Niven, "All the Myriad Ways." In the story, Detective-Lieutenant Gene Trimble investigates a rash of mysterious suicides. Suddenly, all over town, people with no previous history of mental problems are jumping off bridges, blowing their brains out, or even committing mass murder. The mystery deepens when Ambrose Harmon, the billionaire founder of the Crosstime Corporation, jumps off the thirty-sixth floor of his luxury apartment after winning five hundred dollars at a poker table. Rich, powerful, and well-connected, he had everything to live for; his suicide makes no sense. But Trimble eventually discovers a pattern. Twenty percent of the pilots of the Crosstime Corporation have committed suicide; indeed, the suicides started a month after the founding of Crosstime.

Digging deeper, he finds that Harmon had inherited his vast fortune from his grandparents and squandered it backing harebrained causes. He might have lost his entire fortune, but for one gamble that paid off. He had assembled a handful of physicists, engineers, and philosophers to investigate the possibility of parallel time tracks. Eventually, they devised a vehicle that could enter a new time line, and the pilot promptly brought back a new invention from the Confederate States of America. Crosstime then bankrolled hundreds of missions to parallel time lines, where they would discover new inventions that could be brought back and patented. Soon, Crosstime became a billion-dollar corporation, holding the patents to the most important world-class inventions of our time. It looked as if Crosstime would be the most successful corporation of its age, with Harmon in charge.

Each time line, they found, was different. They found the Catholic Empire, Amerindian America, Imperial Russia, and scores of dead, radioactive worlds that had ended in nuclear war. But eventually, they find something deeply disturbing: carbon copies of themselves, living lives almost identical to their own, but with a bizarre twist. In these worlds, no matter what they do, anything can happen: no matter how hard they work, they might realize their most fantastic dreams or live through their most wrenching nightmare. Whatever they do, in some universes they are successful and in others they are complete failures. No matter what they do, there are an infinite number of copies of themselves who make the opposite decision and reap all possible consequences. Why not become a bank robber if, in some universe, you will walk away scot-free?

Trimble thinks, "There was no luck anywhere. Every decision was made both ways. For every wise choice you bled your heart out over, you made all the other choices too. And so it went, all through history." Profound despair overwhelms Trimble as he reaches a soul-wrenching realization: In a universe where everything is possible, nothing makes any moral sense. He falls victim to despair, realizing that we ultimately have no control over our fates, that no matter what decision we make, the outcome does not matter.

Eventually, he decides to follow Harmon's lead. He pulls out a gun and points it at his head. But even as he pulls the trigger, there are an infinite number of universes in which the gun misfires, the bullet hits the ceiling, the bullet kills the detective, and so on. Trimble's ultimate decision is played out in an infinite number of ways in an infinite number of universes.

When we imagine the quantum multiverse, we are faced, as Trimble is in the story, with the possibility that, although our parallel selves living in different quantum universes may have precisely the same genetic code, at crucial junctures of life, our opportunities, our mentors, and our dreams may lead us down different paths, leading to different life histories and different destinies.

One form of this dilemma is actually almost upon us. It's only a matter of time, perhaps a few decades, before the genetic cloning of humans becomes an ordinary fact of life. Although cloning a human being is extremely difficult (in fact, no one has yet fully cloned a primate, let alone a human) and the ethical questions are profoundly disturbing, it is inevitable that at some point it will happen. And when it does, the question arises: do our clones have a soul? Are *we* responsible for our clone's actions? In a quantum universe, we would have an infinite number of quantum clones. Since some of our quantum clones might perform acts of evil, are we then responsible for them? Does our soul suffer for the transgressions of our quantum clones?

There is a resolution to this quantum existential crisis. If we glance across the multiverse of infinite worlds, we may be overwhelmed by the dizzying randomness of fate, but within each world the commonsense rules of causality still hold in the main. In the multiverse theory proposed by physicists, each distinct universe obeys Newtonian-like laws on the macroscopic scale, so we can live our lives comfortably, knowing that our actions have largely predictable consequences. Within each universe, the laws of causality, on average, rigidly apply. In each universe, if we commit a crime, then most likely we will go to jail. We can conduct our affairs blissfully unaware of all the parallel realities that coexist with us.

It reminds me of the apocryphal story that physicists sometimes tell each other. One day, a physicist from Russia was brought to Las Vegas. He was dazzled by all the capitalist opulence and debauchery that sin city had to offer. He went immediately to the gaming tables and placed all his money on the first bet. When he was told that this was a silly gambling strategy, that his strategy flew in the face of the laws of mathematics and probability, he replied, "Yes, all that is true, but in one quantum universe, I shall be rich!" The Russian physicist may have been correct and in some parallel world may be enjoying wealth beyond his imagination. But in this particular universe he lost and left dead broke. And he must suffer the consequences.

WHAT PHYSICISTS THINK ABOUT THE MEANING OF THE UNIVERSE

The debate on the meaning of life was stirred even more by Steven Weinberg's provocative statements in his book *The First Three Minutes.* He writes, "The more the universe seems comprehensible, the more it also seems pointless . . . The effort to understand the universe is one of the very few things that lifts human life a little above the level of farce, and gives it some of the grace of tragedy." Weinberg has confessed that of all the sentences he has written, this one elicited the most heated response. He later created another controversy with his comment, "With or without religion, good people can behave well and bad people can do evil; but for good people to do evil—that takes religion."

Weinberg apparently takes a certain devilish delight in stirring up the pot, poking fun at the pretensions of those who profess some insight into the cosmic meaning of the universe. "For many years I have been a cheerful philistine in philosophical matters," he confesses. Like Shakespeare, he believes that all the world is a stage, "but the tragedy is not in the script; the tragedy is that there is no script."

Weinberg mirrors the words of fellow scientist Richard Dawkins of Oxford, a biologist who proclaims, "In a universe of blind physical forces . . . some people are going to get hurt, and other people are going to get lucky, and you won't find any rhyme or reason in it, nor any justice. The universe that we observe has precisely the properties we should expect if there is, at bottom, no design, no purpose, no evil, and no good, nothing but blind, pitiless indifference."

In essence, Weinberg is laying down a challenge. If people believe that the universe has a point, then what is it? When astronomers peer out into the vastness of the cosmos, with giant stars much larger than our Sun being born and dying in a universe that has been explosively expanding for billions of years, it is hard to see how all this could have been precisely arranged to give a purpose to humanity dwelling on a tiny planet revolving around an obscure star.

Although his statements have generated much heat, very few scientists have risen to confront them. Yet when Alan Lightman and Roberta Brawer interviewed a collection of prominent cosmologists to ask them if they agreed with Weinberg, interestingly, only a handful accepted Weinberg's rather bleak assessment of the universe. One scientist who was firmly in Weinberg's camp was Sandra Faber of the Lick Observatory and the University of California at Santa Cruz, who said, "I don't believe the earth was created for people. It was a planet created by natural processes, and, as part of the further continuation of those natural processes, life and intelligent life appeared. In exactly the same way, I think the universe was created out of some natural process, and our appearance in it was a totally natural result of physical laws in our particular portion of it. Implicit in the question, I think, is that there's some motive power that has a purpose beyond human existence. I don't believe in that. So, I guess ultimately I agree with Weinberg that it's completely pointless from a human perspective."

But a much larger camp of cosmologists thought Weinberg was off base, that the universe did have a point, even if they could not articulate it.

Margaret Geller, a professor at Harvard University, said, "I guess my view of life is that you live your life and it's short. The thing is to have as rich an experience as you possibly can. That's what I'm trying to do. I'm trying to do something creative. I try to educate people."

And a handful of them did indeed see a point to the universe in the handiwork of God. Don Page of the University of Alberta, a former student of Stephen Hawking, said, "Yes, I would say there's definitely a purpose. I don't know what all of the purposes are, but I think one of them was for God to create man to have fellowship with God. A bigger purpose maybe was that God's creation would glorify God." He sees the handiwork of God even in the abstract rules of quantum physics: "In some sense, the physical laws seem to be analogous to the grammar and the language that God chose to use."

Charles Misner of the University of Maryland, one of the early pioneers in analyzing Einstein's general relativity theory, finds common ground with Page: "My feeling is that in religion there are very serious things, like the existence of God and the brotherhood of man, that are serious truths that we will one day learn to appreciate in perhaps a different language on a different scale . . . So I think there are real truths there, and in the sense the majesty of the universe is meaningful, and we do owe honor and awe to its Creator."

The question of the Creator raises the question: can science say anything about the existence of God? The theologian Paul Tillich once said that physicists are the only scientists who can say the word "God" and not blush. Indeed, physicists stand alone among scientists in tackling one of humanity's greatest questions: is there a grand design? And if so, is there a designer? Which is the true path to truth, reason or revelation?

String theory allows us to view the subatomic particles as notes on a vibrating string; the laws of chemistry correspond to the melodies one can play on these strings; the laws of physics correspond to the laws of harmony that govern these strings; the universe is a symphony of strings; and the mind of God can be viewed as cosmic music vibrating through hyperspace. If this analogy is valid, one must ask the next question: is there a composer? Did someone design the theory to allow for the richness of possible universes that we see in string theory? If the universe is like a finely tuned watch, is there a watchmaker?

In this sense, string theory sheds some light on the question: did God have a choice? Whenever Einstein was creating his cosmic theories, he would always ask the question, how would I have designed the universe? He leaned toward the idea that perhaps God had no choice in the matter. String theory seems to vindicate this approach. When we combine relativity with the quantum theory, we find theories that are riddled with hidden but fatal flaws: divergences that blow up and anomalies that spoil the symmetries of the theory. Only by incorporating powerful symmetries can these divergences and anomalies be eliminated, and M-theory possesses the most powerful of these symmetries. Thus, perhaps, there might be a single, unique theory that obeys all the postulates that we demand in a theory.

Einstein, who often wrote at length about the Old One, was asked about the existence of God. To him, there were two types of gods. The first god was the personal god, the god who answered prayers, the god of Abraham, Isaac, Moses, the god that parts the waters and performs miracles. However, this is not the god that most scientists necessarily believe in.

Einstein once wrote that he believed in "Spinoza's God who reveals Himself in the orderly harmony of what exists, not in a God who concerns himself with fates and actions of human beings." The god of Spinoza and Einstein is the god of harmony, the god of reason and logic. Einstein writes, "I cannot imagine a God who rewards and punishes the objects of his creation . . . Neither can I believe that the individual survives the death of his body."

(In Dante's *Inferno*, the First Circle near the entrance to Hell is populated by people of good will and temperament who failed to fully embrace Jesus Christ. In the First Circle, Dante found Plato and Aristotle and other great thinkers and luminaries. As physicist Wilczek remarks, "We suspect that many, perhaps most, modern scientists will find their way to the First Circle.") Mark Twain might also be found in that illustrious First Circle. Twain once defined faith as "believing what any darn fool knows ain't so."

Personally, from a purely scientific point of view, I think that perhaps the strongest argument for the existence of the God of Einstein or Spinoza comes from teleology. If string theory is eventually experimentally confirmed as the theory of everything, then we must ask where the equations themselves came from. If the unified field theory is truly unique, as Einstein believed, then we must ask where this uniqueness came from. Physicists who believe in this God believe that the universe is so beautiful and simple that its ultimate laws could not have been an accident. The universe could have been totally random or made up of lifeless electrons and neutrinos, incapable of creating any life, let alone intelligent life.

If, as I and some other physicists believe, the ultimate laws of reality will be described by a formula perhaps no more than one inch long, then the question is, where did this equation come from?

As Martin Gardner has said, "Why does the apple fall? Because of the law of gravitation. Why the law of gravitation? Because of certain equations that are part of the theory of relativity. Should physicists succeed some day in writing one ultimate equation from which all physical laws can be derived, one could still ask, 'Why that equation?' "

CREATING OUR OWN MEANING

Ultimately, I believe the very existence of a single equation that can describe the entire universe in an orderly, harmonious fashion implies a design of some sort. However, I do not believe that this design gives personal meaning to humanity. No matter how dazzling or elegant the final formulation of physics may be, it will not uplift the spirits of billions and give them emotional fulfillment. No magic formula coming from cosmology and physics will enthrall the masses and enrich their spiritual lives.

For me, the real meaning in life is that we create our own meaning. It is our destiny to carve out our own future, rather than have it handed down from some higher authority. Einstein once confessed that he was powerless to give comfort to the hundreds of well-meaning individuals who wrote stacks of letters pleading with him to reveal the meaning of life. As Alan Guth has said, "It's okay to ask those questions, but one should not expect to get a wiser answer from a physicist. My own emotional feeling is that life has a purpose—ultimately, I'd guess that the purpose it has is the purpose that we've given it and not a purpose that came out of any cosmic design."

I believe that Sigmund Freud, with all his speculations about the dark side of the unconscious mind, came closest to the truth when he said that what gives stability and meaning to our minds is work and love. Work helps to give us a sense of responsibility and purpose, a concrete focus to our labors and dreams. Work not only gives discipline and structure to our lives, it also provides us with a sense of pride, accomplishment, and a framework for fulfillment. And love is an essential ingredient that puts us within the fabric of society. Without love, we are lost, empty, without roots. We become drifters in our own land, unattached to the concerns of others.

Beyond work and love, I would add two other ingredients that give meaning to life. First, to fulfill whatever talents we are born with. However blessed we are by fate with different abilities and strengths, we should try to develop them to the fullest, rather than allow them to atrophy and decay. We all know individuals who did not fulfill the promise they showed in childhood. Many of them became haunted by the image of what they might have become. Instead of blaming fate, I think we should accept ourselves as we are and try to fulfill whatever dreams are within our capability.

Second, we should try to leave the world a better place than when we entered it. As individuals, we can make a difference, whether it is to probe the secrets of Nature, to clean up the environment and work for peace and social justice, or to nurture the inquisitive, vibrant spirit of the young by being a mentor and a guide.

TRANSITION TO TYPE I CIVILIZATION

In Anton Chekhov's play *Three Sisters*, in act 2 Colonel Vershinin proclaims, "In a century or two, or in a millennium, people will live in a new way, a happier way. We won't be there to see it—but it's why we live, why we work. It's why we suffer. We're creating it. That's the purpose of our existence. The only happiness we can know is to work toward that goal."

Personally, rather than be depressed by the sheer enormity of the universe, I am thrilled by the idea of entirely new worlds that exist next to ours. We live in an age when we are just beginning the exploration of the cosmos with our space probes and space telescopes, our theories and equations.

I also feel privileged to be alive at a time when our world is undergoing such heroic strides. We are alive to witness perhaps the greatest transition in human history, the transition to a type I civilization, perhaps the most momentous, but also dangerous, transition in human history.

In the past, our ancestors lived in a harsh, unforgiving world. For most of human history, people lived short, brutish lives, with an average life expectancy of about twenty years. They lived in constant fear of diseases, at the mercy of the fates. Examination of the bones of our ancestors reveals that they are incredibly worn down, a testament to the heavy loads and burdens they carried daily; they also bear the telltale marks of disease and horrible accidents. Even within the last century, our grandparents lived without the benefit of modern sanitation, antibiotics, jet airplanes, computers, or other electronic marvels.

Our grandchildren, however, will live at the dawning of Earth's first planetary civilization. If we don't allow our often brutal instinct for self-destruction to consume us, our grandchildren could live in an age when want, hunger, and disease no longer haunt our destiny. For the first time in human history, we possess both the means for destroying all life on Earth or realizing a paradise on the planet.

As a child, I often wondered what it would be like to live in the far future. Today, I believe that if I could choose to be alive in any particular era of humanity, I would choose this one. We are now at the most exciting time in human history, the cusp of some of the greatest cosmic discoveries and technological advances of all time.

We are making the historic transition from being passive observers to the dance of nature to becoming choreographers of the dance of nature, with the ability to manipulate life, matter, and intelligence. With this awesome power, however, comes great responsibility, to ensure that the fruits of our efforts are used wisely and for the benefit of all humanity.

The generation now alive is perhaps the most important generation of humans ever to walk the Earth. Unlike previous generations, we hold in our hands the future destiny of our species, whether we soar into fulfilling our promise as a type I civilization or fall into the abyss of chaos, pollution, and war. Decisions made by us will reverberate throughout this century. How we resolve global wars, proliferating nuclear weapons, and sectarian and ethnic strife will either lay or destroy the foundations of a type I civilization. Perhaps the purpose and meaning of the current generation are to make sure that the transition to a type I civilization is a smooth one.

The choice is ours. This is the legacy of the generation now alive. This is our destiny.

NOTES

Chapter One: Baby Pictures of the Universe

6   *"a rite of passage for cosmology from speculation . . ."* www.space.com, Feb. 11, 2003.

10   *"What you will hear next is all wrong."* Croswell, p. 181.

10   *"That's a bunch of hooey. It's war—it's war!"* Croswell, p. 173.

12   *"We live in an . . ."* Britt, Robert. www.space.com, Feb. 11, 2003.

12   *"Frankly, we just don't understand it . . ."* www.space.com, Jan. 15, 2002.

28   *"We have laid the cornerstone of a unified coherent theory . . ."* New York Times, Feb. 12, 2003, p. A34.

14   *"No theory as beautiful as this has ever been wrong before."* Lemonick, p. 53.

15   *"Inflation pretty much . . ."* New York Times, Oct. 29, 2002, p. D4.

15   *"What's conventionally called 'the universe' . . ."* Rees, p. 3.

19   *"The universe is behaving like a driver who slows down . . ."* New York Times, Feb. 18, 2003, p. F1.

20   *"Believing as I do . . ."* Rothman, Tony. Discover magazine, July, 1987, p. 87.

21   *"Wormholes, if they exist, would be ideal for rapid space travel . . ."* Hawking, p. 88.

Chapter Two: The Paradoxical Universe

23   *"How do you know? . . ."* Bell, p. 105.

25   *"The Universe is not bounded in any direction . . ."* Silk, p. 9.

26   *"A continual miracle is needed . . ."* Croswell, p. 8.

27   *"How fortunate that the Earth . . ."* Croswell, p. 6.

28   *"Were the succession of stars endless . . ."* Smoot, p. 28.

28   *"When I first read Poe's words I was astounded . . ."* Croswell, p. 10.

29   *"We might have seen . . ."* New York Times, March 10, 2004, p. A1.

30   *"Hubble takes us to within a stone's throw . . ."* New York Times, March 10, 2004, p. A1.

30   *"The misfortune of my poor parents, who for so many years . . ."* Pais2, p. 41.

31   *"Such a principle resulted from a paradox . . ."* Schilpp, p. 53.

33   *If time could change depending on your velocity, Einstein realized . . . The contraction*

of objects moving near the speed of light was actually found by Hendrik Lorentz and George Francis FitzGerald shortly before Einstein, but they did not understand this effect. They tried to analyze the effect in a purely Newtonian framework, assuming the contraction was an electromechanical squeezing of the atoms created by passing through the "ether wind." The power of Einstein's ideas was that he not only got the entire theory of special relativity from one principle (the constancy of the speed of light), he interpreted this as a universal principle of nature that contradicted Newtonian theory. Thus, these distortions were inherent properties of space-time, rather than being electromechanical distortions of matter. The great French mathematician Henri Poincaré perhaps came closest to deriving the same equations as Einstein. But only Einstein had the complete set of equations and the deep physical insight into the problem.

34   "As an older friend, I must advise you against it . . ." Pais2, p. 239.

39   "one of the greatest . . ." Folsing, p. 444.

39   "Not at all . . ." Parker, p. 126.

40   "I feel as if . . ." Brian, p. 102.

42   This is the principle . . . When gas expands, it cools down. In your refrigerator, for example, a pipe connects the inside and outside of the chamber. As gas enters the inside of the refrigerator, it expands, which cools the pipe and the food. As it leaves the inside of the refrigerator, the pipe contracts, so the pipe gets hot. There is also a mechanical pump that drives the gas through the pipe. Thus, the back of the refrigerator gets warm, while the interior gets cold. Stars work in the reverse order. When gravity compresses the star, the star heats up, until fusion temperatures are reached.

Chapter Three: The Big Bang

51   "The evolution of the world can be compared to a display of fireworks . . ." Lemonick, p. 26.

51   "As a scientist, I simply do not believe . . ." Croswell, p. 37.

52   "Ninety percent of Gamow's . . ." Smoot, p. 61.

52   "classes were often suspended when Odessa was bombarded . . ." Gamow1, p. 14.

53   "I think this was the experiment which made me a scientist." Croswell, p. 39.

53   "There was a young fellow from Trinity . . ." Gamow2, p. 100.

55   In typical fashion, Gamow laid out . . . Croswell, p. 40.

56   "Every time you buy a balloon, you are getting atoms . . ." New York Times, April 29, 2003, p. F3.

57   "Extrapolating from the early days of the universe . . ." Gamow1, p. 142.

58   "We expended a hell of a lot of energy giving talks about the work . . ." Croswell, p. 41.

59   "I concluded that, unhappily, I'd been born into a world . . ." Croswell, p. 42.

59   For that impudent act of insubordination . . . Croswell, p. 42.

60   "I think we saw that movie several months before . . ." Croswell, p. 43.

61   *"There is no way in which I coined the phrase to be derogatory . . ."* Croswell, pp. 45–46.

61   *"When I was fifteen, I heard Fred Hoyle give lectures on the BBC . . ."* Croswell, p. 111. Hoyle's fifth and final lecture, however, was the most controversial because he criticized religion. (Hoyle once said, in characteristic bluntness, that the solution to the problem in Northern Ireland was to jail every priest and clergyman. "Not all the religious quarrels I ever saw or read about is worth the death of a single child," he said. Croswell, p. 43.)

63   *"In the excitement of counting . . ."* Gamow1, 127.

69   *"Whether it was the too-great comfort of the Cadillac . . ."* Croswell, p. 63.

69   *"It is widely believed that the existence of the microwave background . . ."* Croswell, pp. 63–64.

71   *"Today's sycophants . . ."* Croswell, p. 101.

72   *He was incensed that he was passed over when the Nobel Prize . . .* Although Zwicky, to his dying day, publicly expressed his bitterness because his scientific discoveries were ignored, Gamow kept quiet in public over being passed over for the Nobel Prize, although he expressed his great disappointment in private letters. Instead, Gamow turned his considerable scientific talents and creativity to DNA research, eventually unlocking one of the secrets of how nature makes amino acids from DNA. Nobel laureate James Watson even acknowledged that contribution by putting Gamow's name in the title of his recent autobiography.

72   *"That became a tag line in my family . . ."* Croswell, p. 91.

74   *"When fossils were found in the rocks . . ."* Scientific American, July 1992, p. 17.

Chapter Four: Inflation and Parallel Universes

85   *"How would you suspend 500,000 pounds of water . . ."* Cole, p. 43.

86   *"Like the unicorn, the monopole has continued to fascinate . . ."* Guth, p. 30.

89   *"I was still worried that some consequence of theory might . . ."* Guth, pp. 186–67.

89   *"Did Steve have any objections to it? . . ."* Guth, p. 191.

90   *"I was in a marginal . . ."* Guth, p. 18.

90   *"This 'inflation' idea sounds crazy . . ."* Kirschner, p. 188.

90   *"a fashion the high-energy physicists have visited on the cosmologists . . ."* Rees1, p. 171.

92   *"I just had the feeling that it was impossible for God . . ."* Croswell, p. 124.

95   *Although we take this for granted, the cancellation . . .* Rees2, p. 100.

95   *There is one apparent exception to this rule . . .* Scientists have looked for antimatter in the universe and have found little (except some streams of antimatter near the Milky Way's core). Since matter and antimatter are virtually indistinguishable, obeying the same laws of physics and chemistry, it is quite difficult to tell them apart. However, one way is to look for characteristic gamma ray emissions of 1.02 million electron volts. This is the fingerprint for the presence of antimatter because this is the minimum

energy released when an electron collides with an antielectron. But when we scan the universe, we see no evidence of large amounts of 1.02-million-electron-volt gamma rays, one indication that antimatter is rare in the universe.

97   *"The secret of nature is symmetry . . ."* Cole, p. 190.

99   *"Everything that happens in our world . . ."* Scientific American, June, 2003, p. 70.

102   *"I'm completely snowed by the cosmic background radiation . . ."* New York Times, July 23, 2002, p. F7.

103   *If a white dwarf star weighs more than 1.4 solar masses . . .* Chandrasekhar's limit can be derived by the following reasoning. On one hand, gravity acts to compress a white dwarf star to incredible densities, which brings the electrons in the star closer and closer together. On the other hand, there is the Pauli exclusion principle, which states that no two electrons can have exactly the same quantum numbers describing its state. This means that two electrons cannot occupy precisely the same point with the same properties, so that there is a net force pushing the electrons apart (in addition to electrostatic repulsion). This means that there is a net pressure pushing outward, preventing the electrons from being crushed further into each other. We can therefore calculate the mass of the white dwarf star when these two forces (one of repulsion and one of attraction) exactly cancel each other, and this is the Chandrasekhar limit of 1.4 solar masses.

For a neutron star, we have gravity crushing a ball of pure neutrons, so there is a new Chandrasekhar limit of roughly 3 solar masses, since the neutrons also repel each other due to this force. But once a neutron star is more massive than its Chandrasekhar limit, then it will collapse into a black hole.

104   *"The Lambda thing has always been a wild-eyed concept . . ."* Croswell, p. 204.

104   *"I was still shaking my head, but we had checked everything . . ."* Croswell, p. 222.

104   *"the strangest experimental finding since I've been in physics."* New York Times, July 23, 2002, p. F7.

Chapter Five: Dimensional Portals and Time Travel

116   *"It would be a true disaster for the theory . . ."* Parker, p. 151.

117   *"The essential result of this investigation is a clear understanding . . ."* Thorne, p. 136

117   *"be a law of Nature to prevent a star from behaving . . ."* Thorne, p. 162.

121   *"Pass through this magic ring and—presto! . . ."* Reesi, p. 84.

122   *"Ten years ago, if you found an object that you thought was a black hole . . ."* Astronomy Magazine, July 1998, p. 44.

125   *"This star was stretched beyond . . ."* Reesi, p. 88.

129   *"This state of affairs seems to imply an absurdity . . ."* Nahin, p. 81.

130   *"Kurt Gödel's essay constitutes, in my opinion, an important contribution . . ."* Nahin, p. 81.

134 *As shown by Jacob Bekenstein and Stephen Hawking* . . . They were among the first to apply quantum mechanics to black hole physics. According to the quantum theory, there is a finite probability that a subatomic particle may tunnel its way out of the black hole's gravitational pull, and hence it should slowly emit radiation. This is an example of tunneling.

136 *"Everything not forbidden is compulsory."* Thorne, p. 137.

139 *"there is not a grain of evidence to suggest that the time machine* . . ." Nahin, p. 521.

139 *"There is no law of physics preventing the appearance of closed timelike curves."* Nahin, p. 522.

139 *"not as a vindication for time travel enthusiasts, but rather* . . ." Nahin, p. 522.

141 *"When I found this solution* . . ." Gott, p. 104.

141 *"To allow time travel to the past, cosmic strings with a mass-per-unit length* . . ." Gott, p. 104.

142 *"A collapsing loop of string large enough to allow you to circle it* . . ." Gott, p. 110.

143 *The sexual paradox.* One well-known example of a sexual paradox was written by the British philosopher Jonathan Harrison in a story published in 1979 in the magazine *Analysis*. The magazine's readers were challenged to make sense of it.

The story begins with a young lady, Jocasta Jones, who one day finds an old deep freezer. Inside the freezer she discovers a handsome young man frozen alive. After thawing him out, she finds out that his name is Dum. Dum tells her he possesses a book that describes how to build a deep freeze that can preserve humans and how to build a time machine. The two fall in love, marry, and soon have a baby boy, whom they call Dee.

Years later, when Dee has grown to be a young man, he follows in his father's footsteps and decides to build a time machine. This time, both Dee and Dum take a trip into the past, taking the book with them. However, the trip ends tragically, and they find themselves stranded in the distant past and running out of food. Realizing that the end is near, Dee does the only thing possible to stay alive, which is to kill his father and eat him. Dee then decides to follow the book's instructions and build a deep freeze. To save himself, he enters the freezer and is frozen in a state of suspended animation.

Many years later, Jocasta Jones finds the freezer and decides to thaw Dee out. To disguise himself, Dee calls himself Dum. They fall in love, and then have a baby, whom they call Dee . . . and so the cycle continues.

The reaction to Harrison's challenge provoked a dozen replies. One reader claimed it was "a story so extravagant in its implications that it will be regarded as a *reductio ad abusurdum* of the one dubious assumption on which this story rests: the possibility of time travel." Notice that the story does not contain a grandfather paradox here, since Dee is fulfilling the past by going back in time to meet his mother. At no point does Dee do anything that makes the present impossible. (There is an information paradox, however, since the book containing the secret of suspended animation and time travel appears from nowhere. But the book itself is not essential to the story.)

Another reader pointed out a strange biological paradox. Since half the DNA of any individual comes from the mother and half from the father, this means that Dee should have half of his DNA from Ms. Jones and half from his father, Dum. However, Dee is Dum. Therefore, Dee and Dum must have the same DNA because they are the same person. But this is impossible since, by the laws of genetics, half their genes come from Ms. Jones. In other words, time travel stories in which a person goes back in time, meets his mother, and fathers himself violate the laws of genetics.

One might think there is a loophole to the sexual paradox. If you are able to become both your father and mother, then all of your DNA comes from yourself. In Robert Heinlein's tale "All You Zombies," a young girl has a sex change operation and goes back twice in time to become her own mother, father, son, and daughter. However, even in this strange tale, there is a subtle violation of the laws of genetics.

In "All You Zombies," a young girl named Jane is raised in an orphanage. One day she meets and falls in love with a handsome stranger. She gives birth to his baby girl, who is mysteriously kidnapped. Jane has complications during childbirth, and doctors are forced to change Jane into a man. Years later, this man meets a time traveler, who takes him back into the past, where he meets Jane as a young girl. They fall in love, and Jane gets pregnant. He then kidnaps his own baby girl and goes further back into the past, dropping the baby Jane off at an orphanage. Then Jane grows up to meet a handsome stranger. This story almost evades the sexual paradox. Half your genes are those of Jane the young girl, and half of your genes are from Jane the handsome stranger. However, a sex change operation cannot change your X chromosome into a Y chromosome, and hence this story also has sex paradox.

144   *"We cannot send a time traveler back to the Garden of Eden . . ."* Hawking, pp. 84–85.

144   *"For example, it can be my will to walk on the ceiling . . ."* Hawking, pp. 84–85.

145   *This eliminates the infinite divergences found by Hawking . . .* Ultimately, to resolve these complex mathematical questions, one must go to a new kind of physics. For example, many physicists, such as Stephen Hawking and Kip Thorne, use what is called the semiclassical approximation—that is, they take a hybrid theory. They assume that the subatomic particles obey the quantum principle, but they allow gravity to be smooth and unquantized (that is, they banish gravitons from their calculations). Since all the divergences and anomalies come from the gravitons, the semiclassical approach does not suffer from infinities. However, one can show mathematically that the semiclassical approach is inconsistent—that is, it ultimately gives wrong answers, so the results from a semiclassical calculation cannot be

trusted, especially in the most interesting areas, such as the center of a black hole, the entrance to a time machine, and the instant of the big bang. Notice that many of the "proofs" stating that time travel is not possible or that you cannot pass through a black hole were done in the semiclassical approximation and hence are not reliable. That is why we have to go to a quantum theory of gravity such as string theory and M-theory.

Chapter Six: Parallel Quantum Universes

150 It was Wheeler who coined . . . Bartusiak, p. 62.

151 "The underlying physical laws necessary for the mathematical theory . . ." Cole, p. 68.

154 "for such an intellect, nothing could be uncertain . . ." Cole, p. 68.

154 "I am a determinist, compelled to act as if free will existed . . ." Brian, p. 185.

156 "Number 1: I calls 'em like I see 'em . . ." Bernstein, p. 96.

156 "Madness is the ability to make fine distinctions . . ." Weinberg2, p. 103.

156 "Is not all of philosophy as if written in honey? . . ." Pais2, p. 318.

156 Physicists also like to tell the apocryphal story supposedly told . . . Barrow1, p. 185.

157 "There was a time when the newspapers said that only twelve men . . ." Barrow3, p. 143.

157 "describes nature as absurd from the point of view of common sense . . ." Greene1, p. 111.

157 "I admit to some discomfort in working all my life in a theoretical framework . . ." Weinberg1, p. 85.

158 "Science cannot solve the ultimate mystery of Nature . . ." Barrow3, p. 378.

159 "It was wonderful for me to be present at the dialogues . . ." Folsing, p. 589.

159 "To Bohr, this was a heavy blow . . ." Folsing, p. 591; Brian, p. 199.

160 "I am convinced that this theory undoubtedly contains . . ." Folsing, p. 591.

160 "Of course, today every rascal thinks he knows the answer . . ." Kowalski, p. 156.

161 "The energy produced . . ." New York Herald Tribune, Sept. 12, 1933.

162 Since there was no stopping the Nazi juggernaut . . . New York Times, Feb. 7, 2002, p. A12.

165 "The average quantum mechanic is no more philosophical . . ." Rees1, p. 244.

165 "was not possible to formulate the laws of quantum mechanics . . ." Crease, p. 67.

165 "Nothing ever becomes real till it is experienced." Barrow1, p. 458.

166 "For me as a human being . . ." Discover magazine, June 2002, p. 48.

169 "There is a universe . . ." Quoted in BBC-TV's Parallel Universes, 2002.

169 "We are haunted by the awareness . . ." Wilczek, pp. 128–29.

169 "Whenever a creature was faced with several possible courses of action . . ." Rees1, p. 246.

171 "Where there's smoke, there's smoke." Bernstein, p. 131.

171 "I am just driven crazy by that question . . ." Bernstein, p. 132.

177 "who know each other . . ."
National Geographic News, www.nationalgeographic.com, Jan. 29, 2003.

178   "Possibly, larger objects . . ."
178   "The key thing for now . . ."

Chapter Seven: M-Theory: The Mother of All Strings

182   "I found a general principle . . ." Nahin, p. 147.
184   "There may be any number of three-dimensional . . ." Wells2, p. 20.
186   "You may be amused to hear . . ." Pais2, p. 179.
186   "I believe I am right . . ." Moore, p. 432.
187   "We in the back are convinced . . ." Kaku2, p. 137.
188   "By rights, twentieth-century physicists . . ." Davies2, p. 102.

191   In an equation barely an inch and a half long, we could summarize all the information contained within string theory. In principle, all of string theory could be summarized in terms of our string field theory. However, the theory was not in its final form, since manifest Lorentz invariance was broken. Later, Witten was able to write down an elegant version of open bosonic string field theory that was covariant. Later, the MIT group, the Kyoto group, and I were able to construct the covariant closed bosonic string theory (which, however, was nonpolynomial and hence difficult to work with). Today, with M-theory, interest has shifted to membranes, but it is not clear if a genuine membrane field theory can be constructed.

192   Similarly, the superstring model of Neveu, Schwarz, and Ramond could only exist in ten dimensions. There are actually several reasons why ten and eleven are preferred numbers in string theory and M-theory. First, if we study the representations of the Lorentz group in increasingly higher dimensions, we find that in general the number of fermions grows exponentially with the dimension, while the number of bosons grows linearly with the dimension. Thus, for only low dimensions can we have a supersymmetric theory with equal numbers of fermions and bosons. If we do a careful analysis of the group theory, we find that we have a perfect balance if we have ten and eleven dimensions (assuming that we have at maximum a particle of spin two, not three or higher). Thus, on purely group theoretic grounds, we can show that ten and eleven are preferred dimensions.

There are other ways to show that ten and eleven are "magic numbers." If we study the higher loop diagrams, we find that in general unitarity is not preserved, which is a disaster for the theory. It means that particles can appear and disappear as if by magic. We find that unitarity is restored for the perturbation theory in these dimensions.

We can also show that in ten and eleven dimensions, "ghost" particles can be made to vanish. These are particles that do not respect the usual conditions for physical particles.

In summary, we can show that in these "magic numbers" we can preserve (a) supersymmetry, (b) finiteness of the perturbation theory, (c) unitarity of the perturbation series, (d) Lorentz invariance, (e) anomaly cancellation.

192   *"Well, John, and how many . . ."* Private communication.

194   *Similar divergences plague any quantum theory of gravity.* When physicists try to solve a complex theory, they often use "perturbation theory," the idea of solving a simpler theory first and then analyzing small deviations from this theory. These tiny deviations, in turn, give us an infinite number of small correction factors to the original, idealized theory. Each correction is usually called a Feynman diagram and can graphically be described by diagrams representing all possible ways in which the various particles can bump into each other.

Historically, physicists were troubled by the fact that the terms in the perturbation theory became infinite, rendering the entire program useless. However, Feynman and his colleagues discovered a series of ingenious tricks and manipulations by which they could brush these infinities under the rug (for which they won the Nobel Prize in 1965).

The problem with quantum gravity is that this set of quantum corrections is actually infinite—each correction factor equals infinity, even if we use the bag of tricks devised by Feynman and his colleagues. We say that quantum gravity is "not renormalizable."

In string theory, this perturbation expansion is actually finite, which is the fundamental reason why we study string theory in the first place. (Technically speaking, an absolutely rigorous proof of this does not exist. However, infinite classes of diagrams can be shown to be finite, and less-than-rigorous mathematical arguments have been given showing that the theory is probably finite to all orders.) However, the perturbation expansion alone cannot represent the universe as we know it, since the perturbation expansion preserves perfect supersymmetry, which we do not see in nature. In the universe, we see that the symmetries are badly broken (for example, we see no experimental evidence of superparticles). Hence, physicists want a "nonperturbative" description of string theory, which is exceedingly difficult. In fact, at present there is no uniform way in which to calculate nonperturbative corrections to a quantum field theory. There are many problems constructing a nonperturbative description. For example, if we wish to increase the strength of the forces in the theory, it means that each term in the perturbation theory gets larger and larger, so that the perturbation theory makes no sense. For example, the sum $1 + 2 + 3 + 4 \ldots$ makes no sense, since each term gets larger and larger. The advantage of M-theory is that, for the first time, we can establish nonperturbative results via duality. This means that the nonperturbative limit of one string theory can be shown to be equivalent to another string theory.

195   *Gradually, they realized the solution might be to abandon the Band-Aid approach and adopt an entirely new theory.* String theory and M-theory represent a radical

new approach to general relativity. While Einstein built up general relativity around the concept of curved space-time, string theory and M-theory are built up around the concept of an extended object, such as a string or membrane, moving in a supersymmetric space. Ultimately, it may be possible to link these two pictures, but at present this is not well understood.

195   *"I'm not one to be modest . . ."* Discover magazine, Aug. 1991, p. 56.

197   *"Music creates order out of chaos . . ."* Barrow2, p. 305.

198   *"Music is the hidden arithmetic exercise of a soul . . ."* Barrow2, p. 205.

198   *"Music and science were [once] identified so profoundly . . ."* Barrow2, p. 205.

203   *This precisely describes the symmetry of the superstring, called supersymmetry.* In the late 1960s, when physicists first began to look for a symmetry that might include all the particles of nature, gravity was pointedly not included. This is because there are two types of symmetries. The ones found in particle physics are those that reshuffle the particles among themselves. But there is also another type of symmetry, which turns space into time, and these space-time symmetries are associated with gravity. Gravity theory is based not on the symmetries of interchanging point particles, but on the symmetries of rotations in four dimensions: the Lorentz group in four dimensions $O(3,1)$.

At this time, Sidney Coleman and Jeffrey Mandula proved a celebrated theorem stating that it was impossible to marry space-time symmetries, which describe gravity, with the symmetries describing the particles. This no-go theorem threw cold water on any attempt to construct a "master symmetry" of the universe. For example, if anyone tried to marry the GUT group $SU(5)$ with the relativity group $O(3,1)$, one found a catastrophe. For example, the masses of the particles would suddenly become continuous rather than discrete. This was disappointing, since it meant that one could not naively include gravity with the other forces by appealing to a higher symmetry. This meant that a unified field theory was probably impossible.

String theory, however, solves all of these thorny mathematical problems with the most powerful symmetry ever found for particle physics: supersymmetry. At present, supersymmetry is the only known way in which to avoid the Coleman-Mandula theorem. (Supersymmetry exploits a tiny but crucial loophole in this theorem. Usually, when we introduce numbers like a or b, we assume that $a \times b = b \times a$. This was tacitly assumed in the Coleman-Mandula theorem. But in supersymmetry, we introduce "supernumbers," such that $a \times b = -b \times a$. These supernumbers have strange properties. For example, if $a \times a = 0$, then a can be nonzero, which sounds absurd for ordinary numbers. If we insert supernumbers into the Coleman-Mandula theorem, we find that it fails.)

205   *Supersymmetry also solves a series of highly technical problems . . .* First, it solves the hierarchy problem, which dooms GUT theory. When constructing unified field theories, we come up with two quite different mass scales. Some

particles, like the proton, have masses like those found in everyday life. Other particles, however, are quite massive and have energies comparable to those found near the big bang, the Planck energy. These two mass scales have to be kept separate. However, when we factor in quantum corrections, we find disaster. Because of quantum fluctuations, these two types of masses begin to mix, because there is finite probability that one set of light particles will turn into the other set of heavy particles, and vice versa. This means that there should be a continuum of particles with masses varying smoothly between everyday masses and the enormous masses found at the big bang, which we clearly do not see in nature. This is where supersymmetry comes in. One can show that the two energy scales do not mix in a supersymmetric theory. There is a beautiful cancellation process that takes place, so that the two scales never interact with each other. Fermion terms cancel precisely against boson terms, yielding finite results. To our knowledge, supersymmetry may be the only solution to the hierarchy problem.

In addition, supersymmetry solves the problem first posed by the Coleman-Mandula theorem of the 1960s, which proved that it was impossible to combine a symmetry group that acted on the quarks, like SU(3), with a symmetry that acted on space-time, as in Einstein's relativity theory. Thus, a unifying symmetry that united both was impossible, according to the theorem. This was discouraging, because it meant that unification was mathematically impossible. However, supersymmetry provides a subtle loophole to this theorem. It is one of the many theoretical breakthroughs of supersymmetry.

217 *"Pure mathematics is, in its way, the poetry of logical ideas."* Cole, p. 174.

217 *"[The universe] cannot be read until we have learnt the language . . ."* Wilzcek, p. 138.

218 *"The discrepancy is not small . . ."* www.edge.org, Feb. 10, 2003.

220 *"There was a lot of excitement when it was first suggested . . ."* www.edge.org, Feb. 10, 2003.

223 *"Maybe the acceleration of the expansion of the universe . . ."* Seife, p. 197.

224 *"That would be like throwing a chair into a black hole . . ."* Astronomy magazine, May 2002, p. 34.

224 *"If you start . . ."* Astronomy magazine, May 2002, p. 34.

224 *"Flat plus flat . . ."* Astronomy magazine, May 2002, p. 34.

224 *"I don't think Paul and Neil come close to proving their case . . ."* Discover magazine, Feb. 2004, p. 41.

224 *"In the long run, I think it's inevitable that string theory and M-theory . . ."* Astronomy magazine, May 2002, p. 39.

225 *"I think it's silly . . ."* Discover magazine, Feb. 2004, p. 41.

229 *"Most physicists want to believe that information is not lost . . ."* Greene1, p. 343.

232 *Maldacena showed that there is a duality between this five-dimensional universe . . .* More precisely, what Maldacena showed was that type II string theory, compactified to a five-dimensional anti–de Sitter space, was dual to a four-

dimensional conformal field theory located on its boundary. The original hope was that a modified version of this bizarre duality could be established between string theory and four-dimensional QCD (quantum chromodynamics), the theory of the strong interactions. If such a duality can be constructed, it would represent a breakthrough, because then one might be able to compute the properties of the strongly interacting particles, such as the proton, directly from string theory. However, at present this hope is not yet fulfilled.

235   "Field theory, with its . . ." Scientific American, Aug. 2003, p. 65.

235   "a final theory . . ." Ibid.

239   "Currently, string theorists are in a position analogous to an Einstein bereft of the equivalence principle . . ." Greene1, p. 376.

Chapter Eight: A Designer Universe?

243   "Without the Moon there would be no moonbeams, no month . . ." Brownlee and Ward, p. 222.

244   "There are worlds infinite in number and different in size . . ." Barrow1, p. 37.

245   "You can think of the star and the large planet as dance partners . . ." www.sciencedaily.com, July 4, 2003.

246   What was so unusual about this planet . . . www.sciencedaily.com, July 4, 2003.

246   "We are working to place all 2,000 of the nearest sun-like stars under survey . . ." www.sciencedaily.com, July 4, 2003.

248   Physicist Don Page has summarized . . . Page, Don. "The Importance of the Anthropic Principle." Pennsylvania State University, 1987.

248   "The exquisite order . . ." Margenau, p. 52.

248   "not just 'any old world,' but it's special and finely tuned for life . . ." Rees2, p. 166.

248   "It is almost irresistible for humans to believe . . ." New York Times, Oct. 29, 2002, p. D4.

249   "I find it hard to believe that anybody would ever use the anthropic principle . . ." Lightman, p. 479.

250   "The apparent fine-tuning on which our existence depends . . ." Rees1, p. 3.

250   Rees points to the fact that . . . Rees2, p. 56.

251   "At one second after the big bang, Omega cannot have differed from unity . . ." Rees2, p. 99.

252   "great gobs of matter would have condensed into huge black holes . . ." Discover magazine, Nov. 2000, p. 68.

253   "If there is a large stock of clothing, you're not surprised . . ." Discover magazine, Nov. 2000, p. 66.

Chapter Nine: Searching for Echoes from the Eleventh Dimension

256   "Other universes can get intoxicating . . ." Croswell, p. 128.

257   Everything from computerized maps inside cars to cruise missiles . . . Bartusiak, p. 55.

257   *But in order to guarantee such incredible accuracy, scientists must calculate slight corrections to Newton's laws due to relativity, which states that radio waves will be slightly shifted in frequency as satellites soar in outer space.* This shift takes places in two ways. Because near-Earth satellites travel at 18,000 miles per hour, special relativity takes over, and time slows down on the satellite. This means that clocks on the satellite appear to slow down a bit compared to clocks on the ground. But because the satellite experiences a weaker gravitational field in outer space, time also speeds up, because of general relativity. Thus, depending on the distance the satellite is from Earth, the satellite's clocks will either slow down (due to special relativity) or speed up (due to general relativity). In fact, at a certain distance from Earth, the two effects will exactly balance out, and the clock on the satellite will run at the same speed as a clock on Earth.

258   *"Every time we have looked at the sky in a new way, we have seen a new universe . . ."* Newsday, Sept. 17, 2002, p. A46.

259   *For their work, they won the Nobel Prize in physics in 1993.* Newsday, Sept. 17, 2002, p. A47.

260   *"Imagine the earth were that smooth. Then the average mountain . . ."* Bartusiak, p. 152.

260   *"Most control systems engineers' jaws drop when they hear . . ."* Bartusiak, pp. 158–59.

260   *"It feels like a rumble . . ."* Bartusiak, p. 154.

261   *Sensitive optical instruments each have their own seismic isolation system . . .* Bartusiak, p. 158.

261   *Altogther, LIGO's final construction cost will be $292 million . . .* Bartusiak, p. 150.

261   *"You go from . . ."* Bartusiak, p. 169.

261   *"People take pleasure in solving these technical challenges . . ."* Bartusiak, p. 170.

261   *With LIGO II, the chances are much better . . .* Bartusiak, p. 171.

262   *If all goes according to plan . . .* The cosmic background radiation measured by the WMAP satellite dates back to 379,000 years after the big bang, because that is when atoms began to condense for the first time after the initial explosion. However, gravity waves that LISA might detect date back to when gravity first began to separate out from the other forces, which took place near the instant of the big bang itself. Hence, some physicists believe that LISA will be able to verify or rule out many of the theories being proposed today, including string theory.

263   *"Half of this deflection is produced by the Newtonian field . . ."* Scientific American, Nov. 2001, p. 66.

264   *"not much hope of observing this phenomenon . . ."* Petters, pp. 7, 11.

264   *Over forty years later, in 1979, the first partial evidence . . .* Scientific American, Nov. 2001, p. 68.

264   *Today, Einstein's rings are an essential weapon . . .* Scientific American, Nov. 2001, p. 68.

264   *Since then, about a hundred galactic arcs . . . Scientific American*, Nov. 2001, p. 70.

266   *In 1998, astronomers at the Harvard-Smithsonian Center for Astrophysics . . . Scientific American*, Nov. 2001, p. 69.

266   *Physicists estimate that a billion dark matter particles . . . Scientific American*, March 2003, p. 54.

267   *So far, experiments with acronyms like UKDMC . . . Scientific American*, March 2003, p. 55.

267   *"If the detectors do register and verify a signal . . ." Scientific American*, March 2003, p. 59.

275   *"So far, Newton is holding his ground . . ."* www.space.com, Feb. 27, 2003.

276   *"Physicists are sure that nature has new tricks up her sleeve . . ." Scientific American*, July 2000, p. 71.

277   *Estimates of the mass of the Higgs boson . . . Scientific American*, June 2003, p. 75.

279   *But the Soviet Union broke apart . . .* In the final days of hearings on the fate of the SSC, a congressman asked the question: what will we find with this machine? Unfortunately, the answer given was the Higgs boson. You could almost hear the jaws hit the floor; $11 billion for just another particle? One of the last questions was asked by Rep. Harris W. Fawell (R-Ill.), who asked, "Will this [machine] make us find God?" Rep. Don Ritter (R-Penn.) then added, "If this machine does that, I am going to come around and support it." (Weinberg1, p. 244). Unfortunately, the congressmen were not given a cogent, persuasive answer by physicists.

As a result of this and other public-relations mistakes, the SSC was canceled. The U.S. Congress had given us a billion dollars to dig the hole for the machine. Then Congress canceled it and gave us a second billion dollars to fill up the hole. The Congress, in its wisdom, had given us $2 billion to dig a hole and then fill it, making it the most expensive hole in history.

(Personally, I think that the poor physicist who had to answer that question about God should have said, "Your honor, we may or may not find God, but our machine will take us the closest that is humanly possible to God, by whatever name you may call the diety. It may reveal the secret of His greatest act, the creation of the universe itself.")

282   *"Although somewhat fanciful, this is my favorite scenario for confirming string theory . . ."* Greene1, p. 224.

282   *Brian Greene lists five possible examples . . .* Greene1, p. 225.

283   *"I am convinced . . ."* Kaku3, p. 699.

Chapter Ten: The End of Everything

289   *The first law states that the total . . .* This law, in turn, means that "perpetual motion machines" which claim to get "something for nothing" are not possible with the known laws of physics.

290   *"The law that entropy always increases . . ."* Barrow1, p. 658.

291   "The Collapse of the Universe: An Eschatological Study." Rees1, p. 194.

292   "Regrettably I have to concur that in this case we have no escape . . ." Rees1, p. 198.

295   Computer simulations done at the University of California at Santa Cruz . . . www.sci-encedaily.com, May 28, 2003; Scientific American, Aug. 2003, p. 84.

296   "As long as people get smarter faster than the Sun gets brighter . . ." Croswell, p. 231.

296   "During the several billion years before the Sun bloats into a red giant . . ." Croswell, p. 232.

296   Because this dwarf star will weigh only 0.55 solar masses . . . Astronomy Magazine, Nov. 2001, p. 40.

297   "Mother Nature wasn't designed to make us happy . . ." www.abcnews.com, Jan. 24, 2003.

298   A mini–black hole the size of a proton might radiate . . . Rees1, p. 182.

299   "And so, finally, after 10^{117} years . . ." Discover magazine, July 1987, p. 90.

301   "Billions of years ago the universe was too hot for life to exist . . ." Scientific American, Nov. 1999, pp. 60–63.

302   "Eternity would be a prison, rather than an endlessly receding horizon . . ." Scientific American, Nov. 1999, pp. 60–63.

Chapter Eleven: Escaping the Universe

306   "Wormholes, extra dimensions, and quantum computers . . ." Rees3, p. 182.

309   The entire population of a type I civilization may be bilingual in this fashion, speaking both a local language and a planetary language. This may also apply to a type I culture. In many third-world countries, an elite that speaks both the local language and English also keeps up with the latest in Western culture and fashion. A type I civilization may then by bicultural, with a planetary culture that spans the entire globe, coexisting with local cultures and customs. So a planetary culture does not necessarily mean the destruction of local cultures.

314   Jun Jugaku of the Research Institute of Civilization in Japan and his colleagues have searched . . . Scientific American, July 2000, p. 40.

315   "Assuming a typical colony spacing of 10 light-years . . ." Scientific American, July 2000, p. 41.

315   However, this does not rule out civilizations that are just beyond us in technology . . . Scientific American, July 2000, p. 40.

316   To prevent the fragmentation of such a Carroll universe . . . Dyson, p. 163.

317   When I reminded him that there are only planets, stars, and galaxies . . . Conceivably, there might be a civilization even higher than type III, which exploits the power of dark energy, which makes up 73 percent of the total matter/energy content of the universe. In the TV series Star Trek, the Q would qualify for such a civilization, since the power of the Q spans the galaxies.

321   "It's quite conceivable that, even if life now exists only here on Earth . . ." Lightman, p. 169.

321   *"If we snuffed ourselves out, we'd be destroying genuine cosmic potentialities . . ."* Lightman, p. 169.

327   *"Does this mean that the laws of physics truly enable us to create a new universe . . ."* Guth, p. 255.

336   *"A future supercivilization might want to lay down . . ."* Gott, p. 126.

339   *"It seems . . . that quantum theory allows time travel on a microscopic basis."* Hawking, p. 104.

340   Each neural connection in the brain would be replaced by a corresponding transistor . . . In principle, this process could be done while you were conscious. As bits of neurons were deleted from your brain, duplicate transistor networks would be created to replace them, placed in the skull of a robot. Since the transistors perform the same function as the deleted neurons, you would be fully conscious during this procedure. Thus, after the operation was finished, you would find yourself in the body of a silicon-and-metal robot.

Chapter Twelve: Beyond the Multiverse

343   *"The question of all questions for humanity . . ."* Kaku2, p. 334.

344   *"I want to know how God created this world . . ."* Calaprice, p. 202.

344   *"Science without religion is lame. But religion without science is blind."* Calaprice, p. 213.

344   *"the extreme difficulty or rather impossibility . . ."* Kowalski, p. 97.

344   *"My theology is simply a muddle."* Ibid.

345   *"Thus is the excellence of God magnified . . ."* Croswell, p. 7.

347   *"The eternal silence of those infinite spaces strikes me with terror . . ."* Smoot, p. 24.

348   *"A man said to the universe . . ."* Barrow1, p. 106.

348   *"If the rate of expansion one second after the big bang . . ."* Kowalski, p. 49.

350   *"There once was a man who said . . ."* Polkinghorne, p. 66.

350   *"Fifty years ago, the universe was generally looked on as a machine . . ."* Kowalski, p. 19.

350   *"It is not only . . ."* Kowalski, p. 50.

351   *"It would be a poor thing . . ."* Kowalski, p. 71.

351   *"The universe, it could be said, exists to celebrate itself and revel in its own beauty . . ."* Kowalski, p. 71.

353   Eventually, he decides to follow Harmon's lead . . . Chown, p. 30.

354   *"The more the universe seems comprehensible, the more it also seems pointless . . ."* Weinberg3, p. 144.

354   *"With or without religion, good people can behave well and bad people can do evil . . ."* Weinberg2, p. 231.

354   *"For many years I have been a cheerful philistine in philosophical matters . . ."* Weinberg2, p. 43.

354   *"but the tragedy is not in the script; the tragedy is that there is no script."* Weinberg2, p. 43.

355 "In a universe of blind physical forces . . . some people are going to get hurt . . ." Kowalski, p. 60.

355 "I don't believe the earth was created for people . . ." Lightman, p. 340.

355 "I guess my view of life. . ." Lightman, p. 377.

356 "Yes, I would say that there's definitely a purpose . . ." Lightman, p. 409.

356 "In some sense, the physical laws seem to be analogous to . . ." Lightman, p. 409.

356 "My feeling is that in religion there are very serious things . . ." Lightman, p. 248.

356 The theologian Paul Tillich once said that physicists are the only scientists . . . Weinberg1, p. 242.

357 "Spinoza's God who reveals Himself in the orderly harmony of what exists . . ." Weinberg1, p. 245.

357 "I cannot imagine a God who rewards and punishes the objects of his creation . . ." Kowalski, p. 24.

357 "We suspect that many, perhaps most, modern scientists . . ." Wilczek, p. 100.

357 Twain once defined faith as . . . Kowalski, p. 168.

358 "Why does the apple fall? . . ." Kowalski, p. 148.

359 "It's okay to ask those questions . . ." Croswell, p. 127.

GLOSSARY

anthropic principle   The principle that the constants of nature are tuned to allow for life and intelligence. The strong anthropic principle concludes that an intelligence of some sort was required to tune the physical constants to allow for intelligence. The weak anthropic principle merely states that the constants of nature must be tuned to allow for intelligence (otherwise we would not be here), but it leaves open the question of what or who did the tuning. Experimentally, we find that, indeed, the constants of nature seem to be finely tuned to allow for life and even consciousness. Some believe that this is the sign of a cosmic creator. Others believe that this is a sign of the multiverse.

antigravity   The opposite of gravity, which would be a repulsive rather than an attractive force. Today, we realize that this antigravity force does exist, probably caused the universe to inflate at the beginning of time, and is causing the universe to accelerate today. This antigravity force, however, is much too small to be measured in the laboratory, so it has no practical implications. Antigravity is also generated by negative matter (which has never been seen in nature).

antimatter   The opposite of matter. Antimatter, first predicted to exist by P. A. M. Dirac, has the opposite charge of ordinary matter, so that antiprotons have negative charge and antielectrons (positrons) have positive charge. When they come in contact, they annihilate each other. So far, antihydrogen is the most complex antiatom produced in the laboratory. It is a mystery why our universe is made mainly of matter rather than antimatter. If the big bang had created equal quantities of both, then they should have annihilated each other, and we would not exist.

atom smasher   The colloquial term for a particle accelerator, a device used to create beams of subatomic energy traveling near the speed of light. The largest particle accelerator is the LHC, to be built near Geneva, Switzerland.

baryon   A particle like the proton or neutron, which obeys the strong interactions. Baryons are a type of hadron (a strongly interacting particle). Baryonic matter, we now realize, makes up only a tiny fraction of the matter in the universe and is dwarfed by dark matter.

big bang   The original explosion that created the universe, sending the galaxies hurtling in all directions. When the universe was created, the temperature was extremely hot, and the density of material was enormous. The big bang took place 13.7 billion years ago, according to the WMAP satellite. The afterglow of the big bang is seen today as the background microwave radiation. There are three experimental "proofs" of the big bang: the redshift of the galaxies, the cosmic background microwave radiation, and nucleosynethsis of the elements.

big crunch   The final collapse of the universe. If the density of matter is large enough (Omega being larger than 1), then there is enough matter in the universe to reverse the original expansion and cause the universe to recollapse. Temperatures rise to infinity at the instant of the big crunch.

big freeze   The end of the universe when it reaches near absolute zero. The big freeze is probably the final state of our universe, because the sum of Omega and Lambda is believed to be 1.0, and hence the universe is in a state of inflation. There is not enough matter and energy to reverse the original expansion of the universe, so it will probably expand forever.

black body radiation   The radiation emitted by a hot object in thermal equilibrium with its environment. If we take an object that is hollow (a black body), heat it up, wait for it to reach thermal equilibrium, and drill a small hole in it, the radiation emitted through the hole will be black body radiation. The Sun, a hot poker, and molten magma all emit approximately a black body radiation. The radiation has a specific frequency dependence that is easily measured by a spectrometer. The microwave background radiation filling up the universe obeys this black body radiation formula, giving concrete evidence for the big bang.

black hole   An object whose escape velocity equals the speed of light. Because the speed of light is the ultimate velocity in the universe, this means that nothing can escape a black hole, once an object has crossed the event horizon. Black holes can be of various sizes. Galactic black holes, lurking in the center of galaxies and quasars, can weight millions to billions of solar masses. Stellar black holes are the remnant of a dying star, perhaps originally up to forty times the mass of our Sun. Both of these black holes have been identified with our instruments. Mini–black holes may also exist, as predicted by theory, but they have not yet been seen in the laboratory.

black hole evaporation   The radiation that tunnels out of a black hole. There is a tiny but calculable probability that radiation will gently seep out of a black hole, which is called evaporation. Eventually, so much of a black hole's energy will leave via quantum evaporation that it will cease to exist. This radiation is too weak to be observed experimentally.

blueshift   The increase in the frequency of starlight because of the Doppler shift. If a yellow star is moving toward you, its light will look slightly bluish. In outer space, blueshifted galaxies are rare. Blueshift can also be created by shrinking the space between two points via gravity or space warps.

boson   A subatomic particle with integral spin, such as the photon or the conjectured graviton. Baryons are unified with fermions via supersymmetry.

brane   Abbreviation for membrane. Branes can be in any dimension up to eleven. They are the basis of M-theory, the leading candidate for a theory of everything. If we take a cross-section of an eleven-dimensional membrane, we obtain a ten-dimensional string. A string is therefore a one-brane.

Calabi-Yau manifold   A six-dimensional space that is found when we take ten-dimensional string theory and roll up or compactify six dimensions into a small ball, leaving a four-dimensional supersymmetric space. Calabi-Yau spaces are multiply connected—that is, they have holes in them, which can determine the number of quark generations that exist in our four-dimensional space. They are important in string theory because many of the features of these manifolds, such as the number of holes they have, can determine the number of quarks there are in our four-dimensional universe.

Casimir effect   Negative energy created by two infinitely long parallel uncharged plates placed next to each other. Virtual particles outside the plates exert more pressure than the virtual particles inside the plates, and hence the plates are attracted to each other. This tiny effect has been measured in the laboratory. The Casimir effect may be used as the energy to drive a time machine or wormhole, if its energy is large enough.

Cepheid variable   A star that varies in brightness at a precise, calculable rate and hence serves as a "standard candle" for distance measurements in astronomy. Cepheid variables were decisive in helping Hubble calculate the distance to the galaxies.

Chandrasekhar limit   1.4 solar masses. Beyond this mass, a white dwarf star's gravity is so immense that it will overcome electron degeneracy pressure and crush the star, creating a supernova. Thus, all white dwarf stars we observe in the universe have mass less than 1.4 solar masses.

Chandra X-ray telescope   The X-ray telescope in outer space that can scan the heavens for X-ray emissions, such as those emitted by a black hole or neutron star.

chaotic inflation   A version of inflation, proposed by Andrei Linde, whereby inflation occurs at random. This means that universes can bud off other universes in a continual, chaotic fashion, creating a multiverse. Chaotic inflation is one way to solve the problem of ending inflation, since we now have the random generation of inflated universes of all types.

classical physics   Physics before the coming of the quantum theory, based on the deterministic theory of Newton. Relativity theory, because it does not incorporate the uncertainty principle, is part of classical physics. Classical physics is deterministic—that is, we can predict the future given the motions of all particles at present.

closed time-like curves   These are paths that go backward in time in Einstein's theory. They are not allowed in special relativity but are allowed in general relativity if we have a large enough concentration of positive or negative energy.

COBE   The Cosmic Observer Background Explorer satellite, which gave perhaps the most conclusive proof of the big bang theory by measuring the black body radiation given off by the original fireball. Its results have since been improved greatly by the WMAP satellite.

coherent radiation   Radiation that is in phase with itself. Coherent radiation, like that found in a laser beam, can be made to interfere with itself, yielding interference patterns that can detect small deviations in motion or position. This is useful in interferometers and gravity wave detectors.

compactification   The process of rolling up or wrapping up unwanted dimensions of space and time. Since string theory exists in ten-dimensional hyperspace, and we live in a four-dimensional world, we must somehow wrap up six of the ten dimensions into a ball so small that even atoms cannot escape into them.

conservation laws   The laws that state that certain quantities never change with time. For example, the conservation of matter and energy posits that the total amount of matter and energy in the universe is a constant.

Copenhagen school   The school founded by Niels Bohr, which states that an observation is necessary in order to "collapse the wave function" to determine the state of an object. Before an observation is made, an object exists in all possible states, even absurd ones. Since we do not observe dead cats and live cats existing simultaneously, Bohr had to assume that there is "wall" separating the subatomic world from the everyday world we observe with our senses. This interpretation has been challenged because it separates the quantum world from the everyday, macroscopic world, while many physicists now believe that the macroscopic world must also obey the quantum theory. Today, because of nanotechnology, scientists can manipulate individual atoms, so we realize that there no "wall" separating the two worlds. Hence, the cat problem resurfaces today.

cosmic microwave background radiation The residual radiation left over from the big bang which is still circulating around the universe, first predicted in 1948 by George Gamow and his group. Its temperature is 2.7 degrees above absolute zero. Its discovery by Penzias and Wilson gave the most convincing "proof" of the big bang. Today, scientists measure tiny deviations within this background radiation to provide evidence for inflation or other theories.

cosmic string A remnant of the big bang. Some gauge theories predict that some relics of the original big bang might still survive in the form of gigantic cosmic strings that are the size of galaxies or larger. The collision of two cosmic strings may allow for a certain form of time travel.

critical density The density of the universe where the expansion of the universe is poised between eternal expansion and recollapse. The critical density, measured in certain units, is Omega = 1 (where Lambda = 0), where the universe is precisely balanced between two alternate futures, the big freeze and the big crunch. Today, the best data from the WMAP satellite indicates that Omega + Lambda = 1, which fits the prediction of the inflation theory.

dark energy The energy of empty space. First introduced by Einstein in 1917 and then discarded, this energy of nothing is now known to be the dominant form of matter/energy in the universe. Its origin is unknown, but it may eventually drive the universe into a big freeze. The amount of dark energy is proportional to the volume of the universe. The latest data shows that 73 percent of the matter/energy of the universe is in the form of dark energy.

dark matter Invisible matter, which has weight but does not interact with light. Dark matter is usually found in a huge halo around galaxies. It outweighs ordinary matter by a factor of 10. Dark matter can be indirectly measured because it bends starlight due to its gravity, somewhat similar to the way glass bends light. Dark matter, according to the latest data, makes up 23 percent of the total matter/energy content of the universe. According to string theory, dark matter may be made of subatomic particles, such as the neutralino, which represent higher vibrations of the superstring.

decoherence   When waves are no longer in phase with each other. Decoherence can be used to explain the Schrödinger cat paradox. In the many worlds interpretation, the wave function of the dead cat and live cat have decohered from each other and hence no longer interact, thus solving the problem of how a cat be simultaneously dead and live. The wave function of the dead cat and the wave function of the live cat both exist simultaneously, but they no longer interact because they have decohered. Decoherence simply explains the cat paradox without any additional assumptions, such as the collapse of the wave function.

de Sitter universe   A cosmological solution of Einstein's equations that expands exponentially. The dominant term is a cosmological constant that creates this exponential expansion. It is believed that the universe was in a de Sitter phase during inflation, and that it has slowly returned to a de Sitter phase within the last 7 billion years, creating an accelerating universe. The origin of this de Sitter expansion is not known.

determinism   The philosophy that everything is predetermined, including the future. According to Newtonian mechanics, if we know the velocity and position of all the particles in the universe, then we can in principle calculate the evolution of the entire universe. The uncertainty principle, however, has proved that determinism is incorrect.

deuterium   The nucleus of heavy hydrogen, consisting of a proton and a neutron. Deuterium in outer space was mainly created by the big bang, not by stars, and its relative abundance allows for the calculation of the early conditions of the big bang. The abundance of deuterium can also be used to disprove the steady state theory.

dimension   A coordinate or parameter by which we measure space and time. Our familiar universe has three dimensions of space (length, width, and depth) and one dimension of time. In string and M-theory, we need ten (eleven) dimensions in which to describe the universe, only four of which can be observed in the laboratory. Perhaps the reason why we don't see these other dimensions is either that they are curled up or that our vibrations are confined to the surface of a membrane.

Doppler effect   The change in frequency of a wave, as an object approaches or moves away from you. If a star moves toward you, the frequency of light increases, so a yellow star appears slightly bluish. If a star moves away from you, the frequency of its light decreases, so a yellow star appears slightly reddish. This change in light frequency can also be created by expanding space itself between two points, as in the expanding universe. By measuring the amount of shift in the frequency, you can calculate the velocity with which a star is moving away from you.

Einstein lenses and rings   The optical distortions of starlight as it passes through intergalactic space due to gravity. Distant galactic clusters often have a ringlike appearance. Einstein lenses can be used to calculate many key measurements, including the presence of dark matter and even the value of Lambda and the Hubble constant.

Einstein-Podolsky-Rosen (EPR) experiment   An experiment devised to disprove the quantum theory but which actually showed that the universe is nonlocal. If an explosion sends two coherent photons in opposite directions, and if spin is conserved, then the spin of one photon is the opposite of the other's spin. Hence, by measuring one spin, you automatically know the other, even though the other particle may be on the other side of the universe. Information has hence spread faster than light. (However, no usable information, such as a message, can be sent in this fashion.)

Einstein-Rosen bridge   A wormhole formed by joining two black hole solutions together. Originally, the solution was meant to represent a subatomic particle, such as the electron, in Einstein's unified field theory. Since then, it has been used to describe space-time near the center of a black hole.

electromagnetic force   The force of electricity and magnetism. When they vibrate in unison, they create a wave that can describe ultraviolet radiation, radio, gamma rays, and so on, which obeys Maxwell's equations. The electromagnetic force is one of the four forces governing the universe.

electron   A negatively charged subatomic particle that surrounds the nucleus of an atom. The number of electrons surrounding the nucleus determines the chemical properties of the atom.

electron degeneracy pressure   In a dying star, this is the repulsive force that prevents electrons or neutrons from completely collapsing. For a white dwarf star, this means that its gravity can overcome this force if its mass is greater than 1.4 solar masses. This force is due to the Pauli exclusion principle, which states that no two electrons can occupy precisely the same quantum state. If gravity is sufficiently large to overcome this force in a white dwarf star, it will collapse and then explode.

electron volt   The energy that an electron accumulates by falling through a potential of one volt. By comparison, chemical reactions normally involve energies measured in electron volts or less, while nuclear reactions may involve hundreds of millions of electron volts. Ordinary chemical reactions involve rearranging the electron shells. Nuclear reactions involve rearranging the shells of the nucleus. Today, our particle accelerators can generate particles with energies in the billions to trillions of electron volts.

entropy   The measure of disorder or chaos. According to the second law of thermodynamics, the total entropy in the universe always increases, which means that everything must eventually run down. Applied to the universe, it means that the universe will tend toward a state of maximum entropy, such as a uniform gas near absolute zero. To reverse the entropy in a small region (such as a refrigerator), the addition of mechanical energy is required. But even for a refrigerator, the total entropy increases (which is why the back of a refrigerator is warm). Some believe that the second law ultimately predicts the death of the universe.

event horizon   The point of no return surrounding a black hole, often called the horizon. It was once believed to be a singularity of infinite gravity, but this was shown to be an artifact of the coordinates used to describe it.

exotic matter   A new form of matter with negative energy. It is different from antimatter, which has positive energy. Negative matter would have antigravity, so it would fall up instead of down. If it exists, it could be used to drive a time machine. However, none has ever been found.

extrasolar planet   A planet orbiting a star other than our own. Over a hundred such planets have now been detected, at a rate of about two a month. Most of them, unfortunately, are Jupiter-like and are not favorable to the creation of life. Within a few decades, satellites will be sent into outer space that will identify Earth-like extrasolar planets.

false vacuum   A vacuum state that does not have the lowest energy. The false vacuum state can be one of perfect symmetry, perhaps at the instant of the big bang, so this symmetry breaks when we descend to a state of lower energy. A state of false vacuum is inherently unstable, and inevitably a transition is made to a true vacuum, which has lower energy. The false vacuum idea is essential to inflationary theory, where the universe began in a state of de Sitter expansion.

fermion   A subatomic particle with half-integral spin, such as the proton, electron, neutron, and quark. Fermions can be unified with bosons via supersymmetry.

fine-tuning   The adjustment of a certain parameter to incredible accuracy. Physicists dislike fine-tuning, considering it artificial and contrived, and try to impose physical principles to eliminate the necessity for fine-tuning. For example, the fine-tuning necessary to explain a flat universe can be explained by inflation, and the fine-tuning necessary to solve the hierarchy problem in GUT theory can be solved using supersymmetry.

flatness problem   The fine-tuning necessary to have a flat universe. In order for Omega to be roughly equal to 1, it must have been fine-tuned to incredible accuracy at the instant of the big bang. Current experiments show that the universe is flat, so either it was fine-tuned at the big bang, or perhaps the universe inflated, which flattened it out.

Friedmann universe   The most general cosmological solution of Einstein's equations based on a uniform, isotropic, homogeneous universe. This is a dynamic solution, where the universe can expand into a big freeze, collapse into a big crunch, or inflate forever, depending on the value of Omega and Lambda.

fusion   The process of combining protons or other light nuclei so they form higher nuclei, releasing energy in the process. The fusion of hydrogen to helium creates the energy of a main sequence star, like our Sun. The fusion of the light elements in the big bang gives us the relative abundance of light elements, like helium.

galaxy   A huge collection of stars, usually containing on the order of 100 billion stars. They come in several varieties, including elliptical, spiral (normal and barred spirals), and irregular. Our galaxy is called the Milky Way galaxy.

general relativity   Einstein's theory of gravity. Instead of being a force, gravity is reduced in Einstein's theory to a byproduct of geometry, so that the curvature of space-time gives the illusion that there is a force of attraction called gravity. It has been verified experimentally to better than 99.7 percent accuracy and predicts the existence of black holes and the expanding universe. The theory, however, must break down at the center of a black hole or the instant of creation, where the theory predicts nonsense. To explain these phenomena, one must resort to a quantum theory.

Goldilocks zone   The narrow band of parameters in which intelligent life is possible. In this band, Earth and the universe are "just right" to create the chemicals that are responsible for intelligent life. Scores of Goldilocks zones have been discovered for the physical constants of the universe, as well as for the properties of Earth.

Grand Unified Theory (GUT)   A theory that unifies the weak, strong, and electromagnetic interactions (without gravity). The symmetry of GUT theories, such as SU(5), mixes the quarks and leptons together. The proton is not stable in these theories and can decay into positrons. GUT theories are inherently unstable (unless one adds supersymmetry). GUT theories also lack gravity. (Adding gravity to GUT theories makes them diverge with infinities.)

grandfather paradox   In time travel stories, this is the paradox that emerges when you alter the past, making the present impossible. If you go back in time and kill your parents before you are born, then your very existence is impossible. This paradox can be resolved either by imposing self-consistency, so you can journey to the past but cannot change it arbitrarily, or by assuming parallel universes.

graviton   A conjectured subatomic particle that is the quanta of gravity. The graviton has spin 2. It is too small to be seen in the laboratory.

gravity wave   A wave of gravity, predicted by Einstein's general relativity theory. This wave has been indirectly measured by looking at the aging of pulsars rotating around each other.

gravity wave detector   A new generation of devices that measure tiny disturbances due to gravity waves via laser beams. Gravity wave detectors like LIGO may soon discover them. Gravity wave detectors can be used to analyze radiation emitted within a trillionth of a second of the big bang. The space-based LISA gravity wave detector may even give the first experimental evidence of string theory or some other theory.

Hawking radiation   The radiation that slowly evaporates from a black hole. This radiation is in the form of black body radiation, with a specific temperature, and is due to the fact that quantum particles can penetrate the gravitational field surrounding a black hole.

heterotic string theory   The most physically realistic string theory. Its symmetry group is E(8) × E(8), which is large enough to incorporate the symmetry of the Standard Model. Via M-theory, the heterotic string can be shown to be equivalent to the other four string theories.

hierarchy problem   The unwanted mixing that takes place between low-energy physics and physics at the Planck length in GUT theories, rendering them useless. The hierarchy problem can be solved by adding supersymmetry.

Higgs field   The field that breaks the symmetry of GUT theories when it makes the transition from the false vacuum to the real vacuum. Higgs fields are the origin of mass in GUT theory and also can be used to drive inflation. Physicists hope that the LHC will finally discover the Higgs field.

horizon   The farthest point you can see. Surrounding a black hole there is a magic sphere, at the Schwarzschild radius, which is the point of no return.

horizon problem   The mystery of why the universe is so uniform no matter where we look. Even regions of the night sky on opposite sides of the horizon are uniform, which is strange because they could not have been in thermal contact at the beginning of time (since light has a finite velocity). This can be explained if the big bang took a tiny uniform patch and then inflated it to the present-day universe.

Hubble's constant   The velocity of a redshifted galaxy divided by its distance. Hubble's constant measures the rate of expansion of the universe, and its inverse correlates roughly to the age of the universe. The lower the Hubble constant, the older the universe. The WMAP satellite has placed the Hubble constant at 71 km/s per million parsecs, or 21.8 km/s per million light-years, ending decades of controversy.

Hubble's law   The farther a galaxy is from Earth, the faster it moves. Discovered by Edwin Hubble in 1929, this observation agrees with Einstein's theory of an expanding universe.

hyperspace   Dimensions higher than four. String theory (M-theory) predicts that there should be ten (eleven) hyperspatial dimensions. At present, there is no experimental data indicating the existence of these higher dimensions, which may be too small to measure.

inflation   The theory which states that the universe underwent an incredible amount of superliminal expansion at the instant of its birth. Inflation can solve the flatness, monopole, and horizon problems.

infrared radiation   Heat radiation, or electromagnetic radiation, that is slightly below visible light in frequency.

interference   The mixing of two waves that are slightly different in phase or frequency, creating a characteristic interference pattern. By analyzing this pattern, one may be able to detect tiny differences between two waves which differ only by an extremely small amount.

interferometry   The process of using the interference of light waves to detect very small differences in the waves from two different sources. Interferometry can be used to measure the presence of gravity waves and other objects that are normally difficult to detect.

isotope   A chemical that has the same number of protons as an element but with a different number of neutrons. Isotopes have the same chemical properties but have different weight.

Kaluza-Klein theory   The theory of Einstein formulated in five dimensions. When reduced down to four dimensions, we find Einstein's usual theory coupled to Maxwell's theory of light. Thus, this was the first nontrivial unification of light with gravitation. Today, Kaluza-Klein theory is incorporated within string theory.

Kerr black hole   An exact solution of Einstein's equations which represents a spinning black hole. The black hole collapses into a ring singularity. Objects falling into the ring experience only a finite force of gravity and may, in principle, fall through to a parallel universe. There are an infinite number of these parallel universes for a Kerr black hole, but you cannot return once you enter one of them. It is still not known how stable the wormhole is at the center of a Kerr black hole. There are severe theoretical and practical problems trying to navigate through a Kerr black hole.

Lambda   The cosmological constant, which measures the amount of dark energy in the universe. At present, the data supports Omega + Lambda = 1, which fits the prediction of inflation for a flat universe. Lambda, which was once thought to be zero, is now known to determine the ultimate destiny of the universe.

laser   A device for creating coherent light radiation. "Laser" stands for Light Amplification through Stimulated Emission of Radiation. In principle, the only limit to the energy contained on a laser beam is the stability of the lasing material and the power source.

lepton   A weakly interacting particle, such as the electron and neutrino, and its higher generations, such as the muon. Physicists believe that all matter consists of hadrons and leptons (strongly and weakly interacting particles).

LHC   The Large Hadron Collider, a particle accelerator for creating energetic beams of protons, based in Geneva, Switzerland. When finally completed, it will collide particles with energies not seen since the big bang. It is hoped that the Higgs particle and sparticles will be found by the LHC after it opens in 2007.

light-year   The distance light travels in one year, or approximately 5.88 trillion miles (9.46 trillion kilometers). The nearest star is about four light-years away, and the Milky Way galaxy is about 100,000 light-years across.

LIGO   The Laser Interferometry Gravitational-Wave Observatory, based in Washington state and Louisiana, is the world's largest gravity wave detector. It went online in 2003.

LISA   The Laser Interferometry Space Antenna is a series of three space satellites using laser beams to measure gravity waves. It may be sensitive enough to confirm or disprove the inflationary theory and possibly even string theory, when it is launched in a few decades.

MACHO   Massive Compact Halo Object. These are dark stars, planets, asteroids, and such which are hard to detect by optical telescopes and may make up a portion of dark matter. The latest data indicates that the bulk of dark matter is nonbaryonic and is not made of MACHOs.

many-worlds theory   The quantum theory which states that all possible quantum universes can exist simultaneously. It solves the Schrödinger cat problem by stating that the universe splits at each quantum juncture, and hence the cat is alive in one universe but dead in another. Recently, an increasing number of physicists have voiced their support for the many-worlds theory.

Maxwell's equation   The fundamental equations for light, first written down by James Clerk Maxwell in the 1860s. These equations show that electric and magnetic fields can turn into each other. Maxwell showed that these fields turn into each other in a wavelike motion, creating an electromagnetic field that travels at the speed of light. Maxwell then made the bold conjecture that this was light.

membrane   An extended surface, in any dimensions. A zero-brane is a point particle. A one-brane is a string. A two-brane is a membrane. Membranes are an essential feature of M-theory. Strings can be viewed as membranes with one dimension compactified.

microwave background radiation   The remnant of the original radiation from the big bang, with a temperature of about 2.7 degrees K. Tiny deviations in this background radiation give scientists valuable data that can verify or rule out many cosmological theories.

monopole   A single pole of magnetism. Usually, magnets have an inseparable pair of north and south poles, so monopoles have never been conclusively seen in the laboratory. Monopoles should have been created in copious quantities at the big bang, but we can find none today, probably because inflation diluted their number.

M-theory   The most advanced version of string theory. M-theory exists in eleven-dimensional hyperspace, where two-branes and five-branes can exist. There are five ways in which M-theory can be reduced down to ten dimensions, thereby giving us the five known superstring theories, which are now revealed to be the same theory. The full equations governing M-theory are totally unknown.

multiply connected space   A space in which a lasso or loop cannot be continuously shrunk down to a point. For example, a loop that winds around the surface of a doughnut hole cannot be contracted to a point, hence a doughnut is multiply connected. Wormholes are examples of multiply connected spaces, since a lasso cannot be contracted around the throat of a wormhole.

multiverse   Multiple universes. Once considered highly speculative, today the concept of the multiverse is considered essential to understanding the early universe. There are several forms of the multiverse which are all intimately related. Any quantum theory has a multiverse of quantum states. Applied to the universe, it means that there must be an infinite number of parallel universes which have decohered from each other. Inflation theory introduces the multiverse to explain the process of how inflation started and then stopped. String theory introduces the multiverse because of its large number of possible solutions. In M-theory, these universes may actually collide with each other. On philosophical grounds, one introduces the multiverse to explain the anthropic principle.

muon   A subatomic particle identical to the electron but with a much larger mass. It belongs to the second redundant generation of particles found in the Standard Model.

negative energy   Energy that is less than zero. Matter has positive energy, gravity has negative energy, and the two can cancel out in many cosmological models. The quantum theory allows for a different kind of negative energy, due to the Casimir effect and other effects, which can be used to drive a wormhole. Negative energy is useful in creating and stabilizing wormholes.

neutrino   A ghostly, almost massless subatomic particle. Neutrinos react very weakly with other particles and may penetrate several light-years of lead without ever interacting with anything. They are emitted in copious quantities from supernovae. The number of neutrinos is so large that they heat up the gas surrounding the collapsing star, thereby creating the explosion of the supernova.

neutron   A neutral subatomic particle which, along with the proton, makes up the nuclei of atoms.

neutron star   A collapsed star consisting of a solid mass of neutrons. It is usually about 10 to 15 miles across. When it spins, it releases energy in an irregular manner, creating a pulsar. It is the remnant of a supernova. If the neutron star is quite large, about 3 solar masses, it might collapse into a black hole.

nucleosynethesis   The creation of higher nuclei from hydrogen, starting from the big bang. In this way, one can obtain the relative abundance of all the elements found in nature. This is one of the three "proofs" of the big bang. The higher elements are cooked in the center of stars. The elements beyond iron are cooked in a supernova explosion.

nucleus   The tiny core of an atom, consisting of protons and neutrons, which is roughly 10^{-13} cm across. The number of protons in a nucleus determines the number of electrons in the shell surrounding the nucleus, which in turn determines the chemical properties of the atom.

Olbers' paradox   The paradox that asks why the night sky is black. If the universe is infinite and uniform, then we must receive light from an infinite number of stars, and hence the sky must be white, which violates observation. This paradox is explained by the big bang and the finite lifetime of stars. The big bang gives a cutoff to the light hitting our eyes from deep space.

Omega   The parameter that measures the average density of matter in the universe. If Lambda = 0, and Omega is less than 1, then the universe will expand forever into a big freeze. If Omega is greater than 1, then there is enough matter to reverse the expansion into a big crunch. If Omega equals 1, then the universe is flat.

perturbation theory   The process by which physicists solve quantum theories by summing over an infinite number of small corrections. Almost all the work in string theory is done via string perturbation theory, but some of the most interesting problems lie beyond the reach of perturbation theory, such as supersymmetry breaking. Thus, we need nonperturbative methods to solve string theory, which at the present time do not really exist in any systematic fashion.

photon   A particle or quantum of light. The photon was first proposed by Einstein to explain the photoelectric effect—that is, the fact that shining light on a metal results in the ejection of electrons.

Planck energy 10^{19} billion electron volts. This might be the energy scale of the big bang, where all the forces were unified into a single superforce.

Planck length 10^{-33} cm. This is the scale found at the big bang in which the gravitational force was as strong as the other forces. At this scale, space-time becomes "foamy," with tiny bubbles and wormholes appearing and disappearing into the vacuum.

powers of ten   Shorthand notation used by scientists to denote very large or very small numbers. Thus, 10^n means 1 followed by n zeros. A thousand is therefore 10^3. Also, 10^{-n} means the inverse of 10^n—that is, 000 . . . 001, where there are $n - 1$ zeros. A thousandth is therefore 10^{-3} or 0.001.

proton   A positively charged subatomic particle which, along with neutrons, makes up the nuclei of atoms. They are stable, but GUT theory predicts that they may decay over a long period of time.

pulsar   A rotating neutron star. Because it is irregular, it resembles a rotating lighthouse beacon, giving the appearance of a blinking star.

quantum fluctuation   Tiny variations from the classical theory of Newton or Einstein, due to the uncertainty principle. The universe itself may have started out as a quantum fluctuation in nothing (hyperspace). Quantum fluctuations in the big bang give us the galactic clusters of today. The problem with quantum gravity, which has prevented a unified field theory for many decades, is that the quantum fluctuations of gravity theory are infinite, which is nonsense. So far, only string theory can banish these infinite quantum fluctuations of gravity.

quantum foam   Tiny, foamlike distortions of space-time at the level of the Planck length. If we could peer into the fabric of space-time at the Planck length, we would see tiny bubbles and wormholes, with a foam-like appearance.

quantum gravity   A form of gravity that obeys the quantum principle. When gravity is quantized, we find a packet of gravity, which is called the graviton. Usually, when gravity is quantized, we find its quantum fluctuations are infinite, which renders the theory useless. At present, string theory is the only candidate which can remove these infinities.

quantum leap   A sudden change in the state of an object that is not allowed classically. Electrons inside an atom make quantum leaps between orbits, releasing or absorbing light in the process. The universe might have made a quantum leap from nothing to our present-day universe.

quantum mechanics   The complete quantum theory proposed in 1925, which replaced the "old quantum theory" of Planck and Einstein. Unlike the old quantum theory, which was a hybrid of old classical concepts and newer quantum ideas, quantum mechanics is based on wave equations and the uncertainty principle and represents a significant break from classical physics. No deviation from quantum mechanics has ever been found in the laboratory. Its most advanced version today is called quantum field theory, which combines special relativity and quantum mechanics. A fully quantum mechanical theory of gravity, however, is exceedingly difficult.

quantum theory   The theory of subatomic physics. It is one of the most successful theories of all time. Quantum theory plus relativity together make up the sum total of all physical knowledge at a fundamental level. Roughly speaking, the quantum theory is based on three principles: (1) energy is found in discrete packets called quanta; (2) matter is based on point particles but the probability of finding them is given by a wave, which obeys the Schrödinger wave equation; (3) a measurement is necessary to collapse the wave and determine the final state of an object. The postulates of the quantum theory are the reverse of the postulates of general relativity, which is deterministic and based on smooth surfaces. Combining relativity and the quantum theory is one of the greatest problems facing physics today.

quark   A subatomic particle that makes up the proton and neutron. Three quarks make up a proton or neutron, and a quark and antiquark pair make up a meson. Quarks in turn are part of the Standard Model.

quasar   Quasi-stellar object. They are huge galaxies that were formed shortly after the big bang. They have huge black holes at their center. The fact that we do not see quasars today was one way to disprove the steady state theory, which says that the universe today is similar to the universe billions of years ago.

red giant   A star that burns helium. After a star like our Sun exhausts its hydrogen fuel, it begins to expand and form a helium-burning red giant star. This means that Earth will ultimately die in fire when our Sun becomes a red giant, about 5 billion years from now.

redshift   The reddening or decrease in frequency of light from distant galaxies due to the Doppler effect, indicating that they are moving away from us. The redshift can also take place via the expansion of empty space, as in the expanding universe.

relativity   The theory of Einstein, both special and general. The first theory is concerned with light and flat, four-dimensional space-time. It is based on the principle that the speed of light is constant in all inertial frames. The second theory deals with gravity and curved space. It is based on the principle that gravitating and accelerating frames are indistinguishable. The combination of relativity with the quantum theory represents the sum total of all physical knowledge.

Schrödinger's cat paradox   The paradox that asks if a cat can be dead and alive at the same time. According to the quantum theory, a cat in a box may be dead and alive simultaneously, at least until we make an observation, which sounds absurd. We must add the wave function of a cat in all possible states (dead, alive, running, sleeping, eating, and so forth) until a measurement is made. There are two main ways to resolve the paradox, either assuming that consciousness determines existence or assuming an infinite number of parallel worlds.

Schwarzschild radius   The radius of the event horizon, or the point of no return for a black hole. For the Sun, the Schwarzschild radius is roughly two miles. Once a star is compressed to within its event horizon, it collapses into a black hole.

simply connected space   A space in which any lasso can be continuously shunk to a point. Flat space is simply connected, while the surface of a doughnut or a wormhole is not.

singularity   A state of infinite gravity. In general relativity, singularities are predicted to exist at the center of black holes and at the instant of creation, under very general conditions. They are thought to represent a breakdown of general relativity, forcing the introduction of a quantum theory of gravity.

special relativity   Einstein's 1905 theory, based on the constancy of the speed of light. Consequences include: time slows down, mass increases, and distances shrink the faster you move. Also, matter and energy are related via $E = mc^2$. One consequence of special relativity is the atomic bomb.

spectrum   The different colors or frequencies found within light. By analyzing the spectrum of starlight, one can determine that stars are mainly made of hydrogen and helium.

standard candle   A source of light that is standardized and the same throughout the universe, which allows scientists to calculate astronomical distances. The fainter a standard candle is, the farther away it is. Once we know the luminosity of a standard candle, we can calculate its distance. The standard candles used today are type Ia supernovae and Cepheid variables.

Standard Model   The most successful quantum theory of the weak, electromagnetic, and strong interactions. It is based on the SU(3) symmetry of quarks, the SU(2) symmetry of electrons and neutrinos, and the U(1) symmetry of light. It contains a large collection of particles: quarks, gluons, leptons, W- and Z-bosons, and Higgs particles. It cannot be the theory of everything because (a) it lacks any mention of gravity; (b) it has nineteen free parameters which have to be fixed by hand; and (c) it has three identical generations of quarks and leptons, which is redundant. The Standard Model can be absorbed into a GUT theory and eventually into string theory, but at present there is no experimental evidence for either.

steady state theory The theory which states that the universe had no beginning but constantly generates new matter as it expands, keeping the same density. This theory has been discredited for various reasons, one being when the microwave background radiation was discovered. Also, it was found that quasars and galaxies have distinct evolutionary phases.

string theory The theory based on tiny vibrating strings, such that each mode of vibration corresponds to a subatomic particle. It is the only theory that can combine gravity with the quantum theory, making it the leading candidate for a theory of everything. It is only mathematically self-consistent in ten dimensions. Its latest version is called M-theory, which is defined in eleven dimensions.

strong nuclear force The force that binds the nucleus together. It is one of the four fundamental forces. Physicists use Quantum Chromodynamics to describe the strong interactions, based on quarks and gluons with SU(3) symmetry.

supernova An exploding star. They are so energetic that they can sometimes outshine a galaxy. There are several types of supernovae, the most interesting being the type Ia supernova. They all can be used as standard candles to measure galactic distances. Type Ia supernovae are caused when an aging white dwarf star steals matter from its companion and is pushed beyond the Chandrasekhar limit, causing it to suddenly collapse and then blow up.

supersymmetry The symmetry that interchanges fermions and bosons. This symmetry solves the hierarchy problem, and it also helps to eliminate any remaining divergences within superstring theory. It means that all the particles in the Standard Model must have partners, called sparticles, which have so far never been seen in the laboratory. Supersymmetry in principle can unify all the particles of the universe into a single object.

symmetry A reshuffling or rearrangement of an object that leaves it invariant, or the same. Snowflakes are invariant under a rotation of a multiple of 60 degrees. Circles are invariant under a rotation of any angle. The quark model remains invariant under a reshuffling of the three quarks, giving SU(3) symmetry.

Strings are invariant under supersymmetry and also under conformal deformations of its surface. Symmetry is crucial in physics because it helps to eliminate many of the divergences found in quantum theory.

symmetry breaking   The breaking of a symmetry found in the quantum theory. It is thought that the universe was in perfect symmetry before the big bang. Since then, the universe has cooled and aged, and hence the four fundamental forces and their symmetries have broken down. Today, the universe is horribly broken, with all the forces split off from each other.

thermodynamics   The physics of heat. There are three laws of thermodynamics: (1) the total amount of matter and energy is conserved; (2) total entropy always increases; and (3) you cannot reach absolute zero. Thermodynamics is essential to understanding how the universe might die.

tunneling   The process by which particles can penetrate barriers that are forbidden by Newtonian mechanics. Tunneling is the reason for radioactive alpha decay and is a by-product of the quantum theory. The universe itself may have been created by tunneling. It has been conjectured that one may be able to tunnel between universes.

type I, II, III civilizations   The classification introduced by Nikolai Kardashev to rank civilizations in outer space by their energy generation. They correspond to civilizations that can harness the power of an entire planet, star, and galaxy, respectively. So far, no evidence has been found for any of them in space. Our own civilization corresponds probably to a type 0.7.

type Ia supernova   A supernova that is often used as a standard candle to measure distances. This supernova takes place in a double star system, where a white dwarf star slowly sucks matter from a companion star, pushing it over the Chandrasekhar limit of 1.4 solar masses, causing it to explode.

uncertainty principle   The principle which states that you cannot know both the location and velocity of a particle with infinite precision. The uncertainty in the position of a particle, multiplied by the uncertainty in its momentum, must be greater than or equal to Planck's constant divided by 2π. The uncertainty principle is the most essential component of the quantum theory, introducing probability into the universe. Because of nanotechnology, physicists can manipulate individual atoms at will and hence test the uncertainty principle in the laboratory.

unified field theory   The theory sought by Einstein that would unify all the forces of nature into a single coherent theory. Today, the leading candidate is string theory or M-theory. Einstein originally believed that his unified field theory could absorb both relativity and the quantum theory into a higher theory that would not require probabilities. String theory, however, is a quantum theory and hence introduces probabilities.

vacuum   Empty space. But empty space, according to the quantum theory, is teeming with virtual subatomic particles, which last only a fraction of a second. The vacuum is also used to describe the lowest energy of a system. The universe, it is believed, went from a state of a false vacuum to the true vacuum of today.

virtual particles   Particles that briefly dart in and out of the vacuum. They violate known conservation laws but only for a short period of time, via the uncertainty principle. The conservation laws then operate as an average in the vacuum. Virtual particles can sometimes become real particles if enough energy is added to the vacuum. On a microscopic scale, these virtual particles may include wormholes and baby universes.

wave function   A wave that accompanies every subatomic particle. It is the mathematical description of the wave of probability locating the position of any particle. Schrödinger was the first to write down the equations for the wave function of an electron. In the quantum theory, matter is composed of point particles, but the probability of finding the particle is given by the wave function. Dirac later proposed a wave equation which included special relativity. Today, all of quantum physics, including string theory, is formulated in terms of these waves.

weak nuclear force   The force within the nucleus that makes possible nuclear decay. This force is not strong enough to hold the nucleus together, hence the nucleus can fall apart. The weak force acts on leptons (electrons and neutrinos) and is carried by the W- and Z-bosons.

white dwarf   A star in its final stages of life, consisting of lower elements such as oxygen, lithium, carbon, and so forth. They are found after a red giant exhausts its helium fuel and collapses. Typically, they are about the size of Earth and weigh no more than 1.4 solar masses (or else they collapse).

WIMP   Weakly interacting massive particle. WIMPs are conjectured to make up most of dark matter in the universe. One leading candidate for the WIMPs are the sparticles predicted by string theory.

wormhole   A passageway between two universes. Mathematicians call these spaces "multiply connected spaces"—spaces in which a lasso may not be shrunk to a point. It is not clear if one may be able to pass through a wormhole without destabilizing it or dying in the attempt.

RECOMMENDED READING

Adams, Douglas. *The Hitchhiker's Guide to the Galaxy*. New York: Pocket Books, 1979.

Adams, Fred, and Greg Laughlin. *The Five Ages of the Universe: Inside the Physics of Eternity*. New York: The Free Press, 1999.

Anderson, Poul. *Tau Zero*. London: Victor Gollancz, 1967.

Asimov, Isaac. *The Gods Themselves*. New York: Bantam Books, 1972.

Barrow, John D. *The Artful Universe*. New York: Oxford University Press, 1995. (referred to as Barrow2)

———. *The Universe That Discovered Itself*. New York: Oxford University Press, 2000. (referred to as Barrow3)

Barrow, John D., and F. Tipler. *The Anthropic Cosmological Principle*. New York: Oxford University Press, 1986. (referred to as Barrow1)

Bartusiak, Marcia. *Einstein's Unfinished Symphony: Listening to the Sounds of Space-time*. New York: Berkley Books, 2000.

Bear, Greg. *Eon*. New York: Tom Doherty Associates Books, 1985.

Bell, E. T. *Men of Mathematics*. New York: Simon and Schuster, 1937.

Bernstein, Jeremy. *Quantum Profiles*. Princeton, N.J.: Princeton University Press, 1991.

Brian, Denis. *Einstein: A Life*. New York: John Wiley, 1996.

Brownlee, Donald, and Peter D. Ward. *Rare Earth*. New York: Springer-Verlag, 2000.

Calaprice, Alice, ed. *The Expanded Quotable Einstein*. Princeton: Princeton University Press, 2000.

Chown, Marcus. *The Universe Next Door: The Making of Tomorrow's Science*. New York: Oxford University Press, 2002.

Cole, K. C. *The Universe in a Teacup*. New York: Harcourt Brace, 1998.

Crease, Robert, and Charles Mann. *The Second Creation: Makers of the Revolution in Twentieth-Century Physics.*. New York: Macmillan, 1986.

Croswell, Ken. *The Universe at Midnight: Observations Illuminating the Cosmos*. New York: The Free Press, 2001.

Davies, Paul. *How to Build a Time Machine*. New York: Penguin Books, 2001. (referred to as Davies1)

Davies, P. C. W., and J. Brown. *Superstrings: A Theory of Everything*. Cambridge, U.K.: Cambridge University Press, 1988. (referred to as Davies2)

Dick, Philip K. *The Man in the High Castle*. New York: Vintage Books, 1990.

Dyson, Freeman. *Imagined Worlds*. Cambridge, Mass.: Harvard University Press, 1998.

Folsing, Albrecht. *Albert Einstein*. New York: Penguin Books, 1997.

Gamow, George. *My World Line: An Informal Biography*. New York: Viking Press, 1970. (referred to as Gamow1)

———. *One, Two, Three . . . Infinity*. New York: Bantam Books, 1961. (referred to as Gamow2)

Goldsmith, Donald. *The Runaway Universe*. Cambridge, Mass.: Perseus Books, 2000.

Goldsmith, Donald, and Neil deGrasse Tyson. *Origins*. New York: W. W. Norton, 2004.

Gott, J. Richard. *Time Travel in Einstein's Universe*. Boston: Houghton Mifflin Co., 2001.

Greene, Brian. *The Elegant Universe: Superstrings, Hidden Dimensions, and the Quest for the Ultimate Theory*. New York: W. W. Norton, 1999. (referred to as Greene1)

———. *The Fabric of the Cosmos*. New York: W. W. Norton, 2004.

Gribbin, John. *In Search of the Big Bang: Quantum Physics and Cosmology*. New York: Bantam Books, 1986.

Guth, Alan. *The Inflationary Universe*. Reading, Penn.: Addison-Wesley, 1997.

Hawking, Stephen W., Kip S. Thorne, Igor Novikov, Timothy Ferris, and Alan Lightman. *The Future of Space-time*. New York: W. W. Norton, 2002.

Kaku, Michio. *Beyond Einstein: The Cosmic Quest for the Theory of the Universe*. New York: Anchor Books, 1995. (referred to as Kaku1)

———. *Hyperspace: A Scientific Odyssey Through Time Warps, and the Tenth Dimension*. New York: Anchor Books, 1994. (referred to as Kaku2)

———. *Quantum Field Theory*. New York: Oxford University Press, 1993. (referred to as Kaku3)

Kirshner, Robert P. *Extravagant Universe: Exploding Stars, Dark Energy, and the Accelerating Universe*. Princeton, N.J.: Princeton University Press, 2002.

Kowalski, Gary. *Science and the Search for God*. New York: Lantern Books, 2003.

Lemonick, Michael D. *Echo of the Big Bang*. Princeton: Princeton University Press, 2003.

Lightman, Alan, and Roberta Brawer. *Origins: The Lives and Worlds of Modern Cosmologists*. Cambridge, Mass.: Harvard University Press, 1990.

Margenau, H., and Varghese, R. A., eds. *Cosmos, Bios, Theos*. La Salle, Ill.: Open Court, 1992.

Nahin, Paul J. *Time Machines: Time Travel in Physics, Metaphysics, and Science Fiction*. New York: Springer-Verlag, 1999.

Niven, Larry. *N-Space*. New York: Tom Doherty Associates Books, 1990.

Pais, A. *Einstein Lived Here*. New York: Oxford University Press, 1994. (referred to as Pais1)

————. *Subtle Is the Lord.* New York: Oxford University Press, 1982. (referred to as Pais2)

Parker, Barry. *Einstein's Brainchild.* Amherst, N.Y.: Prometheus Books, 2000.

Petters, A. O., H. Levine, J. Wambsganss. *Singularity Theory and Gravitational Lensing.* Boston: Birkhauser, 2001.

Polkinghorne, J. C. *The Quantum World.* Princeton, N.J.: Princeton University Press, 1984.

Rees, Martin. *Before the Beginning: Our Universe and Others.* Reading, Mass.: Perseus Books, 1997. (referred to as Rees1)

————. *Just Six Numbers: The Deep Forces that Shape the Universe.* Reading, Mass.: Perseus Books, 2000. (referred to as Rees2)

————. *Our Final Hour.* New York: Perseus Books, 2003. (referred to as Rees3)

Sagan, Carl. *Carl Sagan's Cosmic Connection.* New York: Cambridge University Press, 2000.

Schilpp, Paul Arthur. *Albert Einstein: Philosopher-Scientist.* New York: Tudor Publishing, 1951.

Seife, Charles. *Alpha and Omega: The Search for the Beginning and End of the Universe.* New York: Viking Press, 2003.

Silk, Joseph. *The Big Bang.* New York: W. H. Freeman, 2001.

Smoot, George, and Davidson, Keay. *Wrinkles in Time.* New York: Avon Books, 1993.

Thorne, Kip S. *Black Holes and Time Warps: Einstein's Outrageous Legacy.* New York: W. W. Norton, 1994.

Tyson, Neil deGrasse. *The Sky Is Not the Limit.* New York: Doubleday, 2000.

Weinberg, Steve. *Dreams of a Final Theory: The Search for the Fundamental Laws of Nature.* New York: Pantheon Books, 1992. (referred to as Weinberg1)

————. *Facing Up: Science and Its Cultural Adversaries.* Cambridge, Mass.: Harvard University Press, 2001. (referred to as Weinberg2)

————. *The First Three Minutes: A Modern View of the Origin of the Universe.* New York: Bantam New Age, 1977. (referred to as Weinberg3)

Wells, H. G. *The Invisible Man.* New York: Dover Publications, 1992. (referred to as Wells1)

————. *The Wonderful Visit.* North Yorkshire, U.K.: House of Status, 2002. (referred to as Wells2)

Wilczek, Frank. *Longing for the Harmonies: Themes and Variations from Modern Physics.* New York: W. W. Norton, 1988.

Zee, A. *Einstein's Universe.* New York: Oxford University Press, 1989.

INDEX

Page numbers of illustrations appear in italics.

Abbot, Edwin, 182–83
Adams, Douglas, 147, 348
Adams, Fred, 292
Albrecht, Andreas, 91
Alcubierre, Miguel, 334–35
All the Myriad Ways (Niven), 351–53
"All You Zombies" (Heinlein), 143
alpha particles, 53–54
Alpher, Ralph, 55, 57, 58
Anderson, Poul, 76–78, 292
Andromeda, 47–48, 50, 124
anthropic principle, 240, 242, 247–49
    cosmic accidents and, 246–47,
        348–49
    forms of, 248
antigravity force
    big bang caused by, 19
    dark energy, 12, 37, 41, 317
    Einstein and, 37, 104, 111
    end of the universe and, 288–89
    negative energy and, 131–33
    See also cosmological constant
Arkani–Hamed, N., 219
Asimov, Isaac, 112, 143
Aspect, Alan, 176
asteroids, 295
    impacting earth, 294–95
    1950DA, 295
    1997XF11, 295
As You Like It (Shakespeare), 22, 289, 354
atoms, 17
    absolute zero and, 298
    birth of, 106, 268

Bohr's "wall" separating subatomic
    world from everyday, 156
cosmic accidents and the creation
    of life, 246–47
death of universe and, 298–99
electrons (strings), 17–18
fission and the creation of the
    atomic bomb, 161–63
force holding nucleus together, 53
nanotechnology and, 160
Newton's laws, failure of and,
    147–48
nuclear reactor, first, 162
radioactive decay, cause of, 53–54,
    80
radioactive decay, uranium, 158
resistance by scientists to reality of,
    150–51
Schrödinger wave equation, 151
strong nuclear force, 80, 153, 205–6,
    206, 247
unlocking of secrets of, 150–51
weak nuclear force, 80, 82, 153,
    205–6, *206*, 247
See also quantum theory; subatomic
    particles
atom smashers. See particle
    accelerators

Baade, Walter, 71
Back to the Future (film), 143
Bahcall, John, 6, 12

Baronius, Cardinal, 343
baryonic matter, 72–73
Bear, Greg, 304–5
Bekenstein, Jacob, 134, 231, 233,
     235–36, 298
Bell, John, 175–76
Bell Laboratory, Holmdell, Horn Radio
     Telescope, 68
Bennett, Charles L., 13
Bentley, Richard, 25
Bentley's paradox, 25–26, 36–37, 49
Berkeley, George, 157
Bernstein, Aaron, 31
Betelgeuse, 57, 66, 67
Bethe, Hans, 55
Beyond Einstein and Hyperspace (Kaku),
     xvi
big bang, xv–xvi, 5, 45–75, 105–6
     antigravity force and, 19
     colliding universes and, 222–24
     CP symmetry, 96
     criticism of, 51–52
     evidence (three great "proofs"), 44,
          46, 50, 55–56, 58–59
     false vacuum and origin of, 85–86
     "fossil record" or microwave
          background radiation, 56–58,
          68–70, 74–75, 101–2, 106
     Hubble space telescope photo of
          infant galaxies and, 29–30,
          293–94
     inflationary universe theory and,
          xvi, 13–16, *15*, 42, *43*, 78
     Lemaître and, 51
     light from, 7
     multiple, 5
     named by Hoyle, 61
     Poe and, 51
     quantum fluctuation as cause,
          94, 101
     religious implications, 348
     superatom and, 51
     superforce and, 84
     supersymmetry and, 205–6, *206*

temperature following, 57–58
     what happened before, 16–17
big crunch, 42–43, *42*, 44, 291–92
big freeze, xvi, xvii, 19–20, 41–44, *42*,
     43, 112, 292, 297–98
     escape into hyperspace, 20–21, 112,
          302–3
     Norse legend, 288
     survival by intelligent life, 300–302
Billingsley, Garilynn, 260
black body radiation, 56–57, 74–75
black holes, 20, 64, 67, 111–27
     accretion disk, 123
     colliding, 261–62
     constructing one in slow motion,
          325–27
     death of, 298
     Einstein and, 111–12, 116, 117, 119,
          120, 121
     Einstein-Rosen bridge, 118–22, *120*
     era of, 298
     escape velocity, 123
     event horizon, 117, 121, 123, 134, 225,
          231, 235, 324–25
     frame dragging, 128
     galactic, 122–25, 271 (*see also* Milky
          Way galaxy)
     gamma ray bursters and creation
          of, 125–27
     Hadamard disaster, 116
     Hawking radiation, 228
     information problem, 228–32, 235
     Kerr, 121–22, 324, 326
     M-theory analysis of, 228–30
     mini–black holes, 226–28
     negative energy in, 134
     number in night sky, 125, 324
     observing real and proof of, 122–25,
          257
     poles, 123
     pre–big-bang theory and, 224–25,
          237
     probes sent through, 324–25
     quantum corrections to, 320–21

rotating, 121–22, 123
Schwarzschild radius, 115, 117, 124,
    227
space bent by, 118–22, *120*
as speculation, not real, 115–18
stellar, 123–24
term coined, 150
time bent by, 128–33
tunneling of radiation past (black
    not really black), 134, 228,
    298
universes sprouting from, 253–55
Bohr, Niels, 53, 150, 151, 152–53, 156,
    158–60, 161–63, 170, 187
Boltzmann, Ludwig, 151
Bondi, Hermann, 60
Born, Max, 151–52
Bosma, Albert, 73
Brandenberger, Robert, 222
Braunstein, Samuel, 177, 178
Brawer, Roberta, 355
Brown, Ian, 264
Brownlee, Donald, 243, 244, 297
Bruno, Giordano, 345
Buddhism
    absence of God, 4
    multiverse and Nirvana, 15
    timeless universe, 3, 4–5
Burbidge, Margaret and Geoffrey, 63
Burke, Bernard, 68
Butler, Paul, 246

Calabi-Yau manifold, 207, 217, 282
carbon, formation of, and life, 250–51
carbon nanotube technology, 312
Carroll universe, 316
Casimir, Henrik, 132, 133
Casimir effect, 132, 133, 275–76, 334,
    335, 337
Celestial Mechanics (Laplace), 154
Chandrasekhar, Subrahmanyan, 103
    limit, 326
chaos, 4, 289–90

chaos theory and "butterfly effect,"
    144, 234
Chekhov, Anton, 359–60
China (P'an Ku) creation story, 4, 94
Cicero, 156
civilizations
    current, 308, 309–10, 311, 320
    information classification, 318–19
    miniaturization, 318–19, 339
    Sagan's ranking of advanced, 319–21
    Sagan's refinements of categories,
        308
    transition to type I, 359–61
    type I, 307, 308, 309, 311–13, 318
    type II, 307, 308, 313–14
    type III, 307–8, 315–16, 321–42
    type IV, 317
    waste heat and, 318–19
    written language and, 320
Cline, David B., 267
COBE (Cosmic Background Explorer
    satellite), 7, 74–75, 101–2
comets
    calculating orbit and return, 24
    Halley's, 22–23, 24
    impacting earth, 294–95
computers
    binary, 173
    brain compared to, 318–19
    Browning motion in a fluid, 302
    DNA, 319
    high-speed super, xvi, 5
    molecular transistors, 319
    Moore's law, 172
    quantum, 172–74, 319
Copernican principle, 347–49
Copernicus, Nicolaus, xv, 345
cosmic rays, 226, 227
cosmic strings, 140–42
cosmological constant, 37–38, 51, 86,
    104, 111, 232, 251, 253
    lowered temperature of, 301, 302
    measuring, 265–66
    See also Lambda

cosmological principle, 40–41
cosmology
  big bang, 45–75
  early scientists, xv
  coming of age, 10
  Einstein's theory of relativity and
      second revolution, xv–xvi,
      xvi–xvii
  first revolution in, xv
  golden age, xvii
  high-tech instruments and third
      revolution, xvi, 5, 10
  inflationary universe theory, xvi,
      13–16, 15, 42, 43, 78–108
  as observational science, 54
  religious theories of creation, 4–5
  what it is, xv
  See also big bang; universe; specific
      topics
Crane, Stephen, 348
Crawford, Ian, 315
Cremmer, Eugene, 210
Crick, Francis, 52
Crommelin, Andrew, 39
Croswell, Ken, 43, 256, 296
Curtis, Heber, 47

D (number of spatial dimensions), 252
Dalí, Salvador, 184
Dante, 357
dark energy, 12, 317, 347
  computation of, discrepancy in, 12
  Lambda (energy of space), 41
dark matter, 11–12, 70–74, 347
  capturing a particle, 266–67, 282
  cold, 74
  content of universe, 268, 282
  detecting, with Einstein lenses and
      rings, 264–66
  experiments (listing of), 267
  flatness of universe and, 72–73, 90,
      91
  galaxies analyzed, 270

hot, 74
  in living room, 266–67
  Omega (density of matter) and,
      41–44, 42, 43, 44, 70, 87
  what it is, theories, 74–75, 267–68,
      279
Darwin, Charles, 20
  theory of evolution, 344
Davies, Paul, 133
Dawkins, Richard, 355
  Dead of Night (film), 60
  decoherence, 166–67, 170–71, 174
Democritus, 244–45
  De Revolutionibus Orbium Coelestium
      (Copernicus), 345
  designer universe, 240, 241–55
  de Sitter, Willem, 37–38, 49, 50,
      86
  expansion, 103, 106
  universe, 232
determinism, 154–56
Deutch, David, 173–74
deuterium, 250
Dick, Philip K., 148
Dicke, Robert, 68, 89
dimensional portals (gateways), 21,
      112, 118–21, 185, 305. See also
      wormholes
Dimopoulos, S., 219
Dirac, Paul, 151
Dodgson, Charles (Lewis Carroll), 118,
      316
Doppler effect, 48–49
Droste, Johannes, 115, 116
Duchamp, Marcel, 184
Dvali, G., 219
Dyson, Freeman, 248, 292, 300–301,
      302, 314, 316

Earth
  age of, 11, 50, 60
  age of light from Sun, Moon, and
      stars, 7

cosmic accidents and the creation
of life, 246–47
dark matter wind, 266–67
extinctions, supernovae and, 60,
66–67
fate of, 294–97
Ice Ages, 294
meteor or comet impact, 294–95
"mother" sun of, 67
as oddball of the universe, 40
orbit, moving of, 296
perfect placement ("Goldilocks
zone") and conditions for life,
241–46
Sun swallowing up, 295–96
uniformitarianism vs.
catastrophism, 60
See also life
Eddington, Arthur, 39, 51–52, 117, 186,
290
Egypt, creation story, 4
Ehrenfest, Paul, 159–60, 253
Einstein, Albert, 30–34, 344
advancement of humanity and, 346
antigravity field postulated by, 12
backlash to relativity theory, 39–40
Bentley's paradox and, 36–37
black holes and, 111–12, 116, 117, 119,
120, 121
celebrity of, 39
collapsing universes, 292
cosmic strings and, 140–41
cosmological constant (antigravity
force), 37–38, 51, 86, 104, 111
deflection of starlight, use as
"lens," 263–64
as a determinist, 154–55
development of theory, 31–32
dictum on breaking speed of light,
13
EPR paradox, 174–76
equations, difficulty of, 40, 320
equations as time–reversal
invariant, 323, 329–30
force as the bending of space,
35–36, 38–39, *38*
formula, E=mc², 33–34, 80, 289
girfriend, Mileva Maric, and child,
31
Gödel's solutions and, 129–30
gravity waves and, 258
Lambda, 103–4
mathematical construction and
discovery of nature's laws, 283
meaning of life, unanswerable,
358–59
Mount Wilson observatory visit,
50–51
on the mysterious, 343
nuclear fission and the bomb,
161–62
objective reality of, 154, 156, 238
Omega, value of and, 87
particles and Schwarzschild radius,
325–26
philosophy and, 156
quantum physics and, 158–60,
164–65
reading the "mind of God," 16, 18,
180, 185, 187, 198, 344
space–time and, 33, 34, 35, 97, 130,
135
static universe of, 37, 38, 49
test of theory, solar eclipse 1919,
38–39
theology of and the Old One, 344,
357
"theory of everything" (unified
field theory), 17, 81, 119, 160,
180, 185, 186, 193–94, 198, 227
theory of relativity, xv, xvi–xvii,
33–35, 36, 112, 114–15, 184, 229
time, as relative, 32–33, 128
wave collapse and nature's choice,
167–68
wife of, 51
Einstein lenses and rings, 263–66
Einstein-Rosen bridge, 118–22, *120*

electromagnetism, 79–81, 82, 95, 205–6, *206*, 215–16, 218

electrons, 17–18, 83–84, *84*, 119, 227, 299
  accelerating, 281
  definite state, observation and, 152–53, 156
  EPR paradox, 174–76
  as particle or wave, 151
  partner, selectron, 204
  probability, concept of, and location, 54, 101, 132, 134, 152, 155–56, 158–60, 172–73, 174, 175
  quantum computer and, 172–74
  quantum theory, 147–48
  Schrödinger wave equation, 151–52
  See also quantum theory

elementary particle physics, 79

elements
  birth of heavy, 11, 62–63, 65, 67, 247, 250
  birth of lighter, 55, 65, 106
  on Earth, 11
  5-particle and 8-particle gap, 56, 62, 65
  helium, 11, 55–56, 64–65, 66, 69, 250
  hydrogen, 11, 66, 250
  iron, 62, 67
  lithium and beryllium, 55
  Mendeleev periodic chart, 55
  nucleosynthesis, 55–56, 62–63, 65, 106
  origin of, xvi
  radioactive decay, cause of, 53–54
  supernovae and creation of, 65, 66, 67
  in universe, unknown, dark matter, 11–12, 70
  See also dark matter; helium; hydrogen

End of Eternity, The (Asimov), 143

end of the world (eschatology; death of universe). *See* big freeze; escaping the universe; universe

energy
  accretion disks and, 123
  Casimir effect, 132, 133, 275
  compressed, making black holes and, 227
  content in universe, 94
  dark, 12, 37, 317
  Einstein's relativity theory and, 33
  energy–momentum tensor, 139
  false vacuum, 85–86
  fields, 190–91
  gamma ray bursters and "nuke flashes," 125–27
  multiple universe theory and, 170
  negative, 131–33, 323, 330, 336–38
  negative, problems with, 133–35
  negative, three laws, 337
  Planck's law, 170
  quanta, 153
  thermodynamics and, 289
  vacuum (lowest state), 85, 95, 317
  zero, 94, 290

entropy, 289–90

Eon (Bear), 304–5

Epsilon, 250–51

escaping the universe, 304–42, 346
  computation of conditions of destination, 320–21
  laws of physics and likelihood of, 306–7
  nanobots and, 340–41
  step 1: create and test a theory of everything, 321–23
  step 2: find naturally occurring wormholes and white holes, 323–24
  step 3: send probes through a black hole, 324–25
  step 4: construct a black hole in slow motion, 325–27
  step 5: create a baby universe, 327–30, 329
  step 6: create huge atom smashers, 330–32

step 7: create implosion mechanisms, 332–34

step 8: build a warp drive machine, 334–36

step 9: use negative energy from squeezed states, 336–38

step 10: wait for quantum transitions, 338

step 11: the last hope, 338–41

type II civilization and, 314

wormhole exit, 320

Euler, Leonard, 188

Beta Function, 188

Eureka: A Prose Poem (Poe), 28

Everett, Allen E., 335

Everett, Hugh, III, 168

Faber, Sandra, 355

false vacuum, 85–86, 327–29

Faraday, Michael, 190, 235

Fermi, Enrico, 162

Ferrara, Sergio, 210

Feynman, Richard, 150, 157, 163, 173, 191, 192

sum over paths, 163–65

First Three Minutes, The (Weinberg), 354

Flatland (Abbot), 182–83

Ford, Lawrence, 337

Fowler, William, 63

free will, 154–55

Freedman, Daniel, 210

Freud, Sigmund, 359

Friedmann, Aleksandr, 40–41, 53

expansion, 105, 106

future of the universe and, 43

solution of, three parameters, 41

Fulling, Stephen, 133

functional integrals, 164

galaxies

Abell 2218, 264

Andromeda, 47–48, 50, 124

black holes in, 122–25, 271

catalog, Zwicky's, 71–72

Coma cluster, 70–71

composition of, 55–56

dark matter in, 73, 270

distance, in light-years, 7

distance and speed of expansion, 50

expanding, xv, 12, 19–20, 49–50

galactic arcs and Einstein lenses, 264

Hubble space telescope picture of infant, 29, 293–94

Kant's island universes, 47

M–87, 125

M–100, 45

Milky Way, 9, 47, 72

NGC 4261, 124

number of, 19

red shift, 49–50

RX J1242–11, 125

spiral nebulae, 48

WMAP pictures of, 9, *9*

Galilei, Galileo, xv, 217–18, 343, 345

Gamow, George, xvi, 8, 52–58, 61

alpha-beta-gamma paper, 55

5-particle and 8-particle gap, 56

limerick by, 53

microwave background radiation and, 68–70, 74–75

nucleosynthesis, 55–56, 62, 63

radioactive decay, cause of, 53–54

temperature of universe and, 58, 68, 69

Gardner, Martin, 358

Geller, Margaret, 355–56

Gell-Mann, Murray, 81, 89, 191

Genes, Gamow, and Girls (Watson), 52

Genesis

creation story, 3

multiverse and, 15

repeated occurrence of, 5

Gibbons-Hawking temperature (10^{-29} degrees), 301

Gisin, Nicolas, 178

Glashow, Sheldon, 82, 89, 196
Glenn, John, 311
Global Positioning System (GPS), 257
gluons, 17, 82, 83, *83*, 84, 153, 199, 204, 278
God
    as cosmic consciousness and "invisible hand," 144, 145, 349
    before creation, 5
    Earth's and life's creation and, 241–42, 247–48
    Einstein's, 160, 344, 357
    Newton's watchmaker, 26, 154
    Omega, value of and, 87
    omniscience, 155
    origins of, 3–4
    predetermination, 155
    science and, 344
    scientists on the meaning of the universe and, 356–58
    teleology and, 358
Gödel, Kurt, 129–30
Gods Themselves, The (Asimov), 112–14
Gold, Thomas, 60
Good Will Hunting (film), 202
Goto, Tetsuo, 190
Gott, J. Richard, III, 140–42, 335, 336
gravitino, 210
graviton, 193, 197, 220
gravity
    Bentley's paradox, 25–26
    Einstein's theory of relativity (force as the bending of space), 34–36, 219
    escape velocity, 123
    as fundamental force although weak, 79–81, 95, 218, 251
    high-frequency resonator, to test tiny length scales, 274–76
    infinite, black holes, 115, 116, 119
    infinite, point particles, 201
    leakage into hyperspace proposed, 220–21
    Newton's inverse square law, 274, 276
    Newton's law of, 24–25, 34, 192, 220, 274
    phases of the universe and, 105–6
    Purdue University experiment, atomic level deviations, 275–76
    weakness of, investigation into, 218–21
    Standard Model and, 84
    supergravity, 210–11
gravity wave, 107, 258, 263
gravity wave detectors, xvi, xvii, 5, 16, 226, 258–59
    GEO600, 261
    LIGO, 259–62, 277
    LIGO II, 261–62
    LISA, 226, 262–63, 277
    TAMA, 261
    VIRGO, 261
Green, Mike, 195
Greene, Brian, 239
    five examples of experimental data to confirm string theory, 282
Gross, David, 97, 210
Grossman, Marcel, 31
GUT (grand unified theory), 84–86, 99–101
    era, 105
    flatness problem, 87–88, 90–92
    horizon problem, 88–89, 91–92
    monopole problem, 86, 91–92
    string theory and, 210
    symmetry and, 99–101
Guth, Alan, 13, 15, 79, 85–86, 87–88, 89–91, 94, 102, 169, 224, 249, 327, 359

Hadamard, Jacques, 116
hadrons, 17, 106
Hahn, Otto, 161
Halley, Edmund, 22–23, 24
Halley's comet, 22–23, 24

Harrison, Edward, 28–29
Harrison, Jonathan, 143
Harvey, Jeffrey, 210
Hawking, Stephen, 21
  black holes, tunneling of radiation
    past (black not really black),
    134, 228, 298
  collapsing universes, 292
  information problem, 229–30,
    231–32
  mini–black hole and, 227
  radiation, 228, 230
  religious implications to big bang,
    348
  supergravity and, 311
  time travel questions and
    chronology protection
    hypothesis, 136–40, 339
  wave function of the universe and,
    178–80
Heinlein, Robert, 143
Heisenberg, Werner, 150, 160, 186, 187
  atomic bomb, Bohr meeting, and
    Nazis, 162
  quantum cookbook (principles),
    152–53
  uncertainty principle, 54, 101, 132,
    134, 172–73, 174, 175
helium
  atoms of the big bang, 56
  composition of stars, 56
  creation of, 55–56, 65, 66, 250, 293
  nucleus, 53
  percentage of universe, 55, 64–65,
    69
Helmholtz, Hermann von, 289
Henderson, Linda Dalrymple, 184
Herman, Robert, 57, 58
Higgs boson, 83, *83*, 277–78
higher dimensions, xvi, 181–84, 185,
    202
  measuring the eleventh dimension,
    274–76
  See also hyperspace; multiverse

High-Z Supernova Search Team, 103–4
Hinduism
  Mahapurana, 5
  timeless universe, 4
Hitchhiker's Guide to the Galaxy, The
    (Adams), 146–47, 348
Hogan, Craig, 12
holographic universe, 230–33
Horava, Petr, 215
Horowitz, Gary, 207
Hoyle, C.D., 275
Hoyle, Fred, xvi, 58–65, 250–51
  BBC lectures, 61–62
  big bang christened by, 61
  steady state theory, 59–60, 63–65,
    68–69
Hubble, Edwin, xv, 46–51, 347
  black holes and, 122, 123
  law of, 50–51
  measuring distance to stars, 47–48
  measuring speed of galaxies, 48–51
  mistake in calculations, 50, 59–60
Hubble's constant, 41, 50–51
  measuring, 265
Hubble space telescope, 29–30
  dark matter and, 72
  Einstein ring, 264
  farthest area probed by, 106
  galaxy, M100, 45
  picture of the end of the beginning,
    29–30
  stars in early stages, 293–94
Hubble wars, 50
Hulse, Russell, 258
humanity and man's place in the
    cosmos, question of, 344–45
  Copernican principle vs. anthropic
    principle, 347–49
  historical perspective, 345–47
  indifference of the universe to, 348
Huxley, Thomas M., 343–44
hydrogen, 11
  bomb, 163, 333
  composition of stars, 56

nucleosynthesis and, 55–56, 62–63, 66, 250, 293
spectral lines, 239
hyperspace, 183, 184, 185
    Calabi-Yau manifold, 207
    creatures in, 183–84
    eleven-dimensional, 5, 185, 211–14, 347
    escape to, 20–21, 112, 302–3
    fifth dimensional, 182, 185, 199, 219–20, 232, 233
    Kaluza-Klein higher-dimensional theory, 199–200
    problems with, 198–200
    proof of, 256–57
    strings and antistrings and, 222
    as subatomic, 200
    unified field theory and, 185

Impey, Christopher, 45
Inferno (Dante), 357
inflation (inflationary universe theory), xvi, 13–14
    cause and multiverse, 14–16, *15*, 92–93
    chaotic inflation, 15, 92–93
    colliding universes and, 222–24
    criticism of, 90
    false vacuum and, 85–86, 327
    flatness problem and, 78, 90, 91, 223
    horizon problem and, 89, 223
    Lambda, value of and, 90, 103–4
    Linde and, 165
    M-theory and, 221–26
    Omega, value of and, 90, 102–4
    quantum theory and, 101, 147–48
    shape of the universe and, *42*, 43
    shift in thinking and, 347
    string theory and, 224
    turning off (graceful exit) problem, 91–93, 105–6
    verifying, 257, 262–63
    See also Guth, Alan

interferometers, 5
Internet, 5, 309, 310
    access to Sloan Sky Survey information, 270
    lashing radio telescopes together and, 273
Invisible Man, The (Wells), 181–82

Jacoby, George, 103
Jeans, James, 350
James, Jamie, 198
Jordell Bank Observatory, 264
Julia, Bernard, 210
Jupiter, 243

Kaku, Michio, xvi, 241
    choice of study, 10
    conflicting beliefs, 3
    God and teleology, 358
    London Planetarium talk incident, 317
    M-theory and, 212–13, 238–39
    meaning of life and, 359
    path integral approach and, 164
    Ph.D. thesis, 213–14
    string theory and, 188–89, 191–92, 209–10
    verification of string theory and, 282–83
Kallosh, Renata, 223–24
Kaluza, Theodor, 199–200
Kant, Immanuel, 47
Kardashev, Nikolai, 307, 318, 321
Kelvin, Lord (William Thomson), 29
Kepler, Johannes, xv, 27
Kerr, Roy, 121–22
Kikkawa, Keiji, 190, 191, 209–10, 237
Kirshner, Robert, 90
Kistiakowsky, Vera, 248
Kitt's Peak Observatory, 103–4
Klein, Felix, 199–200
Knox, Ronnie, 350

Koekemoer, Anton, 29
Kofman, Lev, 223–24
Komossa, Stefanie, 125
Kowalski, Gary, 351
Krasnikov, Sergei, 139, 335–36
Krauss, Lawrence, 301, 302

Lambda (energy of space), 41, 103–4,
    251
Lamoreaux, Steven, 132
Landau, Lev, 10
Laplace, Pierre Simon de, 154
Large Hadron Collider (LHC), 226, 227,
    276–80, 330
lasers, xvi, 5, 133
    implosion machine and, 333
    squeezed states and, 133, 336–38
    tabletop accelerators and, 280–82
Laughlin, Greg, 292
Leavitt, Henrietta, 48
Leibniz, Gottfried, 198
Lemaître, Georges, 51, 116
leptons, 17, 82, 99–100, 207
Levy-Leblond, Jean-Marc, 316
Libbrecht, Kenneth, 261
Lick Observatory, 47, 355
    Shane telescope, 272
life
    carbon, formation of, 250–51
    cosmic accidents and the creation
        of, 249–53
    "Goldilocks zone" and conditions
        for life, 241–46, 348
    Ice Age, 294
    leaving the universe, 302–3, 306–7
    man's place in the cosmos, question
        of, 344–45
    meaning of, creating, 358–59
    survival of intelligent, 299–302
    universe, Stelliferous Era, and
        beginning of, 294
light
    bent by black hole, 115

bent by dark matter, 12
bent by Sun (Einstein's theory), 36,
    38–39, 38
    Maxwell's theory, 32
    redshift, 49–50
    speed of (tau zero), 77
    speed of, black holes and, 114
    speed of, and distance to Moon,
        Sun, and stars, 7
    speed of, impossibility of
        surpassing, 13, 88
    speed of, in inertial frames, 34
    velocity of, 32
Lightman, Alan, 355
light-year, 7
LIGO (Laser Interferometer
    Gravitation-Wave
    Observatory), 259–62, 277
LIGO II, 261–62
Linde, Andrei, 14, 15, 91, 92–93, 165–66,
    223–24
LISA (Laser Interferometry Space
    Antenna), 226, 262–63, 277
Livermore National Laboratory, 333
Li-Xin Li, 139
Lorentz-FitzGerald contraction, 33
Lucretius, 25

M-theory, xvi–xvii, 16–18, 185–87,
    207–10, 357
    black hole analysis and, 228–30
    branes and p-branes, 214–15, 221,
        238, 239
    colliding universes and, 222–24
    duality, 215–16
    ekpyrotic universe, 222–23, 226
    eleventh dimension and, 211–14,
        274–76, 347–48
    field theory absent, 214
    holographic universe, 230–33
    inflationary theory and, 221–26
    point particles as "zero-branes,"
        214

quantum theory of gravity and, 283

Randall and, 216–21

size and, 216–17

"smallest distance," 237 (*see also* Planck length)

supermembranes and, 211–14, *213*

symmetries of, 215

T-duality, 236–37

unfinished nature of, 238

as unified field theory, 215, 237–40, 321–22

universe and "three–brane," 214–15, 219

See also string theory

Mach, Ernst, 150

MACHOs (massive compact halo objects), 73, 264–65

Maldacena, Juan, 232

Man in the High Castle, The (Dick), 148, 169

Mandl, Rudi, 263, 264

Maric, Mileva, 31

Martinec, Emil, 210

Matrix, The (film), 233–34

matter

antimatter and, 95–96

content in universe, 94

Einstein's relativity theory and, 33–34

entangled particles, 177

exotic negative, 131–32

quantum theory, wave collapse, and observation, 153, 156, 166, 167, 179, 350

spontaneous breaking (phase transitions), 85, 91, 92, 96

thermodynamics and, 289

See also dark matter; elements

Max, Claire, 272

Maxwell, James Clerk, 32, 97

equations for electricity and magnetism, 215–16

Mayan creation story, 4

McCarthy, Chris, 245

McKellar, Andrew, 69

Meitner, Lise, 161

Melia, Fulvio, 123

Menuhin, Yehudi, 197–98

Mercury (planet), 40, 296

mesons, 17, 81–82, 188

Michell, John, 114

Midi–Pyrenees Observatory, France, 264

Milky Way galaxy, 9, 47

black hole in, 124, 266, 272

center, lack of brightness, 27

dark matter in, 12, 266

Earth in, 244

Einstein lensing of, 265

expansion of the universe and, 19

flatness and dark matter, 72–73

"Great Debate" and, 47

name of, 47

observation of, 272

size, 48

Misner, Charles, 356

Misner space, 136–39, *137*

MIT (Massachusetts Institute of Technology), 13

monopoles, 86, 216, 333

Moon (Earth's), 242–43

Moravec, Hans, 340

Morris, Michael, 131

Morrison, Phillip, 320

Mount Wilson Observatory, xv, 10, 47, 50–51

multiply connected spaces, 118–19. *See also* wormholes

multiverse, xvi

advancement of humanity and, 346

baby universes, 107, 301

budding, 92–93, 222, 328

colliding, 222–24

cosmic accidents and the creation of life, 249–53

creating a baby universe, 327–30, *329*

D (number of spatial dimensions),
 252–53
 Einstein's theory and, 107
 evolution of universes and, 254–55
 existential crisis of, 353–54
 inflationary theory and, 14–16, 15,
  92–95
 lack of spin, 95
 laws of physics and, 240
 M–theory and, 242
 quantum transitions and, 338
 space-time foam and, 134–35
 symmetry breaking and, 99–101
 testing of, 254–55, 279
 time travel and, 145
 what might other universes look
  like, 96–97, 100–101

N (10³⁶), 251
Nambu, Yoichiro, 190
nanotechnology, 160, 275, 311, 319, 339
National Optical Astronomy
  Observatory, Stanford
  University, 264
Neptune, 71, 272
neutralino, 267–68
neutrinos, 17, 74, 80, 82, 83, 83, 247, 282
Neveu, André, 190, 192
Newton, Isaac, 344
 advancement of humanity and, 346
 Bentley's paradox, 25–26
 calculus, 24
 God as watchmaker, universe as
  watch, 26, 154, 248
 Halley and, 23
 inverse square law, 274, 276
 laws of, and cosmology, xv, 155–56
 laws of motion, 26, 123, 154, 234,
  270–71
 Olber's paradox, 26–30
 point particles, gravity of, 201
 Principia Mathematica, 23–24, 25
 time and, 128

universe as static, 25–26, 37, 49
 universal law of gravity, 24–25, 34,
  192, 220, 274
 world view altered by, 24–25
Nielson, Holger, 190
Nietzsche, Friedrich, 311
night sky, why black, 27–30
Niven, Larry, 351–53
Novikov, Igor, 144
nucleosynthesis, 55–56, 62–63

Olbers, Heinrich Wilhelm, 27
Olbers' paradox, 26–30, 49
Omega (density of matter), 41–44, 42,
  43, 44, 251
 dark matter and, 70–74
 fine–tuning problem, 87
 size of and fate of multiverses, 93
 value of, 87–88, 104
Once and Future King, The (White), 136
Oppenheimer, J. Robert, 81, 118, 151
Ostriker, Jeremiah P., 74
Ovrut, Burt, 222–23

Paczynski, Bohdan, 264
Page, Don, 248, 356
parallel worlds
 acceptance of idea, 195–96
 big bang repetition, 5
 gateways to, 112, 119, 185
 inflation and, 76–77, 93
 many worlds solution and, 167–71
 membrane away, 330
 moral implications, 351–54
 quantum computers and, 173–74, 178
 quantum theory and, 148–50
 radio wave analogy, 170
 Rees argument for, 253
 research into, and laws of physics,
  16
 speculation about, current, 5–6
 See also multiverse

particle accelerators, xvii, 81, 82, 106,
    153, 189
  Higgs boson, trying to find, 277–78
  Large Hadron Collider (LHC), 226,
    227, 276–80
  S-matrix, 189
  Superconducting Supercollider
    (SSC), 279–80
  tabletop, 280–82
  Tevatron, 277
  UNK accelerator (Russian), 279–80
path integral approach, 164
Pauli, Wolfgang, 186, 187
Penrose, Roger, 90, 292
  theorem of, 329–30
Penzias, Arno, 68–69, 79
People's Book on Natural Science
    (Bernstein), 31
Perlmutter, Saul, 103
Philosophiae Naturalis Principia
    Mathematica (Newton), 23–24,
    25
photons, 82, 84, 153
  EPR experiment, 176
  teleportation of, 177
  thought experiment, 159–60
Picasso, Pablo, 184
Pierre Auger Cosmic Ray Observatory,
    227–28
Planck, Max, 57, 158
  energy, 206, 221, 226, 278, 315, 330–31
  era, 105
  law, 170
  length, 135, 193, 196, 201, 225, 235,
    236–37, 334
  scale, 222
Planck satellite, 10
planets
  extrasolar, 245–46, 253, 265, 272
  formation of, 65–66, 123
  locating extrasolar, 245
  placement and orbits of, 243–44
Pluto, 243–44
Podolsky, Boris, 174

Poe, Edgar Allan, 28–29, 51
Polkinghorne, John, 165, 248
Polynesian creation story, 4
Poor, Charles Lane, 39–40
Pope, Alexander, 23–24
positronium, 298–99
Primack, Joel, 14, 225

Q (10^{-5}), 251–52
Quantum Chromodynamics, 82
quantum computer, 172–74, 178
quantum entanglement, 174–78
quantum fluctuation
  infinite, problem of, 194
  universe creation, 94, 101, 338
quantum theory, 54, 93, 185–86
  absurdities and successes of, 150–51
  advancement of humanity and, 346
  attempt to reconcile with relativity
    theory, 185–87
  black holes not absolute black and,
    134
  consciousness and, 165, 171, 349–51
  cookbook rules, 152, 165
  decoherence, 166–67, 168, 170–71, 174
  difficulty, 157
  EPR paradox, 174–76
  Feynman's sum over paths, 163–65
  fission and nuclear bomb, 161–63
  gravity and, 194–95
  Heisenberg uncertainty principle,
    54, 101, 132, 134, 172–73, 174, 175,
    290
  inflationary theory and, 101, 147–48
  many worlds solution, 167–71
  observation postulate, 152, 153, 154,
    156–58, 165–66
  paradoxes, 150, 165
  parallel universes and, 93, 101,
    148–50, 163–65
  particle physics and, 93
  postulates of Copenhagen school,
    153, 168, 170

probability, 152, 155–56, 158–60, 165
quanta, 153
Schrödinger's cat, 158–59, 166–67, 170–71, 348, 351
"spooky action-at-a-distance," 175
trees falling in the forest and, 157–58, 349
tunneling, 54, 134
virtual particles in, 132, 135
wave function, 151, 153, 168–69, 179–80
Wigner's friend, 165–66, 349, 351
quarks, 17, 81, 82, 83, 83, 189, 207, 239, 278
antiquarks, 81
GUT theory and, 99–100, 203
symmetry SU(3) and, 203
quark theory, 81–82, 218
quasars, 7, 64, 106, 124, 265–66
Einstein ring and observing, 264, 265
Q0957+561, 264, 265

radiation, background microwave (in space), 258
big bang and, 56–58, 68–70, 101–2, 106
blackness of night sky and, 30, 106
COBE detection of, 7, 74
date of, 293
Earth's masking of, 8
prediction by Gamow, 8
Q (10⁻⁵), 251–52
temperature of, 8, 68, 69
uniformity of, 88–89
WMAP detection of, 6, 8
Ragnarok, 287–88
Ramanujan, 202–3
Ramond, Pierre, 190, 192
Randall, Lisa, 216–21
Reagan, Ronald, 279
Rees, Martin, 15, 249–54
current civilization and, 309

escape from the universe and, 306–7, 321
eschatology and, 291–92
relativity theory, xv, xvi–xvii, 33–35, 36, 112, 184, 185–86
accuracy of, 258–59
attempt to reconcile with quantum theory, 185–87
black holes and, 229
Global Positioning System (GPS) and, 257
Kaluza-Klein higher-dimensional theory, 199–200, 219
Schwarzschild solution, 114–15
Richstone, Douglas, 122
Riess, Adam, 19
Robertson, H. P., 116
Roddenberry, Gene, 335
Rohm, Ryan, 210
Rosen, Nathan, 119, 174, 227
Ross, Hugh, 247
Rothman, Tony, 299
Rubin, Vera, 72–73
Rutherford, Ernest, 161

Sakita, Bunji, 190
Sagan, Carl, 131, 256, 308, 319–21
Sakharov, Andrei, 96
Salam, Abdus, 82
Sandage, Allan, 10
Sanders, Gary, 258
Sargent, Wallace, 61–62
Saulson, Peter, 261
Scherk, Joël, 192–93, 210
Schmidt, Brian P., 103, 104
Schrödinger, Erwin, 150, 151, 160, 186
cat problem, 158–59, 166–67, 170–71, 178–79, 349, 351
wave equation, 151, 153, 168–69, 179
Schwarz, John, 190, 192–93, 195
Schwarzschild, Karl, 114–15
magic sphere of, 115, 116–17

Schwarzschild radius, 115, 116, 227, 325–26
Schwarzschild solution, 114–15
SETI@home, 269
Shakespeare, William, 22, 289, 354
Shapely, Harlow, 47
Sloan Sky Survey, 268–71
Smith, Chris Llewellyn, 276, 277
Smolin, Lee, 254–55
Snyder, Hartland, 118
solipsism, 157
space
    curved, 41–44, *42, 43, 44*, 78, 184, 219
    D (number of spatial dimensions), 252
    fifth dimension and, 182, 185, 199, 219–20 (*see also* hyperspace)
    as finite, 44
    as infinite, 42
    smallest distance, calculating, 134–35
    -time foam, 134–35, 235
    and time warps, xvi, 20–21
    See also hyperspace; relativity theory; universe
space satellites, xvi, xvii, 5, 6
    background radiation, uniformity of, 88
    COBE, 7, 74–75, 101–2
    Global Positioning System (GPS), 257
    photographs of remnants of creation itself, 6
    Planck, 10
    WMAP, 6–9, 11–12, 20, 75
    XMM-Newton satellite, 125
Space Telescope Science Institute, 19, 29, 30
space-time, 33, 34, 35, 97, 130, 135
    dimensions in, 192, 221–22, 232, 347–48
    geometric analog of strings and membranes, 239
    goldfish bowl analogy, 232–33
space travel, 311
    colonies in, 312–13

Mars colony, 312–13
Mars trip, 312
proton-proton fusion, 313
RLVs (reusable launch vehicles), 311
space elevators, 311–12
Spergel, David, 56
spontaneous breaking (phase transitions), 85, 91, 96, 104–7.
    See also symmetry
"standard candle," 47–48
Standard Model, 82–84, *83*, 98–99, 104–7
    big bang and, 205–6, *206*
    generations, 83, *83*
    gravity and, 84, 193–94
    quantum theory and, 153
    string theory and, 206–7, 210, 239
    symmetry and, 98–101, 210
    ugliness of, 82–84, 206
    violations detected, 282
Stanford University, 14
    Linear Accelerator Center (SLAC), 82
Stapledon, Olaf, 169
Starkman, Glenn, 301, 302
Star Maker (Stapledon), 169
stars
    age of oldest, 11
    Betelgeuse, 57, 66, 67
    beyond farthest, 29
    birth of, 65–67, 106–7
    black hole. See black hole
    blue, 72
    bodies of stardust, 67
    brown dwarfs, 72
    Cepheid, 48, 50, 102, 265
    composition of, 55–56, 254–55, 268, 347
    death of, 63, 118, 127
    Degenerate Era, 297–98
    distances to, measuring, 47–51
    double-star (binary) system, 65, 102–3, 258
    energy source for, 33–34
    farthest, 29

flat rotation curve, 72

distance and appearance, 7, 28

"fossil" light of, 7

gamma ray bursters, 125–27

HD 209458, 272

hypernova, 127

lifespan of, 29

light from, cause of, 80

neutron stars, 67, 71, 72, 127, 258, 297

night sky, and Olbers' paradox, 28–29

nucleosynthesis and, 62–63

protostars, 65

PSR 1913+16, 258–59

pulsar, 67, 150

red dwarfs, 297–98

red giants, 57, 296

size and gravity, 218–19

spectrum analysis, 254–55

Spica, 67

"standard candle," 47–48

Stelliferous Era, 293–97

strong nuclear force and, 100

supernovae, 63, 65, 66, 71, 254

supernovae, Ia, 102–3, 265

temperature and color, 57

twinkling, cause of, 271

white dwarf, 66, 102–3, 296, 297

yellow stars, 57

Star Trek, 174, 313–14, 335

steady state theory, 59–60

evidence against, 63–65, 69

Steinhardt, Paul J., 91, 222–23, 224

Stivavelli, Massimo, 30

string and superstring theory, xvi, 16, 17–18, 187–210, 209

bandwagon, scientists jumping on or off, 195–96, 207

black hole thermodynamics and, 229

Calabi-Yau space and, 207, 208, 217

divergences, 201–2, 205, 357

field theory of, 191–92

five examples of experimental data to confirm, 282

five variations, 208, 211–12, 213, 215

God and, 357, 358

gravity and graviton, 193, 194, 197, 220

heterotic SO (32) strings, 210, 215, 216

history of, 187–92

hyperspace and, problems with, 198–200

inflationary theory and, 224–25

Kaku and, 188–89, 191–92, 209–10

mini–black holes and, 227–28

musical analogy, 18, 196–98, 356

Neveu–Scharz–Ramond string, 190, 192

particle transformation and, 196–97

Planck length and, 135, 193, 196, 201, 225

"pre–big bang" theory and, 224–25, 237

spin of particles and, 190

Standard Model and, 206–7, 210

success of, reasons for, 201–3

supersymmetry and, 201, 203–6, 206, 208, 278, 357

ten dimensions and other problems with, 192–95, 215

type I, 209–10, 209

type II, 210, 212

as unified field theory, 187–88, 193, 208, 213

Veneziano model, 188, 189–90, 192, 209–10, 239

verifying, 257, 263, 278–79, 282–83

Strominger, Andrew, 207, 228–29

strong nuclear force, 80, 100, 205–6, 206, 247

subatomic particles, 12, 56, 82–83, 83

baryons, 73

Einstein and, 239

fermions and bosons, 203–5

LHC creation of, 278

mini black holes, 226–28, 278

partners for, 204

physics for, 155–56

predicted by supersymmetry, 267–68

S-matrix, 189, 190

in space, 40

sparticles, 204

spin of, 190

study of, difficulty, 189

tunneling by, 54

types of, 17, 18

as vibrating electrons (strings),
17–18

Sun

age and phase of, 66

bending of starlight around, 38–39,
40

color and temperature of, 57, 66

eclipse of 1919 and testing of
Einstein's theory, 38–39, *38*

fate of, 296

helium in, 64

as power source, 314

swallowing up Earth, 295–96

Sundrum, Raman, 220

supergravity, 210–11, 212, 213

supermembrane theory. *See* M-theory

Supernova Cosmology Project, 103–4

supersymmetry, 201, 203–6, *206*, 211,
267–68

LHC detection of, 276, 278

See also symmetry

Susskind, Leonard, 190

Suzuki, Mahiko, 188, 189

symmetry, 96–101

broken and breaking, 84, 96, 97–98

CP symmetry and big bang, 96

embryo, 98

GUT theory and, 99–101, 203, 205

hidden, 97

Kaluza-Klein theory and, 199–200,
219

quarks and, 203

snowflake, 97, 204

spontaneous breaking, 85, 96

Standard Model and, 98–99

starfish, 98

string theory and, 191–92

supersymmetry, 201, 203–6, *206*, 211,
267–68, 278

unified field theory and, 194–95

universe, origins, 98

Szilard, Leo, 161

Tau Zero (Anderson), 76–78, 292

Taylor, Joseph, Jr., 258

teleportation, 174–78

telescope(s)

Aricebo radio telescope, 269, 315

Chandra X-ray space telescope, 122,
125

compensating for themal
fluctuations, 271–72

discovery of, 345

Horn Radio Telescope, Bell
Laboratory, Holmdell, NJ, 68

Hubble space telescope, 29, 122, 264,
293–94

introduction of, Galileo, xv

lashing radio telescopes together,
273–74

MERLIN radio telescope, 264

Mount Wilson, xv, 10, 47, 50–51

Palomar Sky Survey, 269

search for Type III civilization and,
315–16

Shane telescope, Lick Observatory,
272

Sloan Sky Survey, 268–71

VBLA (very long baseline array), 273

Very Large Array Radio Telescope,
122

W. M. Keck telescope, 272, 274

on WMAP, 8 (*see also* WMAP)

X-ray, xvi

Teller, Edward, 52

Tesla, Nikola, 317

thermodynamics

absolute zero, approaching, and
machines, 299–300

black hole and, 229
First Law of, 289
Second Law of, 289–91
Third Law of, 290
three laws of, 289–91
Thompson, J. J., 39
Thorne, Kip, 131–33, 263
Three Sisters (Chekhov), 359–60
Through the Looking Glass (Carroll), 118,
    121
Tillich, Paul, 356
time
    black holes and, 128–33
    concepts of, 128
    as finite, 44
    as four dimension, 182
    Global Positioning System (GPS)
        and, 257
    infinite nature of, 43
    predicting the future, 154
    as relative, 32–33
    relativity theory and, 32–33, 34, 257
    space-time, 33, 34, 35, 97, 130, 135,
        192, 221–22, 232–33, 239, 347–48
    whirlpools or forks in, 128, 144–45
time travel, xvii, 20, 95, 128–33
    Alcubierre's warp drive and, 335
    bilker's paradox, 142–43
    chaos theory and, 144
    Gödel's universe, 129–30
    Gott time machine, 140–42
    grandfather paradox, 142
    Hawking's investigation of, 136–40,
        145, 339
    information paradox, 142, 229–30
    "invisible hand" interventions,
        144
    laws of physics violated by, 135
    "many worlds theory," 145, 169–70
    Misner space and, 136–39, 137, 145
    sexual paradox, 143
    Thorne time machine, 131–33
    Van Stockum's time machine,
        128–29

Titan (moon of Saturn), 272
Townsend, Paul, 212–13, 215
Turok, Neil, 222–23
Twilight Zone (TV series), 149, 169
Tyron, Edward, 94

Ulam, Stanislaw, 156
unified field theory(ies), 79–92, 358
    Einstein and, 81, 119, 160, 193–94
    finding, to escape the universe,
        321–22
    flatness problem, 87–88, 90–91, 223
    GUT (grand unified theory), 84–86,
        99–101, 105
    hunt for, 185–87, 193–95
    LISA and experimental data on, 262
    M-theory, 215
    mathematical inconsistencies,
        194–95
    Quantum Chromodynamics, 82
    quark theory, 81–82
    Standard Model, 82–84, 83, 98–99,
        104–7, 193–94
    string theory or M-theory, 187–210,
        209
    supersymmetry and, 205
    teleology and, 358
universe
    age, xvi, 7, 8, 10–12, 29, 45–46, 50, 60
    analogy with the Empire State
        Building, 7–8
    "baby pictures" of, 7, 9, 29–30
    Bentley's paradox, 25–26, 36–37
    big bang (origin), xv–xvi, xvii, 5, 7,
        16–17, 28, 44, 50, 56–58, 78, 86,
        94. *See also* inflationary theory
    big crunch, 42–43, 42, 44, 93, 291–92
    big freeze, xvi, xvii, 19–20, 41–44,
        42, 43, 112, 288–89, 292, 297–98,
        300–302
    black holes, 20, 64
    broken symmetry of, 84
    "bud" of, 15, 15

Buddhist and Hindu, timeless concept, 4–5

Chinese (P'an Ku) creation story, 4, 94

closed, 42–43, *44*

composition, xvi, 7, 11–12, 55–56, 65, 69

as computer program, 231, 233–37

continual creation, 5

cosmic music, 18, 356

creatio ex nihilo (creation from nothing), 4, 93–96

dark energy (antigravity field), 12, 37, 41, 317

dark matter in, 11–12, 41, 70, 90

density of, 41

de Sitter, 232

dust clouds, 28

dynamic, xv, 37, 38–39, *38*, 49

Egyptian creation story, 4

ekpyrotic, 222–23, 226

entropy and, 290–91

escape from, 302–3

expanding (and accelerating), xv, xvi, 19–20, 37, 38, 41, 42, 44, 49, 103, 107, 222, 288–89, 293, 301, 302

finite or infinite, question of, 25–26, 27, 37

force (gravity) as the bending of space, 35–36

fundamental forces of (four), 79–81, 105

future and death of, xvi, xvii, 7, 18–20, 40–44, *42, 43, 44,* 287–89

GUT theory and description of, 100

holographic, 230–33

homogeneous, 40

horizon (uniformity) problem, 88–89, 223

Hubble's constant (rate of expansion), 41, 50–51, 265

indifference of, 348

inflation (inflationary universe theory), xvi, 13–16, *15, 42, 43,* 78–108

isotropic, 40

matter/energy content, 94

Mayan creation story, 4

meaning of, scientists on, 354–58

microwave background radiation, 56–58, 68–70, 101–2, 251–52

Olbers' paradox, 26–30

Omega (density of matter in), 41–44, *42, 43, 44,* 70, 87

open, 43, *43*

oscillating, 43, 290–91

participatory, 172, 350

phase transitions, 84–85, 104–7

Polynesian creation story, 4

"pre–big bang" theory and, 224–25, 237

quantum mechanics applied to, 178–80

"real" vs. our perception of, 40

shape of, 41–44, *42, 43, 44,* 78

six numbers that govern (Rees theory), 250–53

size, 47, 48

spin, lack of, 94–95, 129–30

Stage 1: Primordial Era, 293

Stage 2: Stelliferous Era, 293–97

Stage 3: Degenerate Era, 297–98

Stage 4: Black Hole Era, 298

Stage 5: Dark Era, 298–99

static universe, 26, 37, 38, 49

steady state theory, 59–60, 63–65, 69

synthesis of opposing mythologies, 5

temperature, 57–58

theory of general relativity and (*see* relativity theory)

why the night sky is black, 27–30, 106

wormholes and dimensional portals, 21, 112, 114, 118–21, 128, 132–33, 169, 179, 227, 316, 322–23, 340–41

University of Washington, Seattle, 12

Uranus, 71

Vafa, Cumrun, 222, 228–29

Van Nieuwenhuizen, Peter, 210

Van Stockum, W. J., 128–29

Veneziano, Gabrielle, 188
    model, 188, 189–90, 192, 209–10, 239
    "pre–big bang" theory, 224–25, 237

Venus, 242, 296

Virasoro, Miguel, 190

Visser, Matthew, 139–40

W- and Z-bosons, 17, 80, 82, 83, 83, 84, 153, 199, 204

Wald, George, 351

Walsh, Dennis, 264

Ward, Peter, 243, 244

warp drive machine, 334–36

Watson, James, 52

weak nuclear force, 80, 82, 153, 205–6, 206, 247

Wedgwood, Thomas, 57

Weinberg, Steven, 81, 82, 89, 157, 170, 187, 191, 248–49, 354–55
    Weinberg angle, 191

Wells, H. G., 181–82, 183–84, 219

Weyl, Herman, 115

Wheeler, John, 150–51, 156, 161, 162, 163, 164–65, 168, 179, 187, 233, 350
    "It from bit" theory, 171–72

Wheeler-DeWitt equation, 179

White, T. H., 136

white holes, 119, 230, 323–24

Wigner, Eugene, 161, 165
    Wigner's friend, 165–66, 349–50, 351

Wilczek, Frank, 169, 357

Wilkinson, David, 6

Will, Clifford, 257

Wilson, Robert, 68–69, 79

WIMPS (weakly interacting massive particles), 74

Witten, Edward, 104, 188, 196, 197, 207, 211–12, 215, 282

WMAP (Wilkinson microwave anisotropy probe), 6–10, 75
    age of universe, 8, 10–11
    baby picture of universe, 9
    big freeze confirmed by, 20, 292
    cosmological constant, measuring, 266
    dark energy detected by, 12
    dark matter detected by, 12
    data from, 13
    expanding universe and, 288
    Hubble's constant, precise value and, 50
    inflationary universe theory and, 13, 42, 43, 78
    Lambda, value of and, 104
    position, Lagrange point 2, 8
    size, materials, telescopes, 8
    temperature of microwave radiation in space, 8–9, 68–70

Wonderful Visit, The (Wells), 183–84, 219

wormholes, xvi, 21, 112, 114, 128, 227
    atom size, 340–41
    basic questions about, 322–23
    civilization level able to use, 321
    Dodgson and, 118
    Einstein-Rosen bridge, 118–22, 120
    finding, 323–24

many worlds theory and, 169–70,
179, 322
negative energy and, 133, 134
Thorne's time machine and,
132–33
Type III civilization and, 316
See also time travel

Yamasaki, Masami, 237
Yu, L. P., 190

Yurtsever, Ulvi, 131

Zeh, Dieter, 166–67
Zeno, 134–35, 236
Zucker, Michael, 260
Zweig, George, 81
Zwicky, Fritz, 67, 70–72

ALSO BY MICHIO KAKU

BEYOND EINSTEIN
The Cosmic Quest for the Theory of the Universe

Beyond Einstein takes readers on an exciting excursion into the discoveries that have led scientists to superstring theory. This revolutionary breakthrough may well be the fulfillment of Albert Einstein's lifelong dream of a Theory of Everything, uniting the laws of physics into a single description explaining all the known forces in the universe.

Science/0-385-47781-3

HYPERSPACE
A Scientific Odyssey Through Parallel Universes, Time Warps, and the 10th Dimension

Hyperspace is an exploration of the theory of ten-dimensional space, the most exciting development in modern physics. "Taking his material to the limit, Mr. Kaku speculates on how the universe might have begun—perhaps as a tear in the space-time of some other universe, or as a quantum particle that sprang into existence, and then inflated to become a universe. . . . What a wonderful adventure it is, trying to think the unthinkable!" (*The New York Times Book Review*).

Science/0-385-47705-8

VISIONS
How Science Will Revolutionize the 21st Century

In this thrilling tour, Michio Kaku examines the ways the great scientific revolutions of the twentieth century—quantum mechanics, biogenetics, and artificial intelligence—will transform the way we live in the twenty-first century. His unique and compelling vision, based on research already underway at top laboratories around the world, predicts a future in which we are no longer passive bystanders to the dance of the universe, but creative choreographers of matter, life, and intelligence.

Science/0-385-48499-2

ANCHOR BOOKS
Available at your local bookstore, or call toll-free to order:
1-800-793-2665 (credit cards only).